Regenerated cellulose fibres

Other titles in the Woodhead Publishing Limited series on fibres, published in conjunction with The Textile Institute

Series Editor: Professor J E McIntyre

Wool science and technology
Silk, mohair, cashmere, camel hair and related camelid fibres
High performance fibres
Synthetic fibres
Cotton science and technology
Bast and other leaf fibres

Regenerated cellulose fibres

Edited by

Calvin Woodings

The Textile Institute

CRC Press
Boca Raton Boston New York Washington, DC

WOODHEAD PUBLISHING LIMITED
Cambridge England

Published by Woodhead Publishing Limited in association with The Textile Institute

Woodhead Publishing Limited
Abington Hall, Abington
Cambridge CB1 6AH
England
www.woodhead-publishing.com

Published in North and South America by CRC Press LLC,
2000 Corporate Blvd, NW
Boca Raton FL 33431, USA

First published 2001, Woodhead Publishing Ltd and CRC Press LLC

British Library Cataloguing in Publication Data
A catalogue record for this book is available from the British Library.

Library of Congress Cataloging in Publication Data
A catalog record for this book is available from the Library of Congress.

Woodhead Publishing ISBN 1 85573 459 1
CRC Press ISBN 0-8493-1147-0
CRC Press order number: WP1147

Cover design by The ColourStudio
Typeset by Best-Set Typesetter Ltd, Hong Kong
Printed by TJ International Ltd, Cornwall, England

Contents

Preface

Until the early years of the 20th century, textiles were based solely on natural products such as cotton, wool, silk and linen. However a chain of developments commencing in the 1890s led to a global textile industry based largely on fibres manufactured by industrial processes.

These processes can be subdivided into those which use the cell walls of plants *directly* as their starting point, and thermoplastic fibres which start from non-renewable fossil reserves and hence use the plant matter *indirectly*. Synthetic fibres, now economically manufactured from oil reserves valued at little more than the cost of extraction, have come to dominate the manufactured fibre market, and as the 20th century drew to a close, their use overtook that of natural and regenerated fibres combined.

Will this trend continue until natural and regenerated fibres are no more than costly specialities confined to high fashion, 'environmentally friendly' and luxury end-uses? Has cotton usage peaked, like wool and silk in earlier decades, leaving the fibre future to ever-improving synthetics? Will synthetic fibres sustain their cost effectiveness as oil becomes scarcer? If not, will processes such as those described in this book form the basis of a new textile industry based on readily renewable resources, or will fibre-making polymers such as polyesters be producable economically from annual crops? Will thermoplastic hydrophilic polymers ever be synthesised by purpose-designed genetically engineered plants?

These are the big questions facing fibre manufacturers at the turn of the millennium, and this Fibre Series from Woodhead Publishing aims to provide the facts to allow such questions to be debated fully. This book on the regenerated cellulose fibres deals with the original manufactured fibre processes which 'simply' take natural cellulose into solution in order to regenerate it in fibrous form. Cellulosics, including cotton, may be losing market dominance to the cheaper synthetics, but nevertheless remain uniquely capable of providing levels of wearer comfort, absorbency, softness, biodegradability, and environmental compatibility that the synthetics have yet to emulate.

Cellulose will, of course, remain the most abundant biopolymer and will continue to be the driving force of the carbon cycle on which all life depends. Its credentials as a safe, environmentally benign and renewable raw material for many industrial processes will be hard to surpass. Photosynthesis will undoubtedly remain the most efficient polymerisation process on Earth, and it is conceivable that a full understanding of its biochemistry may allow us to engineer organisms to produce modified celluloses and cellulose derivatives directly.

Helping Nature to add to cellulose those properties which the 20th century textile industry has come to regard highly, seems to be at least as worthwhile as trying to make natural processes synthesise the plastic polymers that happen to be the market leaders at the time the fossil reserves, on which they are based, are becoming scarce.

Calvin Woodings

Contributors

Professor John W S Hearle
The Old Vicarage
Church Road
Mellor
Stockport, UK
SK6 5LX

Mr Tim F N Johnson
TFNJ Ltd.
55 Wartnaby Street
Market Harborough
Leicestershire, UK
LE16 9BE

Professor Kenji Kamide
Faculty of Economics
Nara Sangyo University
Tatsuno-kita 3-12-1
Sango Ikoma-gun
Nara 636-8503
Japan

Professor Bruno Lönnberg
Laboratory of Pulping Technology
Åbo Akademi University
Porthansgatan 3
FI-20500 ABO/Turku
Finland

Mr Kazunari Nishiyama
Fibre Division
Asahi Chemical Industries Co Ltd
Dojima 1-2-6 Kita
Osaka 530-8205
Japan

Professor Albin Turbak
Falcon Consultants
7140 Brandon Mill Road
Sandy Springs
Atlanta
Georgia 30328-1923
USA

Mr Patrick A White MBE
Tencel Ltd.
PO Box 5, Spondon
Derby, UK
DE21 7BP

Dr Andrew Wilkes
Acordis UK Ltd
P O Box 5
Spondon
Derby, UK
DE21 7BP

Calvin Woodings
Calvin Woodings Consulting Ltd
The Lodge
Eathorpe
Warwickshire, UK
CV33 9DF

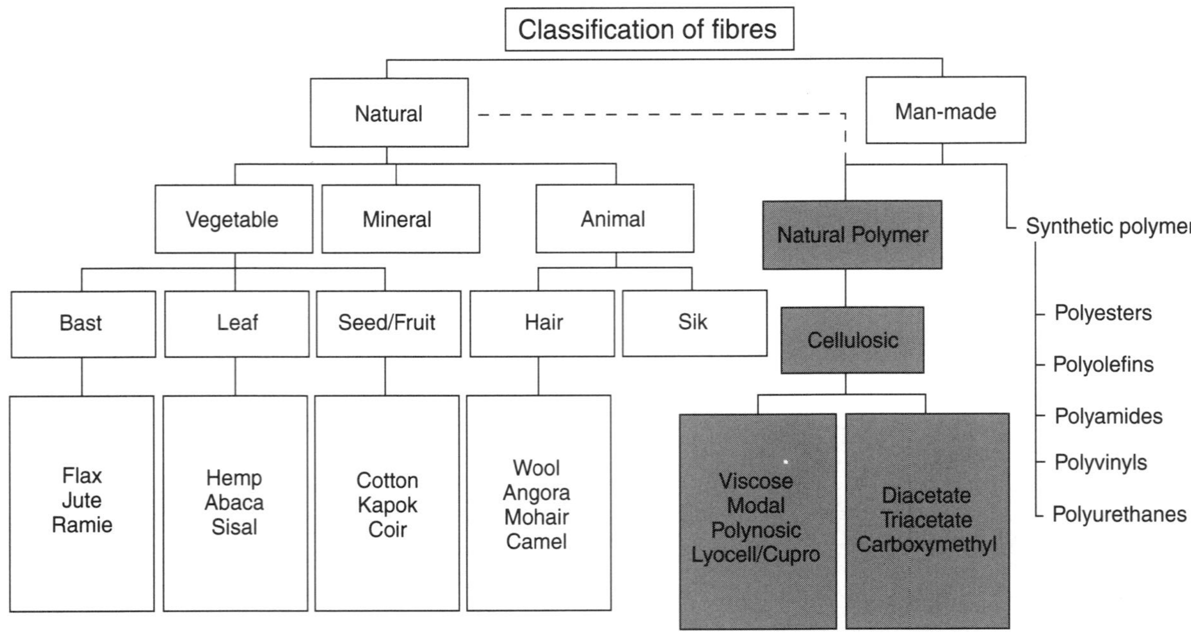

Classification of fibres
Natural
Man-made
Vegetable
Mineral
Animal
Natural Polymer
Synthetic polymer
Bast
Leaf
Seed/Fruit
Hair
Sik
Cellulosic
Polyesters
Polyolefins
Polyamides
Polyvinyls
Polyurethanes
Flax
Jute
Ramie
Hemp
Abaca
Sisal
Cotton
Kapok
Coir
Wool
Angora
Mohair
Camel
Viscose
Modal
Polynosic
Lyocell/Cupro
Diacetate
Triacetate
Carboxymethyl

1

A brief history of regenerated cellulosic fibres

CALVIN WOODINGS

1.1 Introduction

Like so many major new materials, artificial fibres did not arise from the clear identification of a customer need followed by carefully planned research aimed at meeting that need. Visionaries foresaw the potential of artificial fibres but the amateur scientists and professional inventors who made the groundbreaking discoveries were often motivated by products unrelated to today's fibre industry. They worked without knowledge of the underlying chemistry and physics of polymers and progressed towards their goals by trial and error accompanied by careful observation.

Industrially speaking, artificial fibres had their origins in the paper industry, in war materials, and in electricity, that 'internet' of the late 19th century. However the inspiration for the artificial fibre industry is generally credited to Robert Hooke (1635–1703), an English physicist better known for his discovery of the law of elasticity and the development of improved microscopes. In *Micrographia* (Small Drawings, London, 1665) he discussed the possibility of imitating the silkworm by making 'an artificial glutinous composition [and] to find very quick ways of drawing it out into small wires for use'. He also deserves a special mention in the particular context of regenerated cellulose fibres for being the first scientist to use the word 'cell' to describe the honeycomb structure of plant matter.

Renee-Antoine de Réaumur (1683–1747) recorded in 1734 his attempts to force different kinds of varnish through perforated tin cans to form coarse filaments that hardened in warm air. He is therefore credited with the first dry-spinning process even though his extrudate was unusable as fibre. Cellulose (Fig. 1.1) itself was discovered in 1839 by a Frenchman, Anselme Payen (1795–1871) the Professor of Agricultural and Industrial Chemistry at the Central School of Arts and Manufactures in Paris, during an extensive analysis of wood. He also discovered pectin and dextrin and was the first to isolate and concentrate an enzyme – diastase.

More so then than now, silk was *the* luxury fibre and far too expensive

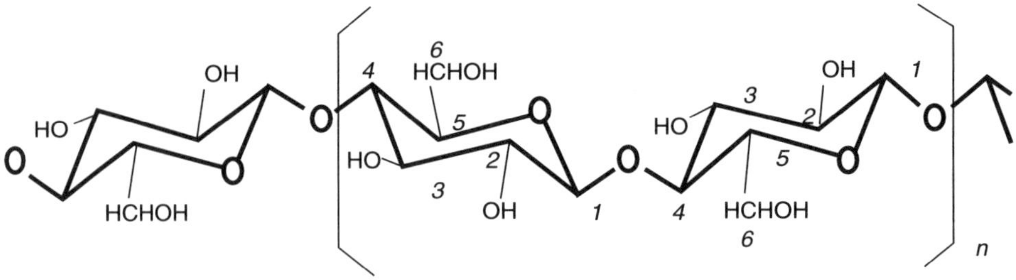

1.1 Structure of cellulose: a linear polymer of two anhydroglucose units with a 1–4 β-glucosidic linkage (see also Fig. 8.2–8.5).

for general use in clothing. Attempts to reduce its cost led to several abortive attempts to extrude the silkworm gum artificially, but the first indication that Hooke's idea might be realisable came in the 1840s when Louis Schwabe, an English silk weaver, developed the precursors of today's spinnerets, the nozzles with fine holes through which liquids could be forced. He extruded molten glass filaments[1] and proceeded to weave fabrics from the resulting continuous filament yarns. At the same time, in apparently unrelated developments, several continental European chemists were working with cotton and the twigs, branches and barks of assorted trees, especially mulberry (the leaves of which are the silkworm's preferred food). Charles Freidrich Schönbein is credited with the accidental discovery in 1846 that nitric acid could nitrate these cellulose sources and result in an explosive substance, 'Schiesswolle' or guncotton. This line of research was to lead to the discovery of dynamite and the founding, by Alfred Nobel in 1867, of the explosives industry. But in the early 1850s, it also led to the first successful attempt to make textile fibres from plant cellulose.

1.2 Cellulose nitrate

George Audemars of Lausanne, Switzerland dissolved the nitrated form of cellulose in alcohol and ether and discovered that fibres were formed as the resulting 'collodion' was drawn into the air. His 1855 patent[2] on *Obtaining and Treating Vegetable Fibres* covers the pulping of the inner bark of mulberry to extract cellulose fibre which 'may be hackled, combed, or carded, and then spun like cotton; or it may be converted into an explosive compound by the action of nitric acid, and then dissolved in a mixture of alcohol and ether, then mixed with an [ether] solution of caoutchouc, and drawn out into fine threads or filaments'.

These soft strong cellulose nitrate fibres could be woven into fabrics but had a very serious drawback that ultimately prevented their widespread use in textiles: they were very flammable.

It was Joseph Swan, the English physicist and chemist, developer of the first electric lamp and inventor of bromide photographic print paper, who learnt how to denitrate the cellulose nitrate using ammonium hydrosulphate. This was part of his quest for a better carbon fibre for lamp filaments and was patented as such in 1883.[3] Swan was nevertheless fully aware of the textile potential of his process, and in 1884, 'some samples of artificial silk . . . the invention of Mr J W Swan' were displayed at a meeting of the Society of Chemical Industry. The first fabrics made from the new artificial silk were also shown at the Exhibition of Inventions in 1885. However with carbon lamp filaments being his main focus (for which he was knighted in 1904) he failed to follow up on the textile possibilities, allowing the Frenchman, Count Louis-Marie-Hilaire Bernigaud, Comte de Chardonnet justly to become regarded as the founder of the regenerated cellulosic fibre industry.

Chardonnet, a scientist and professional inventor, had after all been concentrating on developing artificial silk fibres and textiles, and he did follow through to set up a company to manufacture it. His research evolved a process practically identical to Swan's, albeit slightly later, but he nevertheless perfected the fibres and textiles in time for the Paris Exhibition in 1889. There he attracted the necessary financial backing to produce the first 'Chardonnet silk' from J P Weibel, a French wood-pulp producer. His first factory started up in 1892 in Besançon near to Weibel's pulp and paper mill. His commercial process[4] involved treating mulberry leaves with nitric and sulphuric acids to form cellulose nitrate, which could then be dissolved in ether and alcohol. This *collodion* solution was extruded through holes in a spinneret, but where Swan used a liquid coagulant, Chardonnet used warm air to evaporate the solvent and form solid cellulose nitrate filaments (see Fig. 1.2). Both Swan and Chardonnet denitrated the fibres in a separate step.

British developments began when Freidrich Lehner left the Chardonnet factory in Switzerland and set up his own business, Lehner Artificial Silk Ltd, with British capital in 1892. He wished to exploit his own patented improvements to Chardonnet's process and collaborated with Lister and Co of Bradford Ltd, producing small quantities of cellulose nitrate yarn in 1893. (H G Tetley, the prime mover behind the development of viscose rayon, worked at Lister and Co before joining Samuel Courtauld & Co – see below)

The New Artificial Silk Spinning Co. based at Wolston near Coventry (led by Joseph Cash the Coventry ribbon weaver), also produced cellulose nitrate yarn but with much difficulty. The company went into liquidation in 1900. Their machinery was bought by Glanzstoff (see below).

Although this first artificial fibre process was simple in concept, it proved slow in operation, difficult to scale-up safely, and relatively uneconomic

1.2 The first artificial fibre spinning machine c.1889 developed by Count Hilaire de Chardonnet for cellulose nitrate fibres. The filaments were spun vertically upwards from the nozzles on the floor of the cabinet and collected on reels at the top. (Reproduced from *Courtaulds: An Economic and Social History*, Clarendon Press, 1969.)[11]

compared with later routes. Denitration of the fibres, which was necessary to allow safe use wherever the fabrics might risk ignition, spoilt their strength and appearance. Neverthless, Chardonnet earned and truly deserved his reputation as the 'Father of Rayon'. His process was operated commercially until 1949 when the last factory, bought from the Tubize Co in the USA in 1934 by a Brazilian company, burned down.

1.3 Direct dissolution in cuprammonium hydroxide: cupro

The second artificial silk process to be commercialised was based on a discovery made by the Swiss chemist Matthias Eduard Schweizer in 1857.[5] He found that cotton could be dissolved in a solution of copper salts and ammonia and then regenerated in a coagulating bath. The fibre process was however invented by a French Chemist, Louis-Henri Despeissis,[6] who worked on spinning fibres from Schweizer's solution in 1890. He extruded the *cuprammonium* solution of cellulose into water, with dilute sulphuric acid being used to neutralise the ammonia and precipitate the cellulose fibres.

Despeissis died in 1892 and his patent was allowed to lapse. However in 1891, the German chemist Max Fremery and the Austrian engineer Johan Urban were also using Schweizer's reagent and cotton to make lamp filaments in Oberuch near Aachen. They decided to expand into artificial silk (in German, Glanzstoff) and patented[7] their approach in the name of Dr. Hermann Pauly to avoid attracting the attention of competitors. Pauly, a director of the technical school in Munchen Gladbach, other than lending his name, made no contribution to the development. This patent, essentially a reiteration of the Despeissis process with a practical spinning method added, was however upheld after dispute, thereby allowing Fremery and Urban to begin large scale manufacture as Vereinigte Glanzstoff Fabriken (VGF) in 1899. In 1901, a Dr Edmund Thiele working at J P Bemberg developed a stretch-spinning system. The resulting improved Bemberg® silk went into production in 1908. Its early commercial success owed much to the flammability disadvantages of the Chardonnet process, but it was competition from the viscose process (see below) that led to its decline for all but the finest filament products.

The process is still used today, most notably by Asahi in Japan where sales of artificial silk and medical disposable fabrics still provide a worthwhile income (see Chapter 5). However the relatively high costs associated with the need to use cotton cellulose and copper salts prevented it from reaching the large scale of manufacture achieved by the viscose rayon process. Most producers (Asahi and Bemberg excepted) had abandoned the approach by the outbreak of war in 1914.

1.4 Dissolution via cellulose xanthate: viscose

In 1891 the British chemists Charles Cross, Edward Bevan, and Clayton Beadle, working at Kew in England, discovered that cotton or wood cellulose could be dissolved as cellulose xanthate following treatment with alkali and carbon disulphide:

$$\begin{array}{lclcl} (C_6H_{10}O_5)_n & + \, n\text{NaOH} & \rightarrow & (C_6H_9O_4ONa)_n & + \, nH_2O \\ \text{cellulose} & + \text{ soda} & \rightarrow & \text{sodium cellulosate} & + \text{ water} \end{array} \qquad [1.1]$$

$$\begin{array}{lcl} (C_6H_9O_4ONa)_n \; + \; nCS_2 & \rightarrow & n(\text{S}\overset{\text{SNa}}{\overset{/}{\text{C}}}\text{—OC}_6H_9O_4) \\ \text{sodium cellulose + carbon disulphide} & \rightarrow & \text{sodium cellulose xanthate} \end{array} \qquad [1.2]$$

The treacle-like yellow solution (initially called 'viscous cellulose solution', later contracted to 'viscose') could be coagulated in an ammonium sulphate bath and then converted back to pure white cellulose using dilute sulphuric acid. They patented their process in 1892 without mentioning fibres.

Cross and Bevan had set up a partnership as analytical and consulting chemists to serve the pulp and paper industry in 1881. They collaborated with the Swedish inventor of the sulphite pulping process, C D Eckman, to help the British paper industry develop a wood pulp alternative to the cotton and linen rags that were then the main raw materials of the paper trade. As a natural extension of their work on pulping chemistry, and cotton mercerising* they became interested in dissolving cellulose and the 1892 patent application, entitled *Improvements in Dissolving Cellulose and Allied Compounds*[8], was the result. In 1893 they formed the Viscose Syndicate to grant licences for non-fibre end-uses, and in 1896 the British Viscoid Co Ltd was formed to exploit the process as a route to moulded materials. These companies were later merged to form the Viscose Development Co in 1902. Early licences were granted to papermakers and calico printers for the use of viscose as a coating or size, and to makers of artificial leather and artificial flowers.

In another laboratory at Kew, Charles Henry Stearn and Charles Fred Topham developed the continuous filament spinning process and the machinery needed to wash and collect[9] the viscose yarns. The two had met in about 1874 in Liverpool where Topham was helping his father blow glass for Stearns spectrum tubes and radiometers. At the time Stearn was a cashier in the Liverpool branch of the Bank of England but also an amateur scientist specialising in high vacuum technology. In 1877 Stearn was collaborating with Joseph Swan on the electric lamp, Stearn taking the carbon filaments from Swan's experiments in Newcastle, inserting them in tubes made by Topham in Rock Ferry, and evacuating them at home. In 1889 with a workable lamp filament still eluding Swan and Stearn, Stearn left to direct the Zurich Incandescent Lamp Co of Zurich and London, also at Kew. In 1893, on seeing the viscose patent, he immediately recognised the potential of the process as a cheap route to the now successful Chardonnet silk, and contacted Cross with a view to commencing fibre spinning developments.

His work on spinning equipment culminated in 1898, when, describing himself as an electrician, Stearn patented[10] filament manufacture 'by projecting the [viscose] solution [of cellulose] into a precipitating solution such for instance as alcohol, brine, chloride or sulphate of ammonia or other suitable precipitant.'

In order to finance the development of a commercial spinning process from these ideas, Cross and Stearn set up the Viscose Spinning Syndicate Ltd, incorporated in May 1899. Alfred Nobel (explosives) and Andrew

*John Mercer's 1850s process of using caustic alkali to finish cotton fabric, later (1894) found to be capable of making cotton look like silk if the process was carried out under tension.

Pears (soapmaker, and Beadle's father-in-law) were among the first shareholders. The continental European rights went to a group of German businessmen who had set up Continentale Viskose GmbH in 1897. The Viscose Spinning Syndicate under Cross and Stearn was keen to attract buyers and sell the process, an attitude that inhibited the necessary development of a deeper understanding of the fundamentals of cellulose dissolution. D C Coleman[11] describes the syndicate as 'a curious assemblage of talent and inexperience . . . that managed to be neither a research laboratory nor a production pilot-plant'.

The next few years were fraught with difficulty as the original process was repeatedly shown to be inconsistent and largely uncontrollable. Economics went uncalculated and the consequences of toxic gases and viscose solidifying in sewers were ignored. Filaments produced by spinning into the then favoured alkaline bath were so weak they could only be collected in Topham's specially devised spinning box.[12] Topham's box, later to be regarded as one of the fundamental breakthroughs leading to successful commercialisation of the viscose process, would have appeared unnecessary had acid spinning baths been developed earlier (as they were in Europe). However the box is still relevant to this day as a gentle way of collecting high quality yarn without incurring the strain applied by bobbin systems. His candle filter[13] (to remove particles from the viscose prior to spinning), spinning pump[14] (to allow careful control of filament size), and his development of hot acid fixing were however key, if underrated, process improvements.

The first licensee to discover the inherent unreliability of the early viscose process was Prince Donnersmarck, the majority shareholder of the Continentale Viskose Co who in 1902 decided to set up a plant, using equipment commissioned from Dobson and Barlow, at Settin, near his pulp and paper mill in Altdamm. It started production in August 1903 but despite heavy investment in machines and men failed to exceed 100 kg per day of yarn output over the next two years. His managers became frequent visitors to Kew, but despite free access to the latest technology, they continued to struggle.

An American chemist, Dr Arthur D Little of Boston, had also visited Kew in 1899, and with a Philadelphia businessman, Daniel Spruance, had acquired the US rights to the Cross, Bevan and Beadle patent. They were instrumental in setting up the Cellulose Products Co in the USA to make viscose solutions in 1900, and in 1901 set up the General Artificial Silk Co to spin fibres – acquiring rights to the Stearn spinning patent. Five years of difficulty later, the rights were bought by Silas W Petitt, Spruance's attorney, for $25 000. He dissolved the original companies but continued the project as the Genasco Silk Works until his death in 1908.

The Société Français de la Viscose, who like the pioneering

Donnersmarck, ordered three Dobson and Barlow spinning tables with 50 spindles each, acquired the French rights. Their factory commenced operation in the summer of 1903, but like Donnersmarck, they failed to exceed 100 kg per day output by 1905.

In February 1904, the Kew laboratories were visited by Henry Greenwood Tetley of Samuel Courtauld & Co Ltd. Courtauld were silk weavers looking for new raw materials and new opportunities to grow. The success and profitability of Samuel Courtauld had been built on the 19th century fashion for black silk mourning crepe, and the company was planning its stock market flotation. The visitors to Kew knew that Chardonnet's now established cellulose nitrate process was creating a lucrative market for artificial silk in France. Having also visited Donnersmarck's plant in Germany, they believed that Cross and Bevan's viscose route could make a similar fibre at about a third of the manufacturing costs of the Chardonnet's route. Nevertheless, it took two presentations, the second to a changed Board of Directors after the flotation, before Courtauld was persuaded to acquire the viscose process rights.

On 14 July 1904, the Viscose Spinning Syndicate agreed to sell the viscose process rights and patents to Courtauld & Co for the sum of £25 000. Courtauld took over the Kew laboratories to gain practical experience in the new technology while constructing a production plant. In September 1904 they gained full access to the technology developed by the Société Française de la Viscose, and in December took delivery in Kew of a 12-end Dobson and Barlow viscose yarn pilot line (Fig. 1.3). Siemens Timber Yard by the canal in Foleshill, Coventry, England was acquired, and the Kew equipment transferred to Coventry in August 1905. The first small samples of Coventry viscose filament yarn were shown to the Courtauld Board in November 1905.

While the first fabrics were woven in March 1906 at Courtauld's Halstead Mill in Essex, Coventry's output was largely put to waste. By August 1906, only 25% of the yarn produced was saleable and the Courtauld board were advised that any expansion of the plant would be folly. The problem was largely due to the tender nature of the yarns emerging from the alkaline ammonium sulphate spinning bath but its solution, an acid–salt bath, was emerging from work in the Donnersmarck plant in Germany and at Société Français de la Viscose in France, and, via the technical exchange with the French, at Courtauld & Co in Coventry.

It was Dr Paul Koppe, the technical manager of Donnersmarck's plant, who took out the first patent (May 1904) on what later became known as the Müller spinbath, a mixture of sulphuric acid and another soluble salt. Donnersmarck, a major shareholder in the Viscose Spinning Syndicate, presumably realising this crucial new discovery would devalue the licence being acquired by Courtauld from VSS, had the patent withdrawn. It was

1.3 The first commercial viscose spinning machine (c. 1904) supplied to Samuel Courtauld and Co by Dobson Barlow Ltd. This was a double-sided machine, the other six spinning positions being out of view. A white Topham-Box is shown (out-of-position) on the right. (Reproduced from *Courtaulds: An Economic and Social History*, Clarendon Press, 1969.)[11]

reapplied for in May 1905 (Germany) and April 1906[15] (Britain), this time in the name of Dr Max Müller. By then the alkaline bath process that Donnersmarck had proved unworkable had been safely sold to Courtauld.

Courtauld's first attempt to get the British Müller patent revoked was made jointly with the other European viscose producers and failed in May 1907 in a judgment with curious consequences. The British judge, apparently confused by the complexities of the case, concluded that while the sulphuric acid concentration could vary over the full range claimed in the patent, the salt level always had to be at saturation. So in Britain, lower concentrations of the salt were deemed outside the Müller patent but in the rest of Europe, Donnersmarck's competitors were much more constrained. Courtauld opted for a sulphuric acid/ammonium sulphate/sodium sulphate spinbath with appropiate salt concentrations in mid-1907, and were never challenged by Donnersmarck.

Spinbath evolution in Coventry continued apace with progressive improvements in yarn quality. The addition of glucose[16] in 1907 followed by the progressive removal of the costly ammonium salt in 1909 and the fundamental breakthrough[17] achieved by adding zinc sulphate in 1911 firmly established Courtauld's lead in the new technology. The yield of first quality yarn increased to around 40% in 1907 and 4 years after taking over the Kew operation, early in 1908, the Courtauld viscose business had its first profitable month. By 1911, with the zinc additions to the spinbath, more than 90% of Coventry yarn was meeting the original first quality standard, but by then the standards in use had been re-based to allow further progress.

The acquisition of the rights to the viscose process by Courtauld was to become one of the most profitable investments of all time, and the opening up of the American market was the key to this greater success. Just before he died in 1908, Silas Pettit the owner of the Genasco Silk Works in the USA entered into a royalty agreement to allow Courtauld to sell Coventry yarn in his territory. After his death, his son, John Read Pettit Jnr decided to sell up. He travelled to Coventry in May 1909 and on 3 June reached agreement with Tetley to sell Courtauld the US plant and rights for $150000. The Courtauld board, by then fully convinced of the enormous potential of the process and encouraged by the take-off in US demand for the Coventry yarn, approved the deal on the same day.

On 6 August 1909, the US government applied a duty of 30% to the imported Coventry yarn, but by then Tetley had visited the USA and decided to start production there. Fifty acres of land were purchased in November 1909 and the American Viscose Company registered on 15 March 1910 at Marcus Hook in Pennsylvania. It was set up as a private company with shares bought in cash by Samuel Courtauld and Co, UK. The new US company acquired the rights to the viscose process from Courtauld in exchange for further shares. George Henry Rushbrook, the Courtauld company secretary became the first President. The first yarn was spun on 23 December 1910.

From 1910 to 1920, with the Courtauld viscose patents in force and the production technology developed to provide quality yarn at competitive prices, Tetley's 1904 vision became reality. Donnersmarck's plant in Germany, the originator of the Müller process, failed to achieve Courtauld's level of first quality and was bought out in 1911 by Vereinigte Glanzstoff-Fabriken, the leading producer of cuprammonium silk.

That they too appreciated the superiority of the viscose route, simply underlines the fact that by 1910, viscose was emerging as the winning fibre process. Viscose yarn output may only have been a third of the cellulose nitrate production and a half of the cuprammonium output, but technology and economics were now clearly in its favour. VGF had, in 1908–1910, set up a British cuprammonium plant at Flint on the River Dee to protect their rights to operate the cupro patents in Britain. While at the time Courtauld did not feel this a significant threat, the VGF acquisition of the Donnersmarck viscose process a year later sounded the alarms. In 1911, Courtauld informed VGF they would be prepared to reach a 'general understanding between the makers of cuprammonium and viscose artificial silks.'

The outcome was a consortium of three groups, a German Group (VGF, Austrian Glanzstoff and Donnersmarck, an Anglo–American Group (Samuel Courtauld and Co, The American Viscose Company, and British Glanzstoff at Flint) and a Latin Group (The French, Italian, Belgian, Swiss and Spanish companies). All technology was to be shared, prices fixed, production quotas allotted, and each group would sell yarn only in its own area. While this 'protocol' was signed and put into effect at the end of 1911, it was never converted into a formal contract. It was nevertheless observed in principle if not in detail and provided some marketing consistency and stability at a crucial time. It also allowed faster technical progress by sharing rather than restricting access to innovations, it kept prices high, and it did not prevent the leading exponent, Courtauld, from obtaining even higher prices than its competitors. Only Courtauld had silk weaving know-how allowing them to open up the woven fabric market with a yarn quality the other consortium members could not match (see Chapter 9: Applications development).

The extraordinary success of Courtauld's venture is illustrated in Fig. 1.4 and 1.5.

From 1920 to 1931, after the expiry of the viscose patents, world output increased from 14 000 to 225 000 tonnes per year, as more than 100 companies entered the artificial fibre field. In Europe, VGF (Germany), Enka (Holland), I G Farben (Germany), Snia Viscosa (Italy), Comptoir des Textiles Artificiels (CTA – France), Rhodiaceta (France), Tubize (Belgium) and Chatillon (Italy) were among the new starters.

In the USA the new entrants incuded DuPont (with help from CTA), Tubize, Chatillon, American Enka, The Industrial Fiber Corporation (later The Industrial Rayon Corporation), American Glanzstoff (later North American Rayon) and American Bemberg.

By 1939 Courtauld had six factories in the USA, seven in the UK, one in France, one in Canada, and joint ventures in Germany and Italy.

From the 1920s onwards, sales grew explosively, but the rayon process evolved in a more or less predictable manner. Discoveries enabling the production of stronger yarns led to the development of the tyre-yarn process and, driven by the war and a massive expansion in automobile use in the 1940s and 1950s, this technology boomed.

The introduction of staple fibre, which could be converted on traditional textile spinning equipment, was crucial to continued expansion in the 1930s

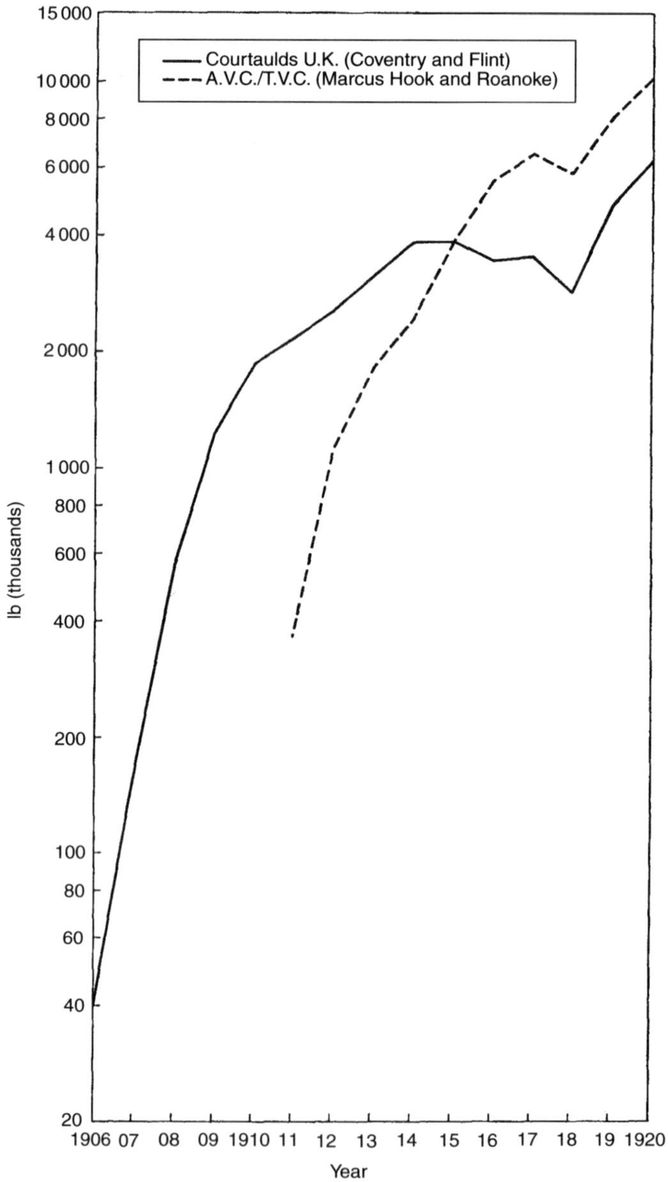

1.4 Growth of Courtauld's viscose filament yarn output: 1906–1920. The 1911 American investment (American Viscose Corporation/The Viscose Company) rapidly overtakes UK output. Despite the viscose success, in 1911 cellulose nitrate was still the world's leading fibre with a 40% share, viscose and cuprammonium yards having 30% each. (Reproduced from *Courtaulds: An Economic and Social History*, Clarendon Press, 1969.)[11]

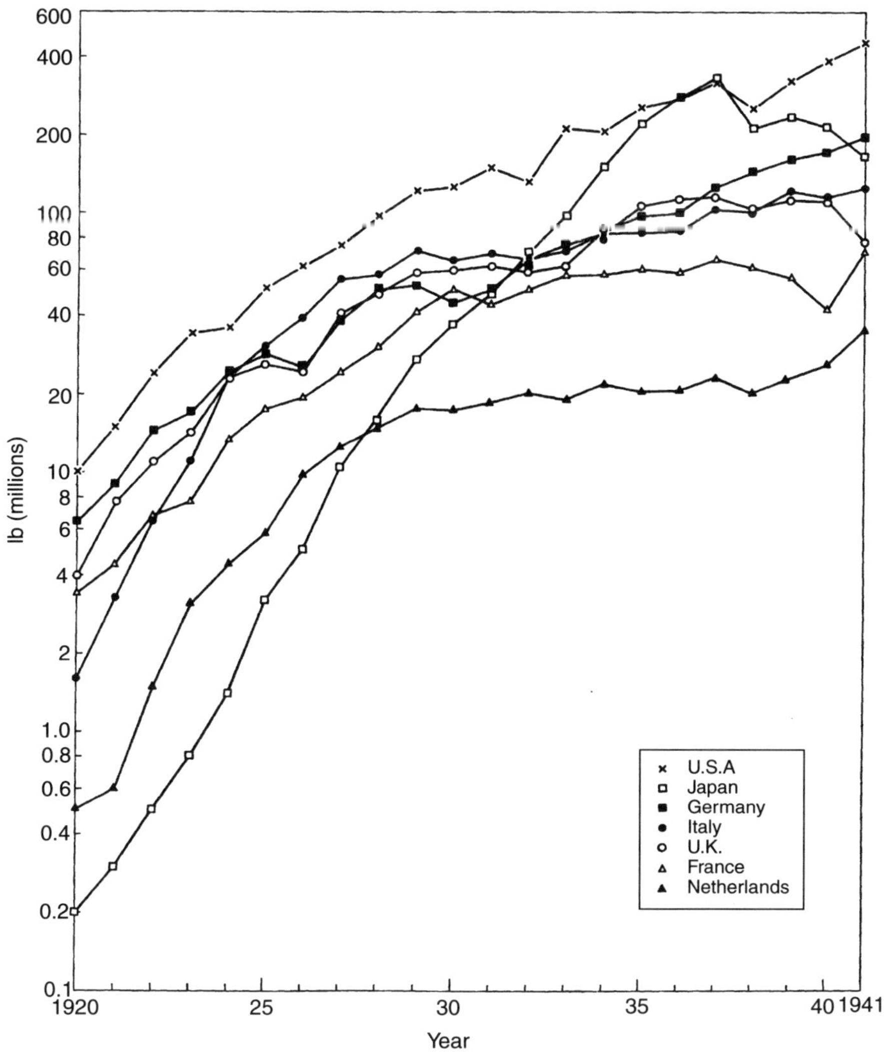

1.5 Regenerated cellulose filament yarns: 1920–1941. Expiry of the key viscose patents allows global growth. (Reproduced from *Courtaulds: An Economic and Social History*, Clarendon Press, 1969.)[11]

but did not involve any startling innovations. However, what had been devised as a route for getting extra value out of yarn waste (chopping it into short lengths and selling it to cotton spinners as a cotton diluent) ultimately outsold the original continuous filament yarns and resulted in many new factories being built in the 1950s and 1960s. Staple fibre is dealt with in Chapter 9: Applications development.

By 1941 as the first synthetic polymers were being converted into the first nylon and polyester fibres, world production of viscose rayon had risen to 1250000 tonnes.[18] It continued to expand into the 1970s recording its highest ever annual output at 3856000 tonnes in 1973. Since then a steady decline has occurred as more and more end-uses switch to the now cheaper synthetic fibres based on oil.

Also in 1973, the century's leading exponent of viscose technology, Courtaulds, who, since the late 1960s had realised that the end was near for viscose, began to explore new ways of converting cellulose into fibre. This is our next subject.

The current position of regenerated fibres in the market and a complete listing of the world's regenerated cellulose fibre plants at the close of the 20th century is given in Chapter 10.

1.5 Direct dissolution in amine oxide: lyocell

Lyocell technology was pioneered in the USA by Eastman Kodak and American Enka, but it was Courtaulds in the UK who persisted with development until a commercially viable fibre process emerged. Furthermore, Courtaulds did it at a time in its history when the very wisdom of continued involvement, not just in cellulosics but in any fibre or textile activity, was being called into question.

As early as the mid-1950s, Courtaulds believed the future of viscose to be so unattractive that it started to divert viscose profits not only into other fibres, but also into totally unrelated businesses.

By the late 1950s, despite accounting for 80–90% of Courtaulds earnings, the reality of viscose's decline was becoming apparent. The usual remedies, reducing costs, improving quality, selling more aggressively and internationally were yielding diminishing returns so the board's reaction was a new strategy[19] involving:

1 developing new internal sources of profit, i.e. utilising the viscose wet-spinning expertise to move into wet-spun acrylic fibres ('Courtelle'), and opening up a vast, and with hind-sight, transient, new market for a coarse and tough viscose in tufted carpets ('Evlan')
2 developing new external sources of profit by acquisitions in 'related but different' products, in reality British Celanese (cellulose acetate fibres and related products) and Pinchin Johnson Paints (later renamed International Paint)
3 developing greater market power by acquiring key elements of the rest of the fibre value-chain – 'Verticalisation' – resulting most notably in the acquisition of the Lancashire Cotton Corporation Ltd and Fine Spinners and Doublers (representing about 35% of the entire Lancashire cotton industry).

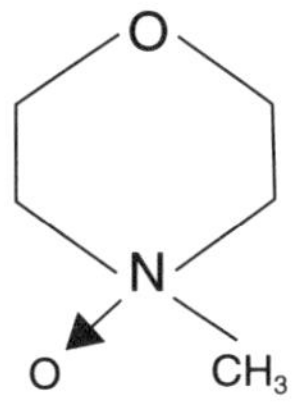

1.6 Structure of *N*-methyl morpholine *N*-oxide.

Of these three, (1) was least favoured with funds. Furthermore, little emphasis was placed on modernisation* of regenerated cellulose fibre production methods to counteract the intertwined problems of synthetic fibre expansions, rising costs and diminishing returns. The demise of viscose thus became a self-fulfilling prophecy. The leading exponent of the technology turned its back on cellulose at a crucial time in its history, a time in fact when new ways of dissolving cellulose were already evident in the research work of other organisations.

Reviews of this early work on direct dissolution are provided by Turbak[20,21] who, in Chapter 7 of this volume, records the efforts to dissolve cellulose directly as a *base* using phosphoric, sulphuric and nitric 'protonic' acids, or using zinc chloride, thiocyanates, iodides, and bromides as Lewis acids. However, despite early promise, the problems of developing fibre production routes using these systems have, with the single exception of the amine–oxide route, so far proved insurmountable.

The amine–oxide solvent, later to become the focus of Courtaulds most costly development project, and the most acrimonious patent battle since the dispute over the Müller bath in 1910, had in fact been discovered back in 1939 by a pair of Swiss chemists, Charles Graenacher and Richard Sallman.[22] But it was not until 1969 that Dee Lynn Johnson of Eastman Kodak described the use of cyclic mono(*N*-methylamine-*N*-oxide) compounds (e.g. NMMO: see Fig. 1.6) as a solvent-size for strengthening paper[23] by partially dissolving the cellulose fibres.

Other Johnson patents[24,25] covered the preparation of cellulose solutions using NMMO and speculated about their use as dialysis membranes, food casings (sausage skins), fibres, films, paper coatings, and nonwoven binders.

NMMO emerged as the best of the amine–oxides and a team at American Enka demonstrated its commercial potential in the late 1970s. In their laboratories in Enka, North Carolina, Neil Franks and Julianna Varga,[26,27] developed a way of making a more concentrated, and hence

* Projects described as 'modernisation' were in fact common, but their scope was restricted largely to taking cost out of viscose dope making, often with losses of quality, by automating the original process.

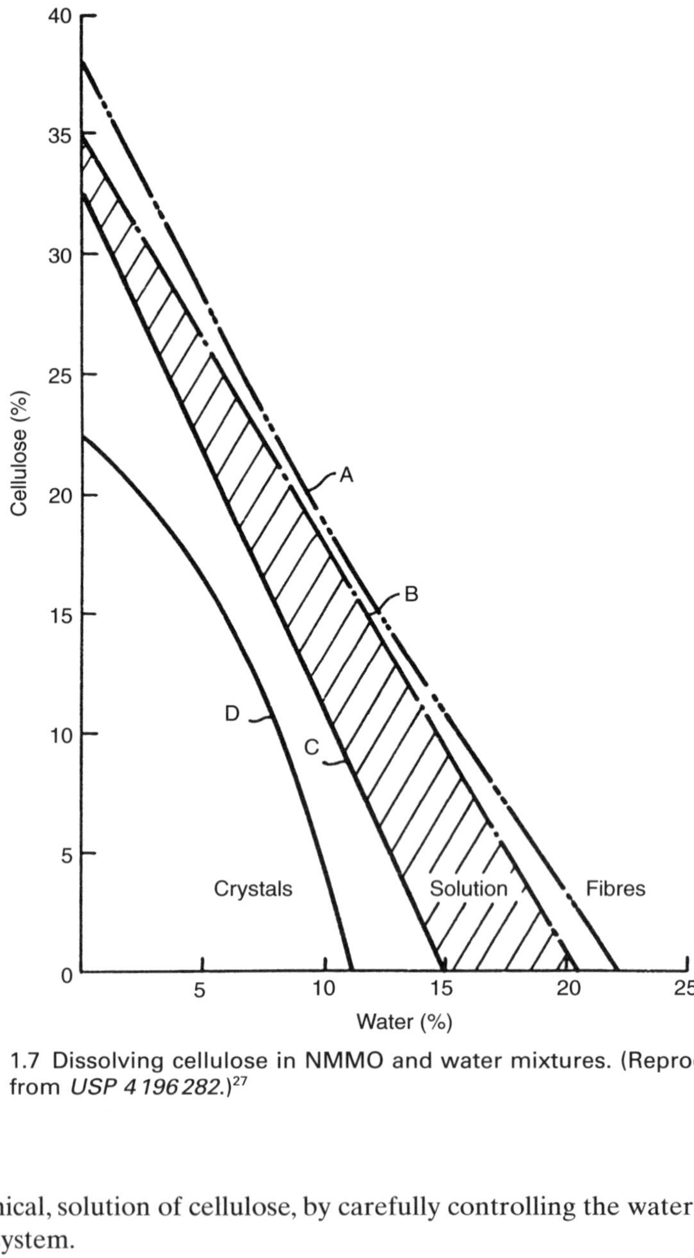

1.7 Dissolving cellulose in NMMO and water mixtures. (Reproduced from *USP 4 196 282*.)[27]

economical, solution of cellulose, by carefully controlling the water content of the system.

Figure 1.7 shows that the concentrations of water and cellulose where complete dissolution of the cellulose occurs (at 95°C), lie between lines B and C. Between lines A and B there can be 95% confidence that the solution would be free from undissolved cellulose fibres, and to the right of line A, undissolved cellulose fibres are bound to be present. Similarly,

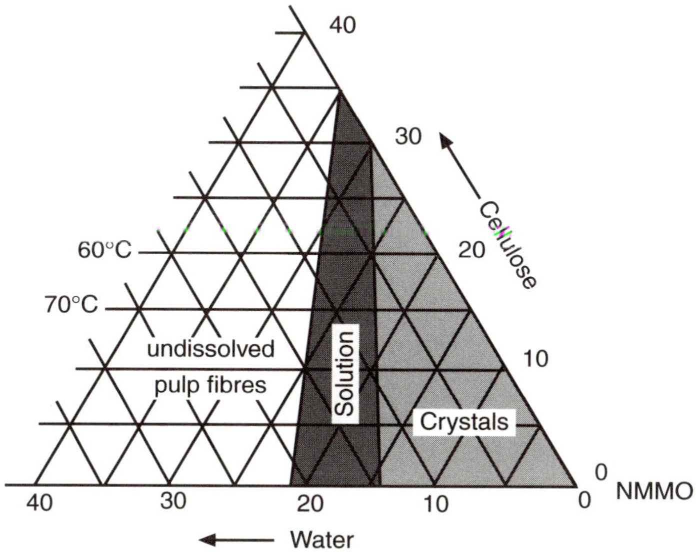

1.8 Ternary diagram showing the effect of temperature on the dissolution of cellulose in NMMO. (Reproduced from Spinnable solutions of cellulose in amine oxides, R N Armstrong, C C McCorsley and J K Varga, *Proceedings of the 5th International Dissolving Pulp Conference*, TAPPI, 1980.)[32]

between C and D there is a 95% chance that crystals of undissolved NMMO will be present, and such crystals will always be present to the left of line D. Figure 1.8 shows the same information in the more familiar ternary diagram form.

Clarence C McCorsley III, also at Enka, developed the key elements of several possible commercial processes. In one, cellulose pulp sheets were soaked in NMMO solution, and after mild heat and vacuum treatment to adjust the water content, the ground-up sheet was fed to an extruder from which fibres could be spun.[28] In another, the solution was made in a large mixer prior to casting it as thick film, freezing it solid, and grinding it up into chips for later extrusion.[29] In a continuous process (Fig. 1.9), a vented extruder fed directly with the ground-up wood pulp and NMMO, mixes the ingredients, creates the solution by removing excess water and volatiles through the vent, and feeds the spinning pumps.[30,31]

Both American Enka[32] and Courtaulds set up pilot plant work in the early 1980s with the objectives of developing the fibre spinning and solvent recovery operations. Courtaulds commercialised first and this, and the continuing development of lyocell is dealt with by White (Chapter 4), who from the outset of practical work in 1979, led the lyocell development effort at Courtaulds.

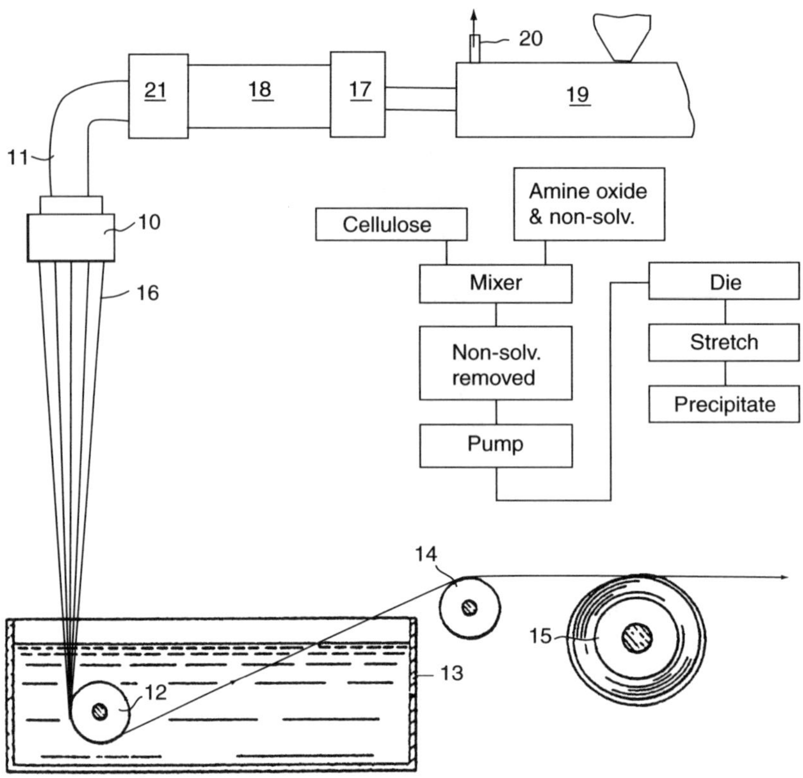

1.9 Clarence McCorsley's lyocell spinning system c. 1979: Dope (containing 15–20% cellulose and at about 115°C) in the extruder barrel (19) is forced through a filter (17), a static mixer (18) to a metering pump (21) and on to the spinneret (10). Filaments (16) pass through an air gap into water in tank 13.

American Enka decided not to commercialise the process and stopped the research in 1981, probably because at that time engineering issues associated with the difficulty of avoiding exothermic reactions looked too hard to resolve economically. However when Courtaulds had demonstrated practical solutions to the many problems discovered during American Enka's early work, they (Enka that is, now part of Akzo Nobel) re-entered the field with the continuous filament version of the lyocell process under their brand name 'Newcell'. The Akzo deal with Courtaulds involved their gaining access to Courtaulds' technology in exchange for granting Courtaulds rights to use some of the key steps in the early patents mentioned above.

Coming right up to date, Akzo Nobel acquired Courtaulds in 1998, and formed Acordis Fibres, bringing together in one company all the key lyocell technology. However, Akzo Nobel had earlier granted a lyocell licence to Lenzing, the Austrian viscose fibre maker, allowing Lenzing to enter the field with a very similar process to Courtaulds. The ensuing patent litigation between Lenzing and Courtaulds was to prove costly to both companies.

Lenzing obtained a patent in the USA for a process, some aspects of which had been operated by Courtaulds for many years, and indeed were used in production at Courtaulds Tencel® plant in Mobile. Courtaulds naturally objected, and applied for summary dismissal of both the US and the subsequent European patent. In Europe, the Munich court would not allow the Lenzing patent to be dismissed summarily and the case went to trial. Courtaulds won, and Lenzing's European patent was disallowed with no right of appeal. In the USA, the Lenzing patent was summarily dismissed, but Lenzing appealed successfully, winning the right to another costly trial. At this point the two companies reached a settlement out of court. The lyocell patent estates of both companies were pooled, to be available royalty free to both companies. It is perhaps worth noting that the settlement only covered patented technology. There was to be no sharing of 'know-how' gained in the operation of the process, which at the time, had only been commercialised by Courtaulds.

1.6 Other routes

Work on other routes to cellulosic fibres has continued, in some cases driven by a desire to utilise the large capital investment in the xanthate route and hence cost less than a completely new fibre process.

The Finnish viscose producer Kemira Oy Saeteri collaborated with Neste Oy on the development of a carbamate derivative route. This system was based on the original work of Hill and Jacobsen[33] who showed that the reaction between cellulose and urea gave a derivative which was easily dissolved in dilute sodium hydroxide:

$$\text{Cell–OH} + \text{NH}_2\text{–CO–NH}_2 \rightarrow \text{Cell–O–NH}_2 + \text{NH}_3 \qquad [1.3]$$

Neste patented an industrial route to a cellulose carbamate pulp[34] which was stable enough to be shipped into rayon plants for dissolution as if it were xanthate. The carbamate solution could be spun into sulphuric acid or sodium carbonate solutions, to give fibres which, when completely regenerated, had similar properties to viscose rayon. When incompletely regenerated they were sufficiently self-bonding for use in papermaking. The process was said to be cheaper than the viscose route and to have a lower environmental impact (Fig. 1.10).[35] It has not been commercialised, so no confirmation of its potential is yet available.

Chen, working on a small scale at Purdue University, claims that

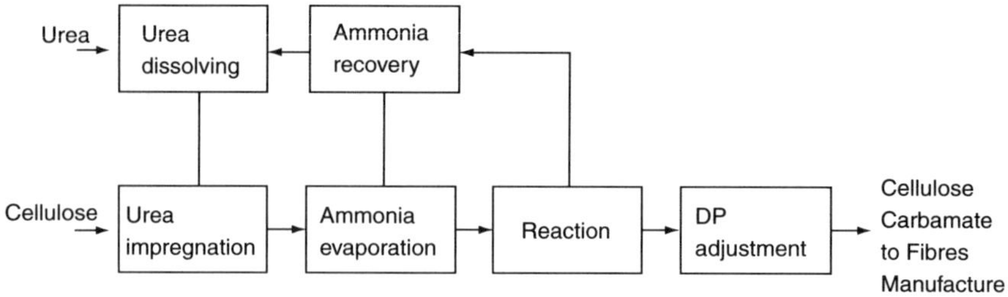

1.10 Manufacture of cellulose carbamate pulp for spinning on viscose equipment. (Reproduced from *Cellulose Carbamate* PIRA International Conference, 1987).[35]

solutions containing 10–15% cellulose in 55–80% aqueous zinc chloride can be spun into alcohol or acetone baths to give fibres with strengths of 1.5–2 g den^{-1}. However, if these fibres were strain dried (i.e. stretched) and rewetted whilst under strain, strengths of 5.2 g den^{-1} were achieved.[36]

Kamide (see also Chapter 5 on cuprammonium rayon) and co-workers at Asahi have been applying the steam explosion[37] treatment to dissolving pulp to make it dissolve directly in sodium hydroxide.[38] In technical papers,[39] they claimed a solution of 5% of steam-exploded cellulose in 9.1% NaOH at 4°C being spun into 20% H_2SO_4 at 5°C. The apparently poor fibre properties (best results being 1.8 g den^{-1} tenacity dry, with 7.3% extension) probably arise because the fibres were syringe extruded at 75 den per fil. Asahi felt at the time that this would be the ultimate process for large scale production of regenerated cellulose fibres but in reality its use appears to be confined to the production of thickeners.

Chanzy, Peguy and co-workers at the Plant Macromolecules Research Centre (CERMAV-CNRS) in Grenoble studied the cellulose/NMMO system in depth;[40] one paper indicates that further strength increases can be obtained by adding ammonium chloride or calcium chloride to the dope.[41]

References

1 H de Leeuw, *Les Soies Artificielles*, Paris, 1932.
2 G Audemars, *Obtaining and Treating Vegetable Fibres*, British Patent 283 (Apr 17, 1855).
3 J W Swan, British Patent 5978 (Dec 31, 1883).
4 A M Chardonnet, French Patent 165 349 (Nov 17th 1884).
5 E Schweitzer, *J Prakt Chem*, 1857 **72** 109.
6 L H Despeissis, French Patent 203 741 (Feb 12th 1890).
7 H Pauly, German Patent 98 642 (1897).

8 C F Cross, E J Bevan and C Beadle, *Improvements in Dissolving Cellulose and Allied Compounds*, British Patent 8700 (May 7th 1892).
9 C F Topham, British Patent 23158 (1900).
10 C H Stearn, British Patent 1020 (Dec 23, 1898).
11 D C Coleman, *Courtaulds. An Economic and Social History: Vol 2*, Clarendon Press, Oxford, 1969.
12 C F Topham, British Patent 23158 (Dec 18th 1900).
13 C F Topham, British Patent 23157 (Dec 18th 1900).
14 C F Topham, British Patent 16605 (1903).
15 Dr Max Müller, British Patent 10094 (April 1906).
16 L P Wilson, for Samuel Courtauld & Co., British Patent 21405 (1907).
17 British Patent 406 (1911).
18 CIRFS: 1997 Information on Man-Made Fibres, International Rayon and Synthetic Fibres Committee, Avenue E Van Niouwen huyse 4, 1160 Bruxelles.
19 D C Coleman, *Courtaulds: An Economic and Social History Vol 3*, Clarendon Press, Oxford, 1969.
20 A F Turbak *et al.*, *A Critical Review of Cellulose Solvent Systems*, ACS Symposium Series 58, 1977.
21 A F Turbak, *TAPPI*, Proceedings of the 1983 International Dissolving and Speciality Pulps Conference, 1983, p 105.
22 C Graenacher and R Sallman (to Society of Chemical Industry, Basle), US Patent 2179181 (Nov 7, 1939).
23 D L Johnson (to Eastman Kodak Company), US Patent 3447956 (June 3, 1969).
24 D L Johnson (to Eastman Kodak Company), US Patent 3447939 (June 3, 1969).
25 D L Johnson (to Eastman Kodak Company), US Patent 3508941 (Apr 28, 1970).
26 N E Franks and J K Varga (to Akzona Inc), US Patent 4145532 (Mar 20, 1979).
27 N E Franks and J K Varga (to Akzona Inc), US Patent 4196282 (Apr 1, 1980).
28 C C McCorsley, Belgian Patent 868735 (1978).
29 C C McCorsley and J K Varga, US patent 4142913 (1979).
30 C C McCorsley, Belgian Patent 875323 (1979).
31 C C McCorsley, Belgian Patent 871428 (1979).
32 R N Armstrong, C C McCorsley and J K Varga, *TAPPI*, Proceedings of the 5th International Dissolving Pulps Conference, 1980.
33 J W Hill and R A Jacobson (to E I du Pont de Nemours), US Patent 2134825 (Nov 1, 1938).
34 K Ekman, O T Turenen and J I Huttunen, Finn Patent 61033 (1982).
35 V Rossi and O T Turenen, *Cellulose Carbamate*, PIRA (Paper Industry Research Association, UK) International Conference, 10–12 Nov 1987.
36 L Fu Chen (to Purdue Research Foundation), US Patent 4999149 (Mar 12, 1991).
37 K Kamide *et al.*, Characterisation of cellulose treated by the steam explosion method, Parts 1–3, *Br Polym J*, 1990 **22** 73–83, 121–128, 201–212.
38 K Kamide and T Yamashiki, *Cellulose Sources and Exploitation*, Ellis Horwood, 1990, Chapter 24.
39 K Kamide *et al.*, *J Appl Polym Sci*, 1992 **44** 691–698.
40 H Chanzy and A Peguy, *J Polym Sci Polym Phys Edn*, 1980 **18** 1137–1144.
41 H Chanzy, M Paillet and R Hagege, *Polymer*, 1990 **31** March 440–405.

2
Industrial cellulose

BRUNO LÖNNBERG

2.1 Wood fibre sources

The main natural fibre source worldwide comprises various softwood and hardwood species. The softwood forest belt of the northern hemisphere covers Russia, the Nordic countries, Canada, and parts of the northern United States, see Fig. 2.1. The main softwood species of these forests in Asia and Europe are larch (*Larix sibirica*) in eastern Siberia, Scots pine (*Pinus silvestris*) and Norway spruce (*Picea abies*). In North America some other species occur, for example white and black spruce (*Picea glauca* and *P. mariana*), balsam fir (*Abies balsamea*), and jack and lodgepole pine (*Pinus banksiana* and *P. contorta*). In the USA hemlocks (*Tsuga* sp.) and Douglas fir (*Pseudotsuga* sp.) also occur, and pines growing in the west and south also deserve mention. These 'Southern pines' comprise among others *Pinus taeda*, *P. elliottii* and *P. palustris*.[1]

The most important hardwood species of the mixed forests south of the true softwood forest belt are birch (*Betula* sp.), aspen (*Populus tremula* and *P. tremuloides*), oak (*Quercus* sp.), gum (*Nyssa* sp.), maple (*Acer* sp.) and beech (*Fagus* sp.).[1]

In addition to these naturally grown wood species, there are some cultivated sotfwood and hardwood species that should be mentioned. The Radiata pine (*Pinus radiata*) is grown in Australia and New Zealand. A number of eucalyptus species are also cultivated in many subtropical areas and play an important rôle in the pulp and paper industry. *Acia* sp. is currently cultivated, and various nonwood fibrous raw materials representing mainly the *Gramineæ* family, for example cereal straws (rice, wheat etc.), sugar cane bagasse and bamboo form a significant fibre source not yet fully utilised.

2.2 Pulping methods

Chemical and mechanical pulping of wood raw materials dominate the paper and board industry worldwide, although other fibrous materials

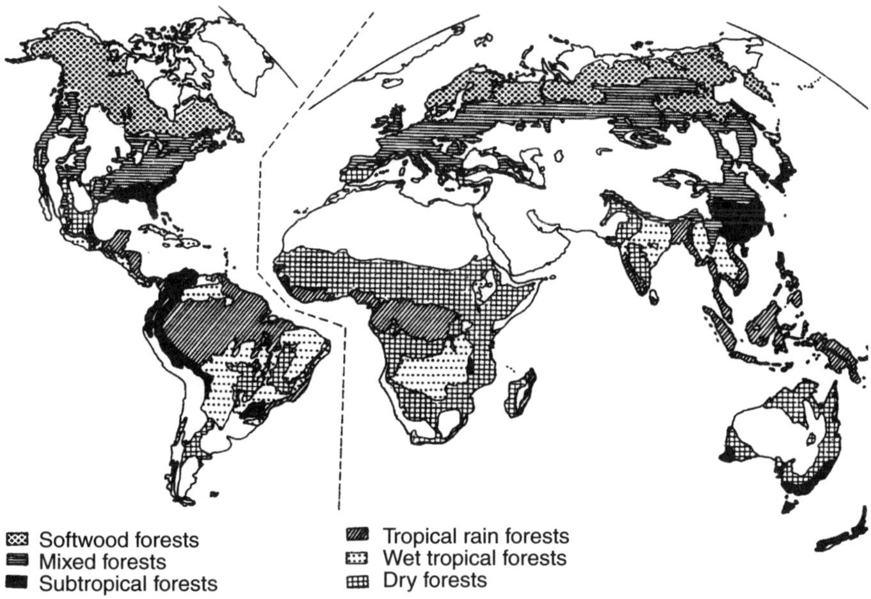

2.1 Global forest belts.[2]

called nonwood are used in certain countries. The most common chemical pulping process is the Kraft pulping process, which is an alkaline process utilising sodium hydroxide and sodium sulphide as active delignification chemicals. The kraft pulping method is able to process a variety of fibre raw materials including both softwood and hardwood raw materials.

Another family of delignification methods comprises various sulphite pulping processes actually based on sulphur dioxide with varying cation, liquor pH and cooking temperature. The acidic calcium bisulphite pulping process spread all over the world, but because its use was limited practically to spruce wood only, and because the chemical regeneration produces insoluble calcium sulphate, its use has been limited. Later, so-called soluble cations, for example magnesium, sodium and ammonia, were introduced to increase the liquor pH from 1–2 typical for the conventional calcium bisulphite process up to 5 for the magnesium bisulphite process, and to even alkaline pH levels for the sodium bisulphite/sulphite processes.

Acidic bisulphite processes and modified two- or three-stage sodium sulphite processes, for example the Rauma process, played an important rôle as dissolving pulp methods for a long period of time, and the acidic bisulphite process is still valid. The main feature of the multistage processes was to alternate between more or less acidic bisulphite/sulphite and alkaline stages. A process could start or end with the alkaline stage, which latter alternative led to a kind of alkaline extraction to diminish the residual hemicelluloses.

2.3 Dissolving pulp processes

2.3.1 Introduction

The pulping process has to dissolve the main bulk of the lignin and modify the residual lignin for successful bleaching. Bleaching then has to remove this residual and increase the pulp brightness and cleanliness. The final result should be a technical cellulose as free as possible from lignin and hemicelluloses as well as extractibles. The α-cellulose content of the final dissolving pulp may normally vary between 90 and 96%, dependent upon the pulping and bleaching processes.

2.3.2 Bisulphite processes

Sulphur dioxide-based pulping processes can be used to produce pure dissolving pulp grades for chemical cellulose. The final result is very much dependent upon liquor pH, bound and total SO_2 contents of the liquor, the time taken to reach cooking temperature and the maximum temperature. The solubility of the cation base determines first of all the pH level, which typically is 1–2 for acidic calcium bisulphite pulping processes, 4.5 for magnesium bisulphite processes and finally up to 14 for sodium sulphite processes because of the high solubility of sodium salts in water. The bound SO_2 must be about 1% and the total SO_2 must be 7–10% of the liquor to guarantee optimum sulphonation and to prevent lignin condensation reactions. Sulphonation begins at temperatures exceeding 110°C, and normally the maximum temperatures are kept at 140–160°C depending on the acidity of the liquors; acidic pH requires lower temperatures, while bisulphite pH allows some higher temperatures.[3,4]

Dissolving pulp grades should practically be lignin free and contain only certain low quantities of hemicelluloses. This can be achieved for spruce and hardwoods under the conditions mentioned above, but as sulphonation of the α-carbon in the side chain of the phenyl propane (lignin) unit forms strongly acidic sulphonates, the pH may be reduced and simultaneously increase the risk of lignin condensation. Condensation is evident particularly when the SO_2 level is low and if phenolic substances other than lignin are available. Pine heartwood contains a type of stilbene called pinosylvine (3,5-dihydroxy-*trans*-stilbene), which is thought to be partly responsible for lignin condensation, and therefore pines are normally not recommended for sulphite pulping processes.[3,5]

2.3.3 Sulphite–soda processes

These processes are worth mention, although they are not in use anymore. These normally two- or three-stage processes were conducted in specific steps for different tasks. Thus the Sivola (sulphite–soda) and the Rauma (bisulphite–sulphite–soda) processes were aimed at performing the delignification in the sulphite stage(s), because the soda stage finally had to control the level of α-cellulose content and viscosity. These levels depended on pH, temperature and time, which preferably were 5–10, 145–180°C and 1–4 h, respectively.[3] The alkaline soda stage was thought to achieve some of the extraction which normally takes place in the alkaline stage of the bleaching sequence. The Rauma process delivered α-cellulose contents up to 96% easily.

2.3.4 Prehydrolysis Kraft processes

Because the Kraft pulping process dominates worldwide as a producer of paper and board pulp grade, it has been interesting to develop the current pulping concept with prehydrolysis before the Kraft cooking stage so as to produce dissolving pulp grades. The prehydrolysis stage is conducted by steaming the wood chips or cooking them in water at 140–170°C, or alternatively by treating the wood with some dilute mineral acid at 110–120°C. The steam or water treatment cleaves acetyl and formyl groups from the wood thus forming acetic and formic acids, which leads to pH 3.5 in the wood. Hence, true acid hydrolysis starts depolymerising the wood components, reducing the wood mass by 5–20% dependent on hydrolysis time and temperature.[3,6] Up to half of the softwood hemicelluloses, mainly glucomannans, dissolve during the hydrolysis, but the lignin remains almost intact, while correspondingly significant amounts of hardwood lignin dissolve. By prolonging the hydrolysis, cellulose may be affected, resulting in a low cellulose yield and a more condensed lignin. This implies difficulties in the subsequent Kraft delignification, because more effective alkalis and higher temperatures would be required.[3] The optimum α-cellulose content (95–96%) for beech (*Fagus silvatica*) was obtained at a wood loss around 20–22% in the prehydrolysis stage.[7] Increased prehydrolysis and Kraft cooking temperatures resulted in shorter residence times in these steps for eucalyptus in a pilot plant producing 11–16 tonnes per day, and simultaneously the viscosity was reduced significantly at a comparable level of α-cellulose content.[8] Further, under comparable overall conditions pine and birch produced the same α-cellulose content (slightly below 96%), while eucalyptus provided α-cellulose slightly above 97%. The viscosities were about the same for the hardwood pulps, but significantly higher than the pine pulp viscosity.[8]

2.3.5 Bleaching technology

2.3.5.1 Introduction

Pulp bleaching aims to remove the residual lignin. This can be done successfully by application of oxidative bleaching agents to depolymerise the highly condensed residual lignin. Moreover, to achieve high final α-cellulose contents, removal of the hemicelluloses* is also required. Further, the bleaching improves pulp purity, that is it decreases the extractives content, and particularly improves the pulp brightness. But bleaching does not significantly affect the cellulose properties, for example degree of polymerisation, viscosity and so on, whose levels are mainly determined in the cooking stage.

2.3.5.2 Elemental chlorine-free bleaching of sulphite pulps

The bleaching sequence is performed in two main stages, an initial stage, where delignification mainly takes place, and a final bleaching stage, where brightness is improved. According to a Nordic Kraft mill survey there are a variety of elemental chlorine-free (ECF) and totally chlorine-free sequences mainly designed for paper pulp grades; the survey emphasises the current development trends in bleaching processes.[11] Delignification in modern bleaching sequences is conducted in the form of ECF processes by application of chlorine dioxide followed by alkaline extraction, which might be strengthened by oxygen and/or hydrogen peroxide. Both oxygen – after being reduced – and hydrogen peroxide produce hydroperoxide ions in an alkaline system, according to the following equilibrium reaction:

$$H_2O_2 + HO^- \rightleftharpoons H_2O + HOO^- \quad [2.1]$$

The hydroperoxide ion is considered to be responsible for a number of lignin depolymerisation and bleaching reactions, which initially involve attachment of hydroperoxide ions to various types of phenoxy radicals thus forming intermediate organic hydroperoxides. Subsequent reactions cause side chain elimination and opening of the aromatic ring to form carboxyl groups.[12] The first type of reaction contributes to delignification and the second type to bleaching and 'hydrophilisation'.

* Chemical pulp fibres dissolve partly when treated in a 10% NaOH solution, but leave a residue (denoted by α-cellulose or R_{10}). The dissolved part (S_{10}) contains low molecular weight cellulose (β-cellulose) and hemicelluloses (γ-cellulose). The last-mentioned can be determined as the substances dissolved in 18% NaOH (S_{18}), and hence the β-cellulose is the difference between the alkaline soluble materials ($S_{10} - S_{18}$).[9,10]

Hydrogen peroxide is subject to decomposition caused by heat, the presence of transition metal ions and radical reactions, where hydroxy and hydroperoxy radicals are involved.[12] Thus, for effective utilisation of the hydrogen peroxide, the temperature should be limited to 110°C and acid washing or chelation should be introduced to eliminate Cu^{2+}, Mn^{2+} and Fe^{2+} particularly.[13] Typical ECF bleaching sequences for full bleaching of softwood pulp, Equation [2.2] and hardwood pulp, Equation [2.3], respectively, may appear as follows:

$$D - Eop - D - Ep - D \tag{2.2}$$

$$D - Ep - D \tag{2.3}$$

where D stands for chlorine dioxide and E for alkaline extraction, while the subindices o and p indicate strengthening by oxygen and hydrogen peroxide, respectively. The short sequence [2.3] might be useful for bleaching dissolving pulps. It has been demonstrated for dissolving pulps made from spruce wood by the acidic bisulphite process that replacement of the conventional bleaching sequence CEHH by D/CEHH (C for chlorine and H for hypochlorite) – where part of the chlorine is substituted by chlorine dioxide – led to some higher α-cellulose content, but a lower filtration value, R_v.[14] Acidic Na bisulphite pulps made from softwood (type A) were bleached by CEH and CDEH, and it was revealed from the compiled results that CEH pulps had more low-molecular cellulose (given by $S_{10} - S_{18}$, which represents substances soluble in 10% and 18% NaOH, respectively), and a lower degree of polymerisation (DP) than the CDEH pulps.[9]

2.3.5.3 *Totally chlorine-free bleaching of sulphite pulps*

Totally chlorine-free (TCF) bleaching is performed without any chlorine chemicals whatsoever. Instead the delignification part of the process may be effected by ozone followed by alkaline extraction as outlined above. The later part of the sequence may also contain ozone or peracid stages in addition to peroxide stages, but since sequences evidently are developed for specific needs, they do not look the same, mainly because the chelation now plays a very important rôle in the bleaching sequences.[11] Acidic bisulphite pulp made from spruce resulted in the same α-content, but with a clearly higher DP and a lower R_v value, when the bleaching was changed from D/CEHH to a pure TCF sequence, EopPEP.[14] Acidic Na bisulphite pulp of softwood (type A) – with an α-cellulose content level around 97% – produced higher viscosity and DP, but definitely lower brightness, when the bleaching system was changed from DPZE (Z for ozone) to a pure TCF system PZE.[9] For beech, acidic bisulphite pulp of κ number 6–7, changing the bleaching process from CEHH to EopPP provided about 1% unit lower α-cellulose content and higher R_v value, while viscosity became slightly higher.[13] Acidic bisulphite pulp of birch wood maintained the α-content, brightness and viscosity roughly at the initial level, when the DEH bleaching sequence was changed to OH, where O stands for oxygen treatment (8 bar oxygen, 6% NaOH, 100°C and 1 h). The pentosans increased, but fortunately the dichloromethane extractives reduced to half of the reference level.[14]

2.3.5.4 Bleaching prehydrolysis Kraft pulps

The conventional CEH bleaching sequence provided a much lower viscosity than the chlorine-free DED sequence, when applied to prehydrolysed beech Kraft pulps (see Fig. 2.2).[16] Simultaneously it was found that the ECF bleaching sequence significantly improved the reactivity of the pulp, expressed as solubility in emulsion xanthation.

Oxygen treatment as a part of the bleaching sequence may reduce the pulp extractives content, as indicated for sulphite pulps.[15] This was also tested on birch prehydrolysis Kraft pulps by application of various sequences. DEoDH was replaced by ODEoDH, which led to a higher α-cellulose content, a halved BOD (biological oxygen demand) discharge and a more than halved extractives content, while replacement of the short DH by ODH, changed the pulp properties only moderately, but still reduced the extractives by 30%.[17]

Ozone displays four mesomeric forms, and it is a dipole and acts as an electrophilic agent. It adds to aliphatic and aromatic double bonds in both free and conjugated phenolic structures, and hence forms polyoxides which after several steps finally end up as carbonyl or carboxyl groups.[12] As a consequence of this, ozone will not only delignify and bleach due to aromatic ring opening, but also attack aliphatic structures. The ozone bleaching stage should preferably be used in the initial stages of the bleaching sequence. This is because ozone, with its high reactivity and relatively low selectivity, affects cellulose fibrils which do not have much protection because of the low residual lignin content at the end of the bulk delignification or at the end of the bleaching sequence.[18]

2.4 Environmental assessment

The acidic bisulphite process is applicable to spruce wood and a number of hardwoods, mainly beech, eucalyptus and birch. However, owing to difficulties separating spruce and pine, and because of the relatively high (neutral) extractives content of many hardwoods, there has been increasing interest in using alkaline pulping processes for production of dissolving

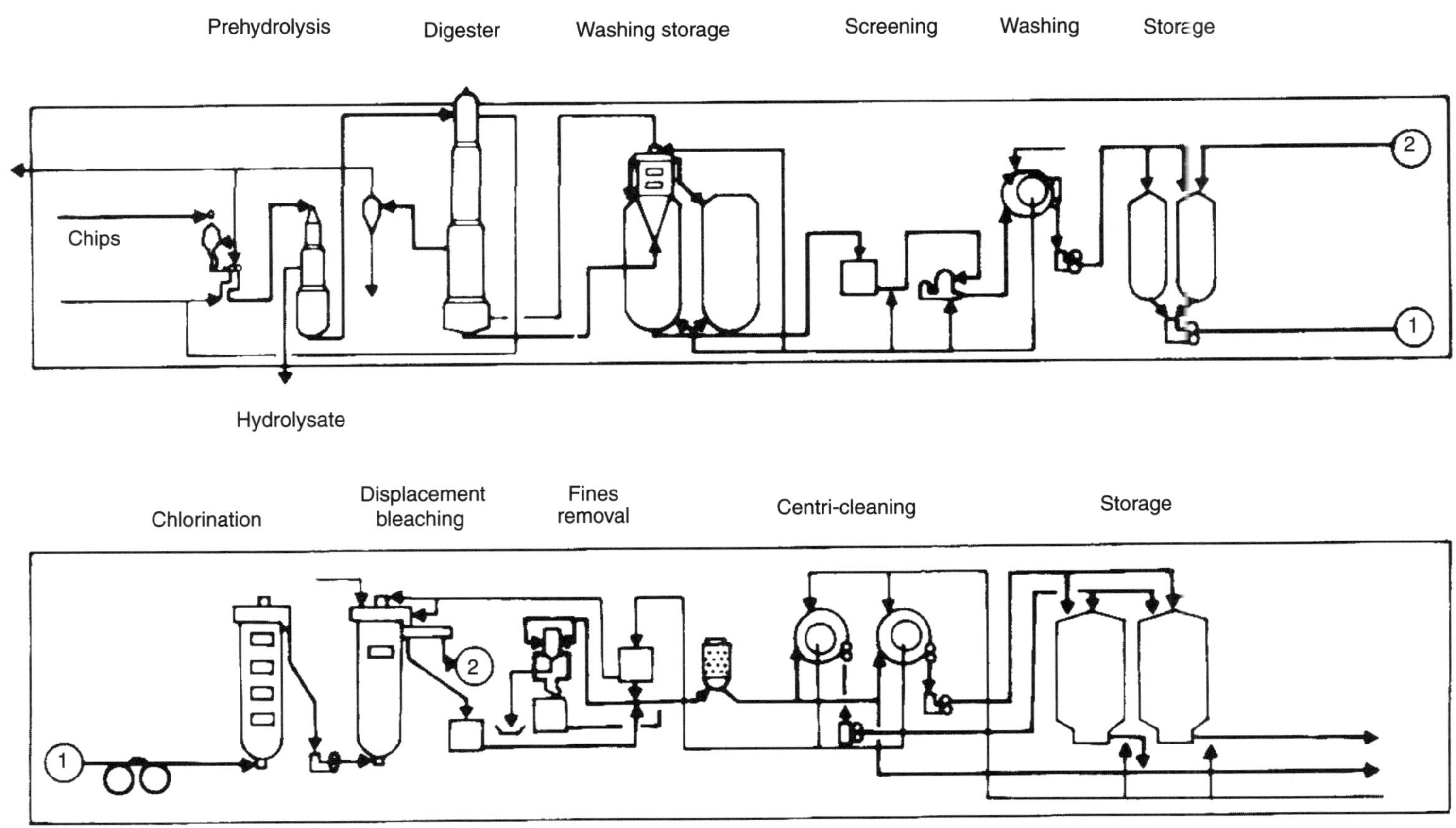

2.2 Typical prehydrolysis Kraft pulping followed by modern displacement bleaching.[4]

Table 2.1 Development of bleaching sequences applied to soft- and hardwood acidic bisulphite and hardwood prehydrolysis Kraft dissolving pulps

Wood material	Pulping method	Bleaching sequence	Year	Reference
Spruce	Acidic bisulphite	(C + D)HEDEH[a]	1983	4
		CEHD	1977	19
		ECHD[b]	1977	19
		EopPEP	1992	14
Softwood (A)	Acidic bisulphite	DPZE	1994	9
		PZE	1994	9
Beech	Acidic bisulphite	CEHH	1992	14
		EopPP	1992	14
Birch	Acidic bisulphite	OH	1976	15
Beech	Prehydrolysis Kraft	CEHED H	1968	7
		DED	1986	17
		ODEoDH	1986	17
		ODH	1986	17
Eucalyptus	Prehydrolysis Kraft	OOAZQP	1999	20

[a] High grade viscose pulp. [b] BOD discharge 60% lower than for CEHD.
Key: C = chlorine, D = chlorine dioxide, H = hypochlorite, E = alkaline extraction, P = hydrogen peroxide, Z = ozone, O = alkaline oxygen, A = acidification and Q = chelation; subindices: o = oxygen enhanced, p = hydrogen enhanced.

pulp grades. Since the mid-1980s tremendous changes have occurred in the wood pulping industry. The entire dissolving pulp process including cooking and bleaching has been developed and often employs a soluble base in acidic cooking to ensure a reasonable recovery of the cooking chemicals. This is not possible for a Ca base and thus Mg or preferably Na or NH_4 bases should be used. The prehydrolysis Kraft process employs a common recovery process which involves black liquor evaporation and incineration in the soda vessel to generate carbonate and sulphide for final regeneration into caustic and sulphide.

The bleaching sequences have been shortened in length, and they have become more or less chlorine-free, and in certain cases employ sequences containing oxygen, ozone and/or hydrogen peroxide stages. This implies that modern dissolving pulp bleaching sequences well may give an absorbable organic halide (AOX)-free discharge. Application of oxygen before the bleaching sequence as a rule reduces the discharges, particularly BOD, in direct proportion to the delignification or treatment in general. Table 2.1 indicates the dramatic changes in bleaching dissolving pulps made from soft- and hardwoods by application of acidic bisulphite and prehydrolysis Kraft methods.

2.5 Dissolving pulp production

2.5.1 Market share

Since the mid-1980s great changes have occurred in wood raw materials and the pulping processes used, as well as in dissolving pulp production. In 1985 the hardwood proportion made up less than 20%, but now is almost 60% of the total usage of dissolving pulp grades. Simultaneously prehydrolysis Kraft pulping increased and now produces about 40% of the total dissolving pulp worldwide.[18] Total production of dissolving pulp was about 4.8 million tonnes in 1978, and in 1998 about 3.5 million tonnes.[18] The main trend is towards a decrease, but after reaching minimum production in 1996 at about 2.8 million tonnes, it seems that some increase will take place.

In 1978 the entire world pulp production including all pulp grades was 127.7 million tonnes.[21] In comparison with world pulp production, the proportion of dissolving grades appeared as 3.7%. In 1998 the corresponding world pulp production was 175.5 million tonnes, and consequently the dissolving pulp proportion was only about 2%.[22] However, the situation is different if the chemical dissolving pulp is compared to the world chemical pulp production of 123.6 million tonnes in 1998, which would provide a share of 2.8% for the dissolving pulp grades.

2.5.2 Pulping economy

The development of the dissolving pulp market will, of course, be dependent on a number of factors, that is, raw material and particularly pulpwood prices, textile fibre demand and prices, production of natural fibres and particularly cotton, and finally production of synthetic fibres, for example polyesters.

The wood raw material costs are projected to be 31% of the total costs including variable, fixed and capital costs for a 500 000 tonnes per annum softwood chemical pulp mill.[23] The unbleached softwood and hardwood Kraft pulp yields are about 47% and 53%, respectively, calculated on the charged wood raw material, and as the high-grade dissolving pulp yield is well below 40%, it is understandable that the wood costs are significant for the dissolving pulp methods. The wood 'working-up' value when used in the pulping process may be computed by Equation [2.4], which gives the wood market price (P_w) in relation to the oil (energy) market price (P_o):[24]

$$\frac{P_w}{P_o} = \eta \frac{P_P}{P_o} + (1-\eta)\frac{H_B}{H_o} \qquad [2.4]$$

where η stands for the pulp yield, P_P for pulp market price, H_B for energy content of the black liquor dry substance, and H_O for the energy content of

Table 2.2 Wood value deficit, expressed by model equation [2.4] as a function of the ratio between the prices of market pulp and fuel oil, P_P/P_O, varying from 2 to 8. The deficit is due to differences in pulp yield, which for hardwood Kraft pulp was put at 53% and for hardwood prehydrolysis Kraft pulp at 35%. Hardwood density was put at 500 kg/m^3

P_P/P_O (tonnes oil/ tonnes pulp)	P_w/P_O (tonnes oil/ tonnes wood)	P_w/P_O (tonnes oil/ m^3 wood)
2	0.27	0.14
4	0.63	0.32
5	0.80	0.40
6	0.99	0.50
8	1.59	0.80

the fuel oil considered. Because the energy values are about constant, 20 and 40 GJ t^{-1} respectively, the ratio $H_B/H_O \approx 0.50$ may be considered to be a material constant. If a reasonable value is taken for the ratio P_P / P_O 5:1, and the hardwood Kraft pulp yield of 53% and the prehydrolysis kraft pulp yield of 35% are considered, the following 'oil equivalent' values would be achieved: 2.88 and 2.08 tonnes oil/tonnes wood for Kraft pulp and prehydrolysis Kraft pulp, respectively. This means that 0.8 tonnes oil/tonnes wood or 0.4 tonnes oil/m^3 wood is the value deficit due to the lower pulp yield, which must be considered in the dissolving pulp production to balance economically. This can be done by a decrease in the wood price or an increase in the dissolving pulp price, see Table 2.2. The use of dissolving pulp may thus be successful only where wood market prices are low and if the demand for pulp pushes the pulp market price up.

2.6 Future of dissolving pulp processes and pulps

As outlined above, the future of dissolving pulp processes and pulp production is not easily predictable; the long-term trend has been decreasing steadily, although there is a slight increase in dissolving pulp production at the moment. Better economy would improve the probability of the dissolving pulp methods surviving, but there is no definite way of making the pulping process cheaper. Better utilisation of the wood raw material and a cheaper pulpwood could improve the situation; the current acidic bisulphite and prehydrolysis Kraft pulps provide a dissolving pulp yield of about 35% only, which actually means that of necessity some of the cellulose material has also been removed during pulping, because the cellulose content of common wood raw materials is about 40%. It will thus be a task for the future to develop methods of producing high-quality celluloses, let's say with an α-cellulose content around 96–98%, but at a significantly higher pulp yield than 35%.

Modern dissolving pulp processes (e.g. Visbatch, which is a displacement method) seem to employ the best of pulping technology including the wood raw material, and are therefore an alternative for the future.[20] It has been demonstrated that acidic bisulphite may produce about 96% cellulose (R_{18}) grades of eucalyptus, while the prehydrolysis Kraft process correspondingly would produce about 98% at the same yield of 35%.[20] Hence, there is a need for an additional alkaline step in the bisulphite process before the bleaching sequence to upgrade the cellulose to high-quality grades for acetates.

The current development of dissolving pulp processing from softwoods to hardwoods, and from acidic bisulphite to alkaline prehydrolysis Kraft methods is well motivated, as hardwoods may produce a high-grade α-cellulose pulp which is also easily bleachable with TCF sequences. The viscose process is still common, but there are new so-called direct spinning methods (Lyocell, Tencel etc.) under development that may change the pre-requisites for the pulping processes and the dissolving pulp properties. Good regenerated cellulose properties, however, require high cellulose activity, high α-cellulose levels, narrowly distributed cellulose DP and hence a well-controlled solution viscosity.

The question is whether other possible dissolving pulp concepts, for example ASAM (methanol containing alkaline sulphite), prehydrolysis soda-AQ, IDE (ethanol containing soda), Milox and so on would provide considerable benefits. This may be clarified only by doing further research work.

Comparison of some of the alternative dissolving pulp technologies with Visbatch would be interesting, because this process is an actual process and it seems to work according to modern displacement pulping principles. The chips are neutralised after the prehydrolysis stage with white and black liquors, which is followed by a white and black liquor hot displacement delivering black liquor to the chemical recovery. This is followed by cooking and finally cold displacement with washing filtrate before the discharge. The hot black liquor coming from this cold displacement is then used in the next cook.[20]

The ASAM cooking concept is an alkaline sulphite process catalysed by anthraquinone and finally strengthened by methanol, whose chemical designation forms the short name. When started with a common prehydrolysis prior to the digestion the process provides high-viscosity dissolving pulps in high yields of about 40%. Beech, however, constitutes an exception, and it seems to require a weak sulphuric acid prehydrolysis to reach the high yield level, although simultaneously the R_{10} level unfortunately reduces below 90%. Compared to Kraft this process requires a lot more total cooking chemicals, 25% on o.d. wood, expressed as NaOH (but in the form of Na_2SO_3 and NaOH).[25]

A modern approach to pulping involves sulphur-free pulping chemicals for several reasons. The base for these processes is the caustic soda which needs to be catalysed by, for example, anthraquinone and further strengthened by simple alcohols.[26,27] The first mentioned method starts with prehydrolysis including methanol, but for quality reasons the subsequent alkaline cooking had to be stopped at κ number 30.[26] The other method called the IDE concept is an alkaline method starting with a strong impregnation followed by delignification and finally extraction, which is a special kind of ethanol washing effectively to remove all residuals.[27] The bleached IDE pulp made from birch wood was tested as such for dissolving, and it was suitable at least for sponges; the pulp had an R_{18} value of about 95% and a high viscosity.[28]

Finally, there are acidic pulping methods aimed at dissolving pulp purposes. Mention should be made of the Milox concept, which is based on formic or acetic acid and hydrogen peroxide involving use of peracids.[29,30]

References

1 S A Rydholm, *Pulping Processes*, John Wiley Interscience, New York, 1965, pp 12–23.

2 Anon, 'Wood: World trends and prospects', *Unasylva*, 1966 **20** 1–2, 40–61.

3 H Sihtola, 'Liukosellut (Dissolving pulps)', in *Puukemia (Wood Chemistry)*, 2nd edition, ed. W Jensen, Turku, The Finnish Paper Engineers' Association and The Finnish Academy of Technical Sciences, 1977, Chapter 14, pp 298–313.

4 R Kekki and H Leppäkoski, 'Liuko- ja erikoissellut (Dissolving and special pulps)', in *Puumassan Valmistus (Production of Wood Pulps)*, 2nd edition, ed. N-E Virkola, Turku, The Finnish Paper Engineers' Association and The Finnish Academy of Technical Sciences, 1983, Chapter 19, pp 1037–1052 (Book II, Part 1).

5 G Pensar, 'Puun uuteaineet (Wood extractives)', in *Puukemia (Wood Chemistry)*, 2nd edition, ed. W Jensen, Turku, The Finnish Paper Engineers' Association and The Finnish Academy of Technical Sciences, 1977, Chapter 9, pp 186–213 (Book I).

6 R P Overend and E Chornet, 'Steam and aqueous pretreatments: Are they prehydrolysis?', *75th Annual Meeting*, Montréal, Canada, Canadian Pulp and Paper Association Technical Section, 1989, Volume B, pp 327–330.

7 M V Tuominen, 'Experience gained in manufacturing viscose pulp by the prehydrolysis-kraft method', *Paper and Timber*, 1968 **50**(9) 517–527.

8 G Annergren, Å Backlund, J Richter and S Rydholm, 'Continuous prehydrolysis-kraft cooking', *Tappi*, 1965 **48**(7) 52A–6A.

9 P C Leithem, Y Huang, E M Wallace and R L Casebier, 'Reduced chlorine usage in the production of sulphite dissolving pulp', *Pulping Conference*, Tappi Press, San Diego, CA, 1994, Book 2, pp 669–713.

10 J F Hinck, R L Casebier and J K Hamilton, 'Dissolving pulp manufacture', in *Pulp and Paper Manufacture*, 3rd edition, Sulfite Science & Technology eds. O V Ingruber, M J Kocurek and A Wong), Montréal, Canada, Canadian Pulp and Paper Association, Technical Section, 1983, Volume 4, Chapter VIII, pp 213–243.

11 J Lund, J Jäkärä, B Lönnberg and J Nyman, 'Current bleaching sequences in Nordic kraft pulp mills', *Pulping Conference*, Tappi Press, Montréal, Canada, 1998, Book 2, pp 711–722.

12 J S Gratzl, 'Abbaureaktionen von Kohlenhydraten und Lignin durch chlorfreie Bleichmittel-Mechanismen sowie Möglichkeiten der Stabilisierung', *Das Papier*, 1987 **41**(3) 120–130.

13 J Colodette, S Rothenberg and C Dence, 'Factors affecting hydrogen peroxide stability in the brightening of mechanical and chemimechanical pulps', *J Pulp Pulp Paper Sci*, 1989 **15**(1) 3–10.

14 K Fischer, K Hubner, H Lindig, D Mauler and I Schmidt, 'Herstellung chlorfrei gebleichter Zellstoffe und deren Reaktivität', *Das Papier*, 1992 **46**(12) 703–709.

15 B Lönnberg, I Palenius and M Ranua, 'The use of an oxygen stage in the production of special-grade pulps', *Tappi*, 1976, **59**(11) 102–105.

16 V Masura, 'Sulfur-free preparation of highly refined dissolving pulp from beechwood', *Nordic Pulp Paper Research J*, 1986 **1**(1) 46–50.

17 H Crönert, 'Erzeugung von Vorhydrolyse-Sulfatzellstoff aus pentosanreichen Hölzern (speziell Birkenholz) – unter besonderer Berücksichtigung der komplexen Verwertung der Holzsubstanz, Teil 1', *Das Papier*, 1986 **40**(12) 619–626.

18 O Kordsachia, R Patt and H Sixta, 'Cellulosegewinnung aus verschiedenen Rohstoffen', *Das Papier*, 1999 **53**(2) 96–108.

19 J Kalish, 'A new future for the sulfite process?', *Pulp Paper Internat*, 1977 **19**(6) 53–55.

20 H Sixta and A Borgards, 'New technology for the production of high-purity dissolving pulps', *Das Papier*, 1999 **53**(4) 220–234.

21 Anon, '1977–78 World pulp, paper & board industry at a glance', *Pulp Paper Internat*, 1979 **21**(8) 22–23.

22 Anon, '1997–98 The world's pulp, paper & board industry: production and trade', *Pulp Paper Internat*, 1999 **41**(7) 12–13.

23 M Diesen, 'Economics of the pulp and paper industry', in *Papermaking Science and Technology*, ed. J Gullichsen, Jyväskylä, The Finnish Paper Engineers' Association and TAPPI, 1998, (Book I), pp 102–124.

24 B Lönnberg, 'The effect of water transport of wood on wood quality', *Seminar on Water Transport of Wood*, Finland, National Board of Vocational Education, 1983, pp 223–226.

25 R Patt, M Zimmermann, O Kordsachia and J F Hinck, 'Optimization of the prehydrolysis and pulping steps in ASAM dissolving pulp production', *Pulping Conference*, Tappi Press, San Diego, CA, 1994, Book 1, pp 141–46.

26 U Neumann, O Kordsachia and R Patt, 'Aufschluss von Kiefernholz im Natronlauge/AQ/Methanol-System', *Das Papier*, 1997 **51**(11) 573–579.

27 M Backman, B Lönnberg, K Ebeling, K Henricson and T Laxén, 'Impregnation – depolymerization – extraction pulping', *Paper and Timber*, 1994 **76**(10) 644–648.

28 S Ciovica, B Lönnberg and K Lönnqvist, 'Dissolving pulp by the IDE pulping concept', *Cellulose Chem Technol*, 1998 **32**(3–4) 279–290.

29 K Poppius-Levlin, R Mustonen, T Huovila and J Sundquist, 'MILOX pulping with acetic acid/peroxyacetic acid', *Paper and Timber*, 1991 **73**(2) 154–158.
30 J Puls, B Saake, J C Parajo, S Abad, H Sixta, H Harms, H P Fink, P Weigel, H Al Ghatta and A Glionna, 'Comparative production of dissolving pulps by Acetosolv-, Formacell- and Milox-pulping', *10th International Symposium on Wood and Pulping Chemistry*, Yokohama, Japan, 1999, Volume 2, pp 280–283.

3
The viscose process

ANDREW G WILKES

3.1 Introduction

The first patent on the viscose process was granted to Cross and Bevan in England in 1893. By 1908 the fibre spun from viscose dope had been accepted as a key component of the burgeoning textile industry (see also Chapter 1). Viscose (or rayon) still enjoys the unique position of being the most versatile of all artificial fibres. This has resulted from an ability to engineer the fibre chemically and structurally in ways that take advantage of the properties of the cellulose from which it is made.

Over the past 100 or so years the viscose process has undergone many refinements. However, the basic chemistry is still the same. Through this route short-fibre cellulose (wood pulp) is converted in a series of controlled and coordinated steps to a spinnable solution (dope) and then into longer filaments which may be precisely controlled in terms of length, denier, physical properties and cross-sectional shape.

The pulp is first steeped in an aqueous solution of sodium hydroxide (17–19%) which causes the fibres to swell and converts the cellulose to sodium cellulosate, commonly called alkali cellulose or white crumb. After steeping, the swollen mass is pressed to obtain a precise ratio of alkali to cellulose, and then can be shredded to provide adequate surface areas for uniform reaction in subsequent process steps. The alkali cellulose is aged under controlled conditions of time and temperature to depolymerise the cellulose by oxidation to the desired degree of polymerisation (DP) prior to reacting with carbon disulphide to form sodium cellulose xanthate. The xanthate, which is a yellow to orange crumb, is dissolved in dilute sodium hydroxide to yield a viscous orange-coloured solution called viscose. The solution is filtered, deaerated, and ripened to the desired coagulation point (called salt index) appropriate for spinning (see Fig. 3.1).

The rayon filaments are formed when the viscose solution is extruded through the very small holes of a spinneret into a spin bath consisting basically of sulphuric acid, sodium sulphate, zinc sulphate and water. The spin

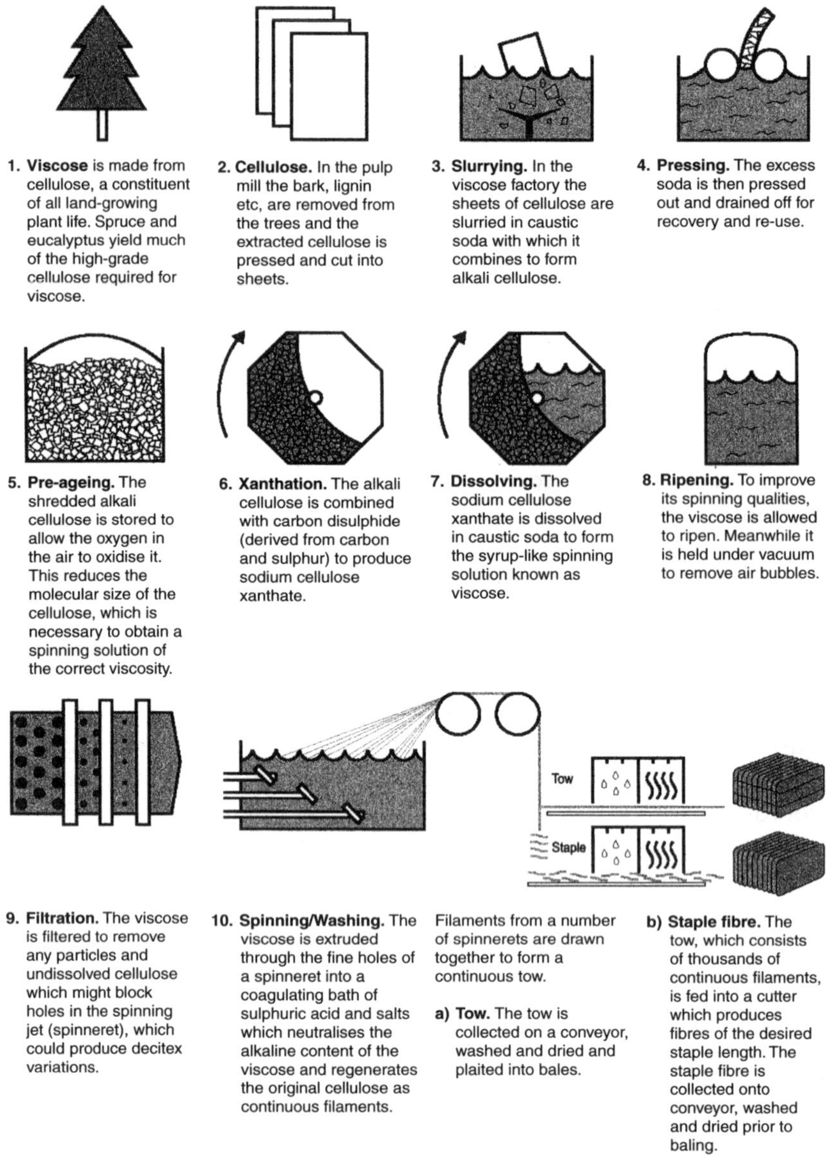

3.1 The viscose process. (Reproduced by courtesy of Acordis.)

bath often also contains a low level of surfactant. Coagulation of the filaments occurs immediately upon neutralising and acidifying the cellulose xanthate followed by controlled stretching and decomposition of the cellulose xanthate to cellulose. These latter steps are important for obtaining the desired tenacity and other properties of the fibre. Finally, the newly formed

Table 3.1 Common grades of dissolving pulp

Pulp manufacturer	Wood type	Cooking process
IP	Southern hardwood/northern softwood blends	Kraft
Tembec	Mixed softwoods	Sulphite
Riocell	Eucalyptus	Kraft
Borregaard	Spruce	Suphite
Modo	Spruce	Sulphite
Lenzing	Beech	Sulphite
Saiccor	Eucalyptus/acacia	Sulphite
Bacell	Eucalyptus	Kraft

filaments are washed free of acid, chemically treated (desulphurised), and bleached prior to final washing and applying a processing finish. This is done either in the form of continuous filament (yarns or tow) or as cut staple, prior to drying and packaging.

3.2 Viscose making

3.2.1 Pulp

A varicty of dissolving-grade wood pulps can be used as the cellulose source in viscose rayon processes, as shown in Table 3.1. A minority of viscose is also made from cotton linters.

For commercial reasons, many viscose plants have optimised their process conditions for pulp furnishes containing predominantly one or two types of pulp. This may be a combination of short and long fibre grades. Wood pulp is usually supplied as either bales of pulp sheets or flock (compressed bales of pulp fibres).

A typical pulp used for viscose manufacture is made by SAPPI Saiccor. This is predominantly a eucalyptus pulp (short fibre hardwood), with perhaps some acacia (Wattle) also present in the blend. SAPPI Saiccor (Fig. 3.2–3.4) use the acid sulphite process for manufacturing pulp, and bleach with an elemental chlorine-free (ECF) sequence, having replaced their chlorine gas delignification stage first with oxygen, and then with chlorine dioxide treatments. The replacement of chlorine bleaching has been a recent trend in the pulp industry brought about by environmental pressures (see also Chapter 2). Other market pulp manufacturers use a range of wood types, and some also employ the alkaline Kraft process. For dissolving pulp,

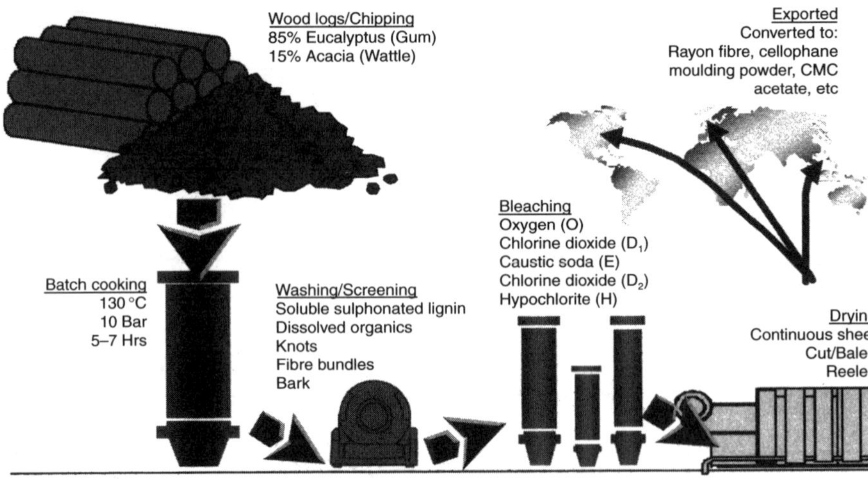

3.2 Typical sulphite pulping operation – overview.

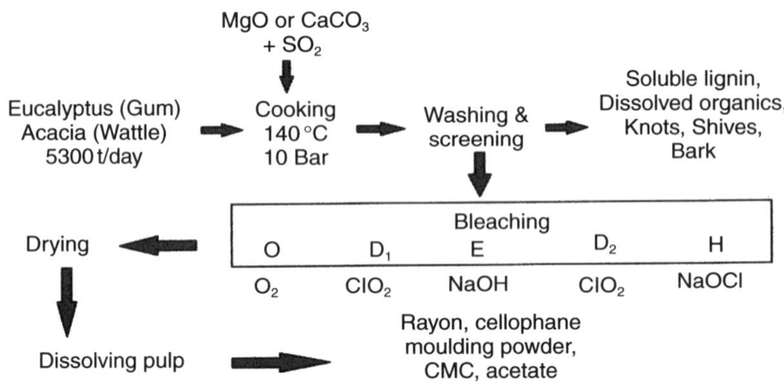

3.3 Process overview.

the Kraft process must be preceded by a prehydrolysis acid treatment stage to remove the primary wall from the fibres in the wood.

Different pulps require different steeping conditions in order to achieve optimal viscose quality. This issue becomes particularly important when market pulps are used in blends in the viscose factory. Identifying the best compromise between optimum steeping conditions and effecting an intimate blend of the pulps is critical in achieving acceptable and stable viscose

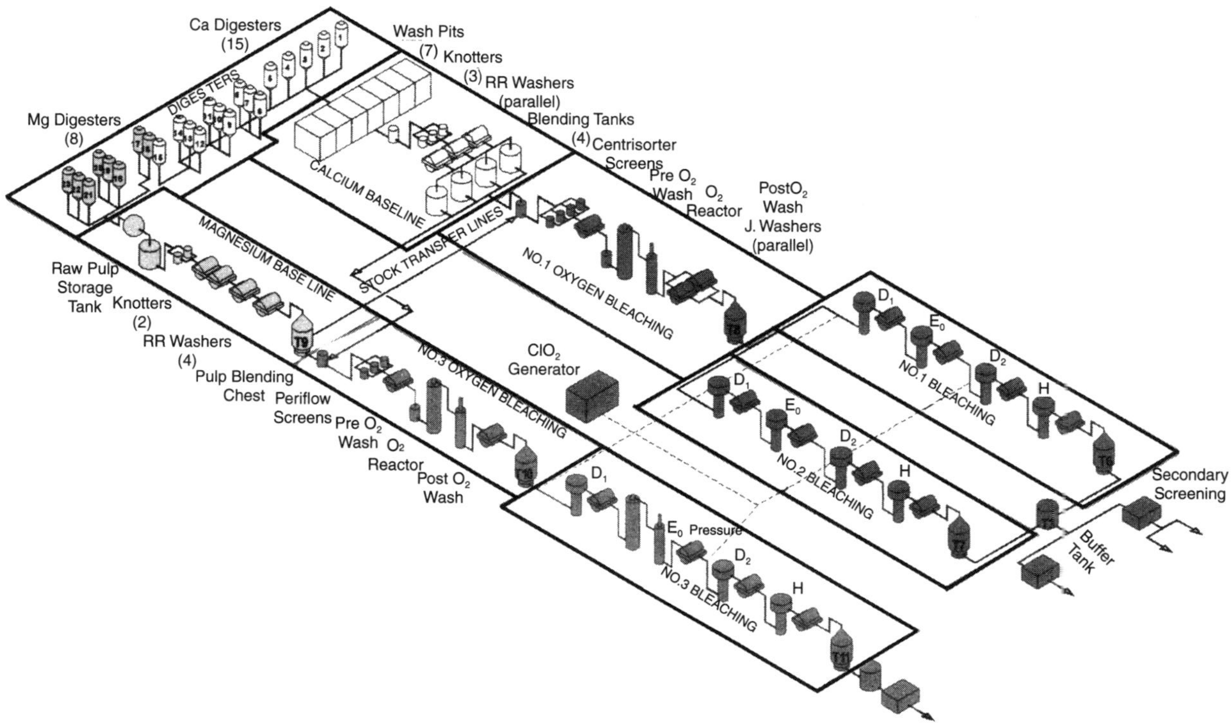

3.4 Process overview.

quality. Such differences are largely associated with the different fibre dimensions of the pulp fibres derived from different wood sources, which result in different reaction rates in the viscose process. Hardwood (short fibre) and softwood (long fibre) blends are generally not recommended for this reason, although the introduction of a small percentage of softwood is often made to improve slurry drainage.

Many dissolving pulps which are to be used in the viscose process now contain a small amount of synthetic resin (surfactant) which is added prior to drying at the pulp mill. This material aids soda penetration at steeping and improves the swelling of the pulp.

Many pulp properties have a significant effect on the viscose process and subsequent fibre properties. These include the degree of polymerisation, the oxidation state of the cellulose, the level of lignin, the soda solubility and the residual level of pulp impurities like iron and silica.

3.2.2 Steeping

Historically, sheet steeping was the norm in the viscose industry, whereby sheets of dissolving pulp or cotton linters were placed in a rack and submerged in a tank of soda (steeping lye). Now, virtually all viscose production utilises slurry steeping, where the pulp is added to a vigorously agitated tank of soda. The process can be operated either continuously or batchwise. Because of viscosity effects, the slurry is relatively dilute, the consistency being typically less than 6% cellulose.

The objective of the steeping process is to convert the cellulose to its alkoxide derivative (alkcell) in as effective a manner as possible. However, the extent of conversion is hampered by the compact nature of the cellulose sheets, and by the accessibility of the soda into the discrete cellulose fibres. For effective xanthate formation later in the process, it is important that the cellulose is converted to its alkoxide form in both its amorphous and crystalline regions. A further objective is to remove the undesirable short-chain materials present in the pulp (hemicellulose and γ-cellulose), as these materials will otherwise consume CS_2 at xanthation and potentially deteriorate fibre quality.

Hence steeping conditions are optimised to ensure good formation of the alkoxide, and dissolving out of the hemi- and short-chain celluloses. Two main parameters are used to achieve optimum steeping, soda temperature and soda concentration.

A higher soda temperature dissolves out more of the short-chain material, but reduces the swelling of the pulp. Hence there is an optimum temperature, usually around 45–55°C. Some depolymerisation also takes place during steeping, and the extent of this will increase with increasing temperature.

For soda concentration, pulp swelling increases as the concentration is reduced, to a maximum at around 11%. A high degree of swelling is desirable because it opens up the pulp structure to enable short-chain material to be leached out, and to enable the soda to penetrate effectively into the fibre structure. However, higher concentrations are necessary to ensure that the conversion to the alkoxide derivative proceeds to an acceptable extent. In practice, around 17–19% soda is typical.

Typically, the residence time through steeping is around 5–10 min. After the steeping vessel (normally referred to as a pulper), the slurry is pumped into a retention tank to aid blending and ensure complete soda penetration. A typical residence time in the retention tank would be 10–20 min. The primary function of the retention tank is to provide a buffer volume to ensure a continuous supply of slurry to the presses.

3.2.3 Double steeping

Double steeping is a technique which was developed to reduce CS_2 consumption and by-product formation. Essentially, as the name suggests, there are two steeping stages:

Stage 1: steep as normal in 17–19% soda, followed by relatively light pressing, and no shredding;

Stage 2: steep in lower concentration soda, say 11–13%, followed by pressing and shredding as normal.

The first steep swells the pulp and converts the cellulose to the alkoxide derivative to the required extent. The second steep swells the pulp further still, allowing extra hemicellulose to be removed. Additionally, the excess soda remaining after pressing has a lower NaOH content than if the steep were performed with just normal concentration steep soda. Hence following the second steep the alkcell going to the churn has lower concentrations of hemi and NaOH than normal, resulting in lower CS_2 consumption.

Despite the theoretical advantages of this process, double steeping has found only limited commercial application. Attempts to achieve reductions in CS_2 using this technique have largely proved unsuccessful to date.

3.2.4 Pressing and soda recovery

The slurry presses remove excess soda from the alkcell slurry. The excess is returned to the steep soda circuit for re-use. Essentially the slurry is pumped between two nip rollers with a narrow gap of the order of 5–15 mm. The extent to which the soda is squeezed out will be determined by the roll speed (which essentially governs the residence time of the slurry between the rollers), and by the gap between the rollers. The roll material is sintered metal, which allows the expressed soda to pass through and away for recovery. The recovered soda (press soda) is always of lower concentration than the steep soda because some of the NaOH combines with the cellulose. Hence strong soda is added to the steep soda circuit to maintain the concentration at target levels.

Other control parameters of importance at pressing are:

- slurry surge pressure, with high pressures giving high levels of pressing, but potentially excessive fibre levels in the press soda, and clumping of the cellulose fibres leading to uneven alkcell
- nip pressure, which increases as pressing becomes more difficult. High nip pressure reduces the press roll life.

Highly swelled pulp makes pressing difficult. Hence, as with the steeping stage of the process, there will be a trade-off required between achieving increased swelling to ensure good accessibility and hemi removal, and achieving acceptable press throughputs.

3.2.5 Hemi and low-molecular weight cellulose

The press soda contains a relatively high level of fines – usually fibres or pieces of fibre from the pulp. These are removed usually by Waco and plate and frame filters in series, which filter the fibres out from a sidestream off the soda system.

For normal staple fibre production, hemi levels in the steep soda circuit are allowed to reach a natural equilibrium. However, where fibre tensile properties are critical, it is often valuable and sometimes vital to reduce the hemicellulose at steeping down to 1.0–1.5%. This is usually achieved by employing a dialysis (or reverse osmosis) system, where the NaOH and hemi/low-molecular weight cellulose are separated using a membrane. This is usually performed on a sidestream from the steep soda circuit, with the flow through the dialysis plant being set to achieve the desired hemi level.

Following dialysis, the 'clean' soda is returned to the steep soda circuit. The 'dirty' or hemi-rich soda is used elsewhere, often in a less critical viscose process or for xanthate dissolving. The key here is that although the hemi that has been removed by dialysis may well be returned into the viscose at the mixing stage, it is absent during the critical xanthation stage.

3.2.6 Shredding

Following pressing, the alkcell composition is typically 30–36% cellulose and 13–17% soda. Although now present as a 'crumb', the alkcell is relatively dense. To assist the subsequent mercerising (pre-ageing) and xanthation reactions, shredding is usually performed at this stage to open up the alkcell and facilitate the penetration of oxygen and CS_2 into the alkcell. Shredding reduces the alkcell density typically from $250–450\,g\,l^{-1}$ down to $80–170\,g\,l^{-1}$.

Not all viscose operations utilise shredding but this is by far the normal approach.

3.2.7 Mercerising (pre-ageing)

As received, pulp typically has a DP of 750–850. For commercial viscose processes, this must be reduced to the point at which viscose dope of an acceptable viscosity will result, while maintaining the final fibre DP high enough to achieve acceptable fibre tensile properties.

For regular staple production, the DP of the alkcell going to xanthation needs to be around 270–350. The reduction in DP is achieved in all commercial viscose operations by oxidative depolymerisation, using time and temperature to control to the required DP. Some manufacturers employ catalysts (e.g. manganese) or enriched oxidation to speed up this process.

The alkcell is often stored in temperature and humidity controlled silos while the DP reduction takes place. Typical mercerising times range from 0.5–5 hours at temperatures of 40–60°C. A variety of devices are used to mix the alkcell including rotating horizontal drums and moving belts. In all cases, cellulose chain length is reduced by a combination of free radical and alkaline degradation. Alkaline degradation tends to occur either at the chain ends ('nibbling', which gives very little overall reduction in mean DP) or at reactive sites such as carbonyl groups on the chain backbone. Such sites arise naturally in cellulose, and also as a result of bleaching in the pulp process. Hence pulp with a high copper number (a measure of the number of carbonyl groups on cellulose) will mercerise faster than will pulp with a low copper number.

Free radical depolymerisation occurs randomly along the cellulose chain, although the dependence upon oxygen to initiate the sequence means that accessibility is an issue, with the crystalline areas potentially undergoing less degradation than the more accessible amorphous regions.

3.2.8 Irradiative depolymerisation

The reduction of pulp DP down to the levels required in alkcell at xanthation can be achieved by irradiative treatment, as opposed to the usual chemical mercerising. Electron beam and gamma irradiation both enable the required DP reductions to be achieved, and in addition these treatments have been found to 'activate' the pulp, making it more reactive in the viscose process. Practically, this means that lower levels of CS_2 and NaOH are required to achieve a homogeneous viscose than is the case if regular non-irradiated pulp is used. Additionally, the irradiated pulp appears to give different rheological properties to the viscose dope, enabling much higher cellulose-in-viscose levels to be used whilst maintaining acceptable filtration and spinning performance.

The benefits associated with irradiation appear to stem from the ability of the electron or gamma beams to penetrate all regions of the pulp, crystalline and amorphous, equally, thereby giving a completely homogeneous depolymerisation. Additionally there is disruption of the crystalline lattices. These aspects facilitate much more even derivatisation, and therefore the resulting xanthate is more readily brought into solution than would otherwise be the case.

Electron (e-beam) processing is the treatment route that has found most favour. A number of industrial-scale trials have been performed and the results achieved have been very interesting. Further development work is required still to characterise the e-beam process fully, before any manufacturer would be in a position to commit to commercial operation using e-beam treated pulp. However, this is expected to occur by the end of 2002.

3.2.9 Alkcell transportation and cooling

Alkcell can be conveyed from the presses to mercerising and ultimately to xanthation by a mixture of platform conveyors, belts and/or airvey systems. All of these, but particularly the airvey system, allow the alkcell to cool from the steeping temperature to closer to the target xanthation temperature of 28–35°C. However, it is important to ensure that the alkcell, as it enters the xanthation vessel, has a minimal temperature difference relative to the xanthation vessel, to avoid condensation. Any alkcell which comes into contact with water will revert substantially to cellulose from its alkoxide form, and thereafter be unreactive at xanthation. Unreacted cellulose following xanthation will cause potentially extreme filtration problems, and may necessitate discarding of the viscose.

Some plants employ a forced cooling system for alkcell, in which a semi-fluidised bed of alkcell is passed continuously through a temperature and humidity controlled atmosphere. Other processes rely on the alkcell transportation to cool the alkcell.

Metal detectors are often employed to screen out extraneous metallic objects from the alkcell prior to addition to the churns. Metal items in the churn can cause spark formation which could in turn lead to an explosion when the churn is vented to atmosphere upon completion of the xanthation reaction.

3.2.10 Xanthation

During xanthation, the mercerised alkcell reacts with CS_2 vapour to produce sodium cellulose xanthate. It is this derivative which is soluble in dilute caustic soda, and when so dissolved forms viscose dope. The reaction is performed under vacuum to ensure vaporisation of the CS_2 (CS_2 is a liquid at room temperature). Mass transfer rates are key to the xanthation reaction, as the reaction is heterogeneous. Commercially, three types of reaction vessel are used:

1 Wet churns, where the alkcell is slowly agitated as the CS_2 is added. Upon completion of the xanthation (seen as a vacuum 'regain' as the residual CS_2 vapour is consumed by the alkcell), the dissolving or mix soda is added directly to the churn. This is a batch process.
2 Dry churns, where the xanthate crumb is discharged into a separate dissolving vessel upon completion of the reaction. This is also a batch process.
3 Continuous belt xanthators, where the xanthate is conveyed through the reacting vessel, again under reduced pressure, as the CS_2 is added continuously. At the end of the conveyor, the formed xanthate drops into the mix soda and is pumped out to dissolving.

The time for complete xanthation depends on temperature and target CS_2 level, and typically lies between ½ and 1½ hours. Xanthation vessels are often jacketed to ensure constant temperature (typically 25–37°C).

During xanthation, CS_2 reacts principally with the C2 and C3 hydroxy substituents on the glucose rings in cellulose, as these positions are kinetically favoured over the C6 position. However, the availability of free NaOH and hemicelluloses/low molecular weight cellulose encourages a number of CS_2 – consuming side reactions to take place. Hence minimising the amount of these materials in the alkcell entering the churn is important in achieving low consumption of CS_2 and NaOH.

Sodium cellulose xanthate formation occurs as follows, Equation [3.1]:

$$\text{cellulose–O}^-\text{Na}^+ + CS_2 \rightarrow \text{cellulose–OCS}_2^-\text{Na}^+ \qquad [3.1]$$

By-product formation occurs by a much more complex series of reactions which may be summarised simply as follows, Equations [3.2] and [3.3]:

$$3CS_2 + 6NaOH \rightarrow 2Na_2CS_3 + Na_2CO_3 + 3H_2O \qquad [3.2]$$

$$Na_2CS_3 + 6NaOH \rightarrow 3Na_2S + Na_2CO_3 + 3H_2O \qquad [3.3]$$

By-product formation is encouraged at higher temperatures, but dropping the xanthation temperature to achieve better CS_2 usage efficiency also extends the xanthation time; hence a compromise is necessary between productivity and chemical usage.

3.2.11 Mixing (dissolving) of the xanthate

To form the viscose solution, xanthate must be dissolved in dilute sodium hydroxide solution of the required concentration to give the final target viscose composition in terms of percentage cellulose and soda in viscose. For a typical alkcell composition, the concentration of dissolving soda is normally 1.0–2.0% NaOH. Low temperatures are better, as the xanthate has greater solubility in NaOH at lower temperatures, and although most operations use 8–12°C, forced cooling of the dissolving soda down to 0–5°C will allow better dissolving and can enable CS_2 usage to be reduced somewhat. The xanthate is also more soluble in higher NaOH concentrations, and it is therefore doubly advantageous to press off as much soda as possible from the alkcell prior to the churns and use this to increase concentration of soda at mixing.

The mix soda is dosed into wet churns in two to three stages following the attainment of regain. In some operations the remaining churn vacuum is vented prior to addition, and for others venting occurs after addition. To minimise the risk of explosion at venting (as air mixes with the CS_2 vapour), nitrogen purging of the churns is sometimes used. Some operations employ a staged addition of mixing liquors of varying concentrations, emptying the churn between each addition to ensure a good degree of churn cleanliness for the next batch.

3.2.12 Viscose ageing

Viscose dope must be aged before spinning can take place to allow for the distribution of CS_2 evenly on the cellulose chains. Even distribution is vital if stable spinning and good fibre properties are to be achieved.

Sodium cellulose xanthate is an unstable compound. Although in the churn (or continuous belt xanthator), the xanthate substituted at C2 or C3 is kinetically favoured, this derivative is thermodynamically unstable when compared with the C6 derivative. Consequently, as time goes by the proportion of C6 derivative increases relative to the C2 and C3 substituents. The shift toward a greater proportion of C6 is due to a combination of 'trans-xanthation' (CS_2 coming off the C2 and C3 positions and re-xanthating at C6), and losses of CS_2 from the C2 and C3 positions to form more stable 'by-products' with hemi and soda. Clearly, for the trans-xanthation mechanism, CS_2 is also able to rexanthate a previously unxanthated part of the chain, and this also helps to make the cellulose better solvated.

The degree of xanthation and the evenness of xanthation are measured by determining the concentration of salt solution required to precipitate the xanthate from viscose. The salt figure test uses sodium chloride solution, the Hottenroth test uses ammonium chloride.

3.2.13 Filtration

Regardless of how well the xanthate is brought into solution, there will always be particulate material in the viscose. This type of impurity needs to be substantially removed prior to spinning to prevent blockage of the holes in the spinning jet. Historically, cloth filters were employed in filter presses to remove particulates, with typically three stages of filtration, each stage consisting of a number of plate and frame (P&F) units in parallel. Once blinded by particulates, cloths would be removed manually and washed for re-use or discarded. It is believed to be important to establish a reasonable residence time for the viscose between each stage of filtration in order to achieve maximum particulate removal efficiencies.

The later generations of P&F filters employ a backwash facility, where cloths can be regenerated *in situ* by taking the filter off-line and backflushing with caustic soda. The backwash soda would then itself be filtered to concentrate the waste particulate material into a single stream for disposal, with the 'clean' backwash soda then being re-used elsewhere in the process (e.g. for mixing, washing, scrubbing, effluent neutralisation, etc.). A range of 'cloths' can be used for P&F filters based on cellulosic and/or synthetic materials.

Most modern viscose plants now favour the use of automatic mechanical filters. These essentially consist of sintered metal screens with hole sizes in the 10–20 μm range. The filters backflush automatically on the basis of pressure build-up, with the reject material itself then being filtered, as with the P&F filters, to concentrate in a waste stream for disposal. In comparison to P&F units, this type of filter (e.g. Lenzing KKF, Brunswick) has a very low labour requirement, is compact, and is not prone to leaks. However, capital cost is higher than for P&Fs.

The two types of filter have very different performance characteristics. P&F filters will stop most particulates within the design specification limit of the filter medium until the filter 'blinds', at which point particulates will be pushed through the cloth and efficiency will start to fall. Additionally, P&F cloths are prone to releasing any deposited particulates if the system pressure or flow changes suddenly (this is referred to as 'shocking' the system). Mechanical filters can also allow certain materials to pass through, particularly fibres which in cross-section are smaller than the sinter hole size.

In addition to the 2 or 3 stages of viscose filtration in the 'cave', further filtration is sometimes effected at the spinning machines, either by central filters on each spinning line, by candle filters on each spinning rounder arm and/or by filter cloths in each jet assembly.

3.2.14 Deaeration

To ensure continuity at spinning, the viscose must be deaerated to remove any dispersed air or other gases that might otherwise cause small bubbles to form as the viscose is extruded into filament form through the jet. Traditional deaeration techniques are employed (cone, film and tank deaerators) where a vacuum is applied while the viscose is passed over a surface to maximise its surface : volume ratio. Some CS_2 and water is lost from the viscose at deaeration.

3.2.15 Additives

Spinning aids, modifiers and pigments can be added to the viscose shortly before spinning. A range of surface active materials can be used to aid spinning performance. Largely these are fatty acid or fatty amine ethoxylates. Most only need a short residence time in viscose, and are added via a barrel mixer around 10–15 min before spinning. Titanium dioxide can also be added at this point to produce dull or matt fibres. The TiO_2 is usually either dispersed in dilute soda to form a fine slurry, or dispersed in viscose to produce an 'injector viscose' which itself is then metered into the main viscose.

3.3 Spinning

3.3.1 Jets and extrusion

Most jets used to spin viscose fibres are fabricated from a platinum/gold alloy, the technologies used historically have, however, been very diverse. Two types of spinning jet are usually available, known as 'full-face' and 'cluster'. With 'full-face' jets the holes are effectively punched directly in the jet cap. With cluster jets, which are becoming the norm, assemblies of discrete smaller spinerets or thimbles are housed in a single larger plate. The gaps between the thimbles provide channels for spinbath liquor to access the forming filaments.

In cluster plates, the thimbles may be recessed to avoid damage. However, the recess has on occasion been implicated in poor spinning performance, as this region can suffer from poor acid access and flow vortex formation. Therefore some manufacturers have adopted spinning configurations where the thimbles are protruding. Typical cluster jets for spinning fine dtex fibre contain 30000–50000 holes. Each hole is typically 40–60 μm in diameter.

Each hole in the jet usually has an entry cone on its viscose feed side, typically at an angle of 20° to the hole axis. This is to allow for smooth, non-turbulent flow of the viscose through this region. The ratio hole length to diameter (*L*/*D* ratio) is normally 1.20–1.80, with the larger ratios (longer capillaries) in theory giving better spinning performance by virtue of reduced turbulence and increased shear.

3.3.2 Spinning chemistry and filament regeneration

The sequence of physical and chemical transformations taking place at spinning is extremely complex and still not fully characterised. The main reaction is clearly reformation of cellulose from sodium cellulose xanthate by the action of sulphuric acid, Equation [3.4]:

$$\begin{aligned} &2\ \text{cellulose–O}\ CS_2\ Na + H_2SO_4 \\ &\quad \rightarrow 2\ \text{cellulose–OH} + Na_2SO_4 + 2CS_2 \end{aligned} \qquad [3.4]$$

However, this follows the transient formation of the more stable zinc cellulose xanthate in certain regions, with additional more complex species also being formed if viscose modifiers are used. Zinc sulphate is soluble up to around pH 8.5, so this material penetrates into the forming filaments ahead of neutralisation (reaction 3.4 above). However, after the initial formation of zinc cellulose xanthate (followed rapidly by regeneration to cellulose) in the outer regions of the filaments, further penetration is believed to be impeded by the formation of other zinc species such as $Zn(OH)_2$ and ZnS. Hence the formation of a transient region of zinc cellulose xanthate is 'overtaken' by the penetration of acid.

The region in which zinc cellulose xanthate has formed is believed to be responsible for the skin effect which can be seen in rayon fibre cross-sections. The skin region retains dye because of a greater degree of crystal structure relative to the core. The transient zinc complex is believed to aid the formation of greater crystallinity by delaying regeneration relative to sodium cellulose xanthate (the zinc variant is more stable), and possibly also by cross-linking between neighbouring cellulose chains, allowing greater orientation to be achieved.

Because of the effect of zinc on crystallinity, which ultimately translates to improved tensile strength, the production of high strength rayon usually involves high levels of zinc in the spinbath. Most plants use between 0.5 and 1.5% zinc sulphate in the spinbath for regular staple production. Below 0.5%, a significant deterioration in fibre tensiles results which can to some extent be recovered by raising sodium sulphate concentration and reducing the acid concentration. At very low levels of zinc in the spinbath, the filament cross-section becomes very irregular, with only a small amount of skin being apparent.

Sodium sulphate in the spinbath causes the filament to shrink and precipitate by a 'salting out' mechanism, with the high concentration salt drawing out the water from the viscose filament (85% of viscose is water).

Temperature and immersion depth affect the extent of regeneration, the latter only to a small degree owing to the amount of spinbath liquor usually carried forward with the tow.

3.3.3 Hydrogen sulphide formation at spinning

The by-products formed between CS_2 and NaOH degrade to CS_2, H_2S and Na_2SO_4 at spinning. The percentage of CS_2 used at xanthation which is converted to H_2S is referred to as the A-factor, and is dependent principally on the age of the viscose. For low CS_2 viscoses, there is a requirement to spin at a low age in order to maintain the spinning viscose salt figure at an acceptably high level. Hence low CS_2 viscose gives a low level of conversion of CS_2 to H_2S by virtue of the necessarily lower age.

The by-product formation reactions shown earlier support this empirical observation, with the sodium trithiocarbonate which is formed initially reacting with further NaOH to produce sodium sulphide. At regeneration, sodium sulphide forms H_2S and Na_2SO_4. Sodium trithiocarbonate, on the other hand, releases both CS_2 and H_2S, Equations [3.5] and [3.6]:

$$Na_2CS_3 + H_2SO_4 \rightarrow CS_2 + H_2S + Na_2SO_4 \qquad [3.5]$$

$$Na_2S + H_2SO_4 \rightarrow H_2S + Na_2SO_4 \qquad [3.6]$$

Hence as more trithiocarbonate is converted to sulphide as the viscose ages, the greater will be the amount of H_2S formed at spinning.

3.3.4 Modifiers

Retarding the rate of regeneration helps to achieve improved fibre tensile properties because it allows extra orientating stretch to be applied. As explained earlier, zinc helps to achieve this by forming a more stable intermediate species. Certain surface active materials, particularly amines and polyglycols can further retard regeneration. A number of mechanisms have been proposed to explain how these materials function.

For primary and secondary amines formation of a more stable intermediate complex with free CS_2, zinc, and cellulose xanthate is thought to be the main effect. The complex may additionally act as a semi-permeable membrane, retarding penetration of acid into the filament, or may allow the zinc to penetrate further into the filament by being soluble at higher pHs. For tertiary amines, which include most commercial modifiers, the materials possibly undergo xanthation, and then act in the same way as the primary and secondary amines in slowing regeneration.

On the other hand polyglycol materials can become protonated upon contact with the spinbath liquor, thereby forming a positively charged barrier to impede further acid penetration into the filament. Additionally, they can complex with zinc to give a more stable zinc cellulose xanthate and/or a less pH-sensitive zinc species.

Modifiers have also found application in the production of certain speciality modified cross-section fibres where their use is essential if a symmetrical and resilient fibre cross-section with clearly defined limbs is to be achieved.

3.3.5 Stretching

To achieve acceptable tensile properties particularly for textile end-uses, the fibre must be stretched during or very soon after extrusion. Cellulose is not thermoplastic like polyester, and therefore cannot be heat stretched at a later stage in the process. However, at the very early stages in regeneration the structure is not fixed within the filaments, and therefore the cellulose chains which will ultimately constitute the fibre can be aligned by stretching.

The stretch may be applied in a number of ways. The filaments may be run over godet rollers at a slower speed than the final traction units. Some stretch may also be applied in the spinbath where a significant speed differential exists between extrusion velocity and take-up speed at the godet. More commonly two stages of stretch are applied: pin stretching immediately after the spinbath, followed immediately by roll stretching (akin to godet stretching) of the whole tow. Although pin stretching has limited flexibility in terms of making changes to the degree of stretch applied relative to godets, the process is considerably simpler and enables the majority of stretch to be applied as early as possible.

Some plants employ hot stretch systems where the secondary stretch occurs in hot dilute sulphuric acid. This enables the final fixing of the fibre structure to take place while the fibre is held under tension in its state of maximum orientation.

3.3.6 Effects of process variables on fibre properties

This is a complex area where only a cursory treatment is possible.

3.3.6.1 Viscose dope viscosity

Higher dope viscosities ('ballfalls') generally result in more stable extrusion, and consequently lower trash figures. Low ballfalls, particularly less than 30 ballfall seconds (as measured at 18°C), can give rise to dead viscose (DV) and stuck filament (SF) trash* problems.

3.3.6.2 Salt figure

High salt figure viscose requires more time for complete regeneration than is the case with lower salt figure. Consequently, this can enable increased stretch to be applied, with correspondingly higher tenacity and extension. However, stuck filament trash can be formed as a result of neighbouring poorly regenerated filaments sticking together just after extrusion.

3.3.6.3 Spinbath acid

Reduced acid levels allow more stretch to be applied and greater tenacity/extension to be realised. However, high DV and SF trash can result from low acid spinning. Low acid will result in greater crimp formation and potentially higher fibre cohesion.

3.3.6.4 Spinbath zinc

Reduced levels of zinc will adversely affect tenacity, significantly so below 0.5%.

3.3.6.5 Sodium sulphate

High levels of sulphate should in principle help tenacity, although with normal levels of zinc in the spinbath and at normal acid levels the effect is likely to be small.

3.3.6.6 Cutting and washing

Many companies now employ some form of tow washing prior to cutting. Most systems enable a relatively high degree of washing to be achieved using a relatively small volume of wash liquor, and this means that the recovery of chemicals from this wash liquor becomes more economically viable.

Maurer cutters are the industry norm. These provide a relatively clean cut end, but with bigger tows they can result in poor fibre openness, with

* Dead viscose and stuck filaments are respectively lumps and sticks of hard cellulose that will disrupt web and yarn making processes. Together they are known as trash.

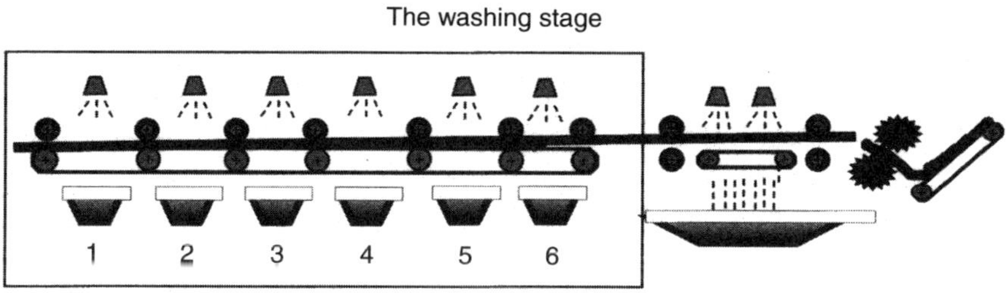

1. Acid water wash
2. Na_2S/NaOH wash, desulphurisation
3. Water wash
4. Bleaching (Peroxide or NaClO)
5. Acid water wash
6. Water wash

3.5 The washing stage.

high levels of compressed fibre 'chips' being carried down to the wash conveyor.

Washing is an essential component of the viscose fibre production process (see Fig. 3.5). After regeneration the fibre is contaminated principally with sulphuric acid, zinc sulphate, sodium sulphate, hydrogen sulphide and carbon sulphide as well as various by products from the spinning reaction including sulphur and polysulphides. To remove these contaminants the fibre must be thoroughly washed. Washing of viscose fibre requires an interchange of liquor and is not determined by residence time. Washing is usually accomplished by having a multistage countercurrent wash machine. This may be enclosed and of a moving rail, flat bed or tank immersion type. Fibre is usually washed as staple but may be kept as tow for all or part of the washing cycle. The flowrate of wash liquor has to be about 10 times the fibre rate to ensure good displacement. The presentation of the fibre blanket has to be good enough to ensure effective and consistent penetration of liquor into the fibre.

The first washing stage may actually be the stretching zone following spinning if this stretching is carried out in hot water or hot dilute acid. Although the primary purpose of this step is to align the molecules, some washing and further regeneration occurs. The next stage is usually termed the hot water wash. It is carried out at around 90°C and its purpose is to complete the regeneration reaction and to vaporise the H_2S and CS_2. In addition it has the secondary objective of washing away all the soluble contaminants, by countercurrent washing. This can be efficiently achieved on modern lines, however, some of the sulphur contaminants are insoluble and remain in the fibre. The effluent from this section is sent for heat recovery and in most cases for zinc recovery.

The next stage is the desulphurisation stage, which is carried out in a recirculating liquor stream of NaOH/NaSH at around 60°C and pH 11–12. The intention here is to dissolve any sulphur or residual polysulphides, so that they can be washed in the next stage. Also any acid is neutralised and any zinc is converted to a less soluble form. The fibre at this stage will swell and washing becomes harder.

Following this the sulphide wash is performed. This is a simple hot water countercurrent washing stage performed at around 70°C which will remove the alkaline soluble contaminants. The highly alkaline effluent from this section is usually sent for heat recovery and then to effluent treatment.

After the sulphide wash most fibres are bleached using either hydrogen peroxide or ozone for 'totally chlorine-free' products or sodium hypochlorite for 'elementally chlorine-free' products. Bleaching usually consists of a recirculating bath containing the aqueous bleaching agent. However, bleach can also be applied with the finish. Bleaching gives the fibre a bright appearance and is also an important purifying stage. After bleaching, residual bleach is removed usually by a single-stage post-bleach section with clean ambient-temperature water. The effluent from this stage may be used to feed a single prebleach section.

After post-bleach washing, the fibre is exposed to a final fresh water wash at around 50°C. This should remove all the contaminants to leave a pure product. Half way through this section a small amount of acid may be added to the wash liquor. This will correct the fibre pH and convert any residual zinc to a soluble form. At this stage the fibre is ready for finishing with a processing lubricant.

3.3.6.7 Finishing

In almost all commercial cases viscose fibres are finished with a processing lubricant prior to drying and baling. The choice of the lubricant depends on the end-use requirements for the fibre but in fact finishing is also necessary for subsequent smooth processing through the baling stage.

Lubrication during down stream processing must be provided because the virgin surfaces of the fibres give such high frictional forces that moving contact with another surface will abrade the fibre and ultimately break the filaments. An antistat may also be added to control static generation during processing which can make the fibres difficult to handle. Modern processing equipment is putting increasing demands on the performance of fibre lubricants and these chemicals are now engineered to work effectively at very high speeds and temperatures. Finishes must also be capable of being handled easily and applied consistently, usually in aqueous systems. Lubricant chemicals must also have very high levels of safety and environmental approval. This is particularly true for sensitive medical and hygiene

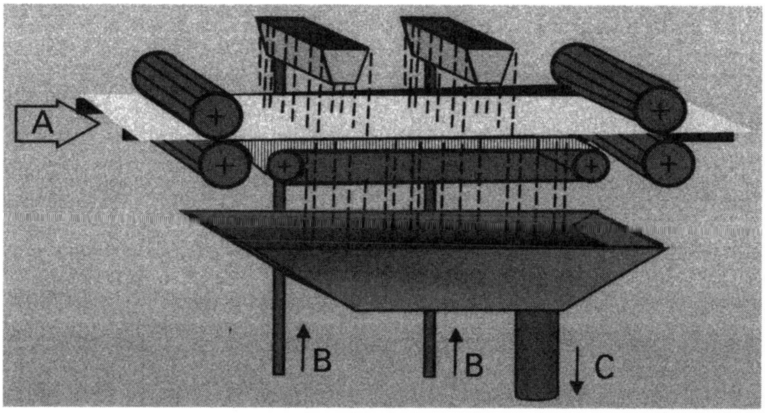

3.6 The finishing stage. (A) Fibre blanket from wash machine; (B) Soft finish inlet; (C) Soft finish regulation to make-up.

applications where a wide range of regulatory criteria must be met. In the case of viscose, lubricant materials are usually chosen to have the minimum impact on the hydrophilicity, softness and lustre of the fibre. However, hydrophobic treatments for viscose are available.

The most commonly used lubricants on viscose fibres are mixtures of fatty acids, salts of fatty acids and ethoxylated fatty acids and ethers. The antistatic agents of choice are either quaternary salts of fatty acids or phosphates. These finish mixtures once selected may be applied to the fibre by a variety of methods including full or part bath immersion, from a 'dripbar' or as a fine spray and via a kiss roller or padding device. Full bath immersion usually leads to the most consistent application but this also depends on the method of drying, the fibre moisture, the contact time between bath and finish and the length of storage time in the bale (Fig. 3.6).

In terms of finish level analysis on viscose fibre, wet chemical test methods are being replaced by advanced spectroscopic methods including IR, near-IR and NMR. These techniques are beginning to be used on-line with good success.

Figure 3.7 shows viscose SEMs.

3.4 Chemical recovery and environmental control

3.4.1 Viscose process chemicals

As outlined earlier, the soda pressed out of the alkcell slurry is returned to the main steep soda circuit. There are off-takes from this circuit for other

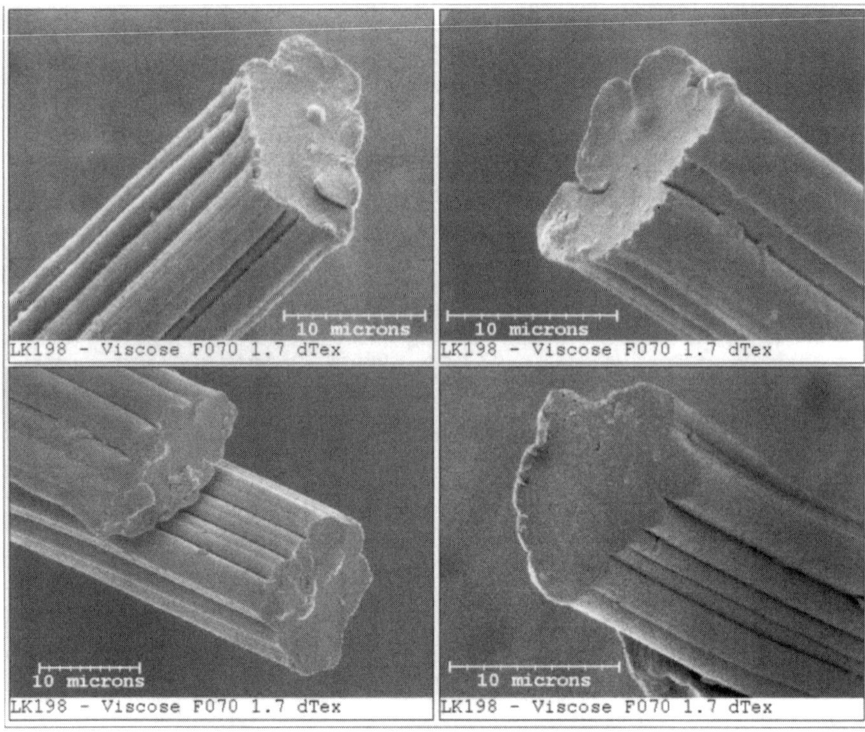

3.7 Viscose fibre scanning electron microscope (SEM) micrographs.

process areas which require soda (filter backwash, xanthate dissolving, H_2S scrubbing, fibre washing, effluent neutralisation being examples), but otherwise the system is operated largely as a closed cycle. The exception is the soda in viscose, which at spinning is neutralised to Na_2SO_4 and must be removed from the spinbath system in order to maintain an acceptable equilibrium level (see later).

Viscose waste arising at the filters is concentrated by further filtration, and then either fed to the waste water treatment plant (WWTP) or incinerated. Alkcell waste is either landfilled, incinerated or fed to the WWTP. In emergency situations, volumes of viscose or steeping soda will be discharged to drain, to be dealt with by the WWTP.

3.4.2 Degassing

Following spinning, the spinbath liquor contains appreciable amounts of dissolved CS_2 and H_2S. These components need to be removed by degassing prior to evaporation, not just to facilitate their recovery, but also because the presence of appreciable quantities of CS_2 and H_2S will foul the evaporators and present safety problems. Some operations use vacuum degassers, where others rely on the high level of liquor turbulence in the spent tank system plus air sparges to effect removal of CS_2 and H_2S. Some plants have no degassers on the main liquor system, but do degas the evaporator feed in a packed tower with air.

3.4.3 Evaporation

A side-stream of spent spinbath acid is taken for evaporation, the volume flow being set to achieve the desired level of water removal and (following crystallisation) the required Na_2SO_4 concentration. Evaporation is effected by passing the recovered spinbath liquor through bundles of tubes which are heated on the outside by steam. Multistage evaporators are used to maximise heat efficiency, whereby the hot vapours from later stages will be used to heat the early stages or to preheat the incoming process liquor.

3.4.4 Crystallisation and anhydrous sodium sulphate production

As with evaporation, crystallisation is effectively a heat transfer process where the hot saturated spinbath liquor is cooled in heat exchange with chilled water. Many crystallisers are available including plastic films, where the chilled water flows down one side of a plastic sheet, and the acid down the other. The precipitated crystals of hydrated Na_2SO_4 (Glauber salt) may then be removed by rotary vacuum filters. Vacuum crystallisers are also available, with separation of the Glauber salt usually done by centrifuge.

Historically, Glauber salt was disposed of via the process effluent. However, most viscose plants now heat the Glauber salt to convert it into saleable anhydrous sodium sulphate (ASS) for use, for example, in detergent manufacture.

3.4.5 Spinbath clarification

Various particulate materials accumulate in the spinbath circuit over time. The extruded viscose contains particulate material, often dirt from the pulp or other process chemicals, but also potentially resins from the pulp, silica, and so on. At spinning, a range of insoluble by-products are formed, particularly sulphur, polysulphides and zinc sulphide, and these materials can form deposits on process pipework, and more critically on the spinning jets.

To avoid this type of deposition, surface-active additives are often added to the liquor at parts per million (ppm) levels, and these materials effectively disperse the particulates in the spinbath. Although the dispersion of such materials in the spinbath does not cause any problems with spinning or fibre quality, particulates can cause problems at evaporation (tube fouling and loss of efficiency) and in ASS production (product contamination). Hence there is a need to remove particulates.

Removal can be effected by filtration, sand filters being the norm, or by flocculation followed by froth flotation for example. As with evaporation, clarification/filtration is performed on a degassed sidestream from the main spinbath circuit.

3.4.6 Recovery of acid and zinc

The majority of acid and zinc employed during spinning remains in the spinbath circuit. However, a significant quantity is carried forward by the advancing tow to later process stages. For both environmental and cost reasons it is important to recover as much as possible from these later process stages. However, the later stages of the washing process result in more and more dilute waste streams, and these become increasingly costly to evaporate in order to re-use the acid/zinc concentrate in the spinbath circuit. Consequently, recovery of the chemicals from the wash liquors is usually only economically feasible for the initial wash stages.

3.4.7 Effluent treatment

As outlined in the previous section, there are some effluent streams at most factories for which chemical recovery is impractical: this is particularly true of the later wash stages. Where evaporation capacity is limited, there will be a corresponding limit on the amount of acid that can be recovered for re-use; beyond this limit, the process liquors must be discharged as effluent. Additionally, there are sometimes circumstances where a bulk discharge of material must be made in an emergency. For example, a 'bad batch' of viscose which is likely to cause the immediate blockage of filtration and loss of spinning, will need to be discharged to effluent.

The viscose plants employ a range of effluent treatment facilities, ranging from just buffering and pH correction through to chemical treatment and biological aerobic digestion. Some viscose manufacturers use a biological treatment plant to reduce the effluent COD (chemical oxygen demand) to acceptable levels. The waste sludge arising from the biological processes is usually separated by filtration, dried, then either landfilled or incinerated.

3.4.8 CS_2 and H_2S recovery

Increasingly stringent environmental legislation has forced the viscose industry to employ more and more effective systems for recovering CS_2 and H_2S. However, certainly for CS_2 there is clearly a cost benefit to the process in re-using this expensive raw material.

As with evaporating spinning liquors for acid recovery, the key to achieving an efficient level of recovery is to establish as high a concentration of CS_2 in the waste air stream as possible. CS_2 is a highly flammable material, however, with a wide explosive range (10–50%), and hence any streams falling in this range must be made inert by replacing air with steam or nitrogen as the ventilation gas.

Condensation is the most widely used means of recovering CS_2 from the more concentrated process exhaust gases. Adsorption is commonly employed for the lower concentration streams, but absorption is also used. Condensation is insensitive to the presence of H_2S, but H_2S must be removed prior to adsorption to prevent fouling of the activated carbon. H_2S is commonly scrubbed using either NaOH (to give NaSH and/or Na_2S), or by a redox-based catalytic oxidation process such as the Ferrox process (based on ferric hydroxide scrubbers).

For some plants, it is attractive to take the combined foul air stream and catalytically convert the CS_2 and H_2S to sulphuric acid. The acid thus produced can then be re-used in spinbath liquor make-up.

Biological scrubbing of CS_2 is a new technique which has found favour in some quarters. However, this approach generates large volumes of dilute sulphuric acid which, given existing limits on evaporative capacity for spinbath liquor recovery, could only be neutralised and disposed of as effluent.

4
Lyocell: the production process and market development

PATRICK WHITE MBE

4.1 Overview

Lyocell is the first in a new generation of cellulosic fibres. The development of lyocell was driven by the desire for a cellulosic fibre which exhibited an improved cost/performance profile compared to viscose rayon. The other main driving force was the continuing demands for industrial processes to become more environmentally responsible and utilise renewable resources as their raw materials. The resultant lyocell fibre meets both demands.

Lyocell was originally conceived as a textile fibre. The first commercial samples were produced in 1984 and fibre production has been increasing rapidly ever since. Fabrics made from lyocell can be engineered to produce a wide range of drapes (how the fabric hangs), handles (how the fabric feels) and unique aesthetic effects. It is very versatile and can be fabricated into a wide range of different fabric weights from women's lightweight blouse fabric through to men's suiting.

Other end-uses, such as nonwoven fabrics and papers, are being developed. These non-textile end-uses will become progressively more important as the special properties of lyocell fibres enables products with enhanced performance characteristics to be developed.

Lyocell is a 100% cellulosic fibre derived from wood-pulp produced from sustainable managed forests. The wood-pulp is dissolved in a solution of hot *N*-methyl morpholine oxide (abbreviated to NMMO or amine oxide in this text). The solution is then extruded (spun) into fibres and the solvent extracted as the fibres pass through a washing process. The manufacturing process is designed to recover >99% of the solvent, helping minimise the effluent. The solvent itself is non-toxic and all the effluent produced is non-hazardous.

The direct dissolution of the cellulose in an organic solvent without the formation of an intermediate compound differentiates the new generation of cellulosic fibres, including lyocell, from other cellulosic fibres such as viscose. This has led to the new generic name 'lyocell' being accepted for labelling purposes.

Lyocell has all the benefits of being a cellulosic fibre, in that it is fully biodegradable, it is absorbent and the handle can be changed significantly by the use of enzymes or chemical finishing techniques. It has a relatively high strength in both the wet and dry state which allows for the production of finer yarns and lighter fabrics. The high strength also facilitates its use in various mechanical and chemical finishing treatments both under conventional and extreme conditions. The physical characteristics of lyocell also result in its excellent blending characteristics with fibres such as linen, cashmere, silk and wool.

In common with other highly oriented cellulosic fibres, such as cotton, cuprammonium and polynosic rayon, lyocell fibrillates but its ease of fibrillation is greater. Fibrillation becomes apparent when the fibre is abraded in the wet state and surface fibrils (small fibre-like structures) peel away from the main body of the fibre but remain attached. The fibrillation behaviour of the fibres can be exploited by using a variety of different mechanical, chemical and enzyme treatments to produce a vast range of fabric aesthetics. The control of the fibre's fibrillation behaviour, both to increase and decrease fibrillation, is a major area of continuing research.

4.2 Amine oxide technology – timeline

1939	Patent appears describing the dissolution of cellulose in amine oxide.
1966–1968	D L Johnson of Eastman Kodak Inc. publishes a series of papers discussing a range of compounds, including cellulose, which dissolve in amine oxide.
1969–1979	American Enka/Akzona Inc. work on spinning fibre from a solution of cellulose in amine oxide but decide not to scale up production.
1979	Courtaulds start research on the new cellulosic fibre, which is to become Tencel®.
1983	Pilot plant tow/staple line built in Coventry, UK.
1984	First commercial staple samples produced.
1988	Small commercial plant set up at Grimsby, UK.
1989	BISFA (Bureau International pour la Standardisation de la Rayonne et des Fibres Synthétiques) agrees to new generic name – lyocell.
1992	Full-scale production plant set up at Mobile, USA.
1996	Second stage expansion takes place in Mobile.

1997 Lenzing starts a production plant in Austria.
1998 Courtaulds* starts a new plant in Grimsby, UK.

Lyocell was first made in the laboratory and then on the pilot plant at Coventry during the early 1980s. In 1988 a semi-commercial plant (capacity 30 tonnes/week) was started up at Grimsby, this plant is known as S25 because originally there were 25 spinning ends. Courtaulds invested around £6m in this plant primarily to develop the lyocell process technology. Although the factory is capable of operating at a profit it was never intended that it would recoup the capital investment. The operation of S25 enabled the manufacturing process to be proven and developed and provided enough fibre to initiate full-scale market development. S25 now concentrates on process and product development together with some more speciality lyocell products.

In May/June 1992 the first full-scale commercial factory (SL1 – Spinning Line 1) was commissioned in Mobile, Alabama. Courtaulds invested some £67m in this venture. The process and market scale-up represented a not inconsiderable risk to Courtaulds. However, the success of producing and selling fibre from SL1 convinced the board to invest heavily in the future of lyocell.

Acordis's principal competitor in making lyocell is Lenzing of Austria who commenced production in 1997.

4.2.1 Expansion of production capacity

Courtaulds initial target was to have 150 000 tonnes/year production capacity on stream by the end of 1999. If we consider that SL1 can produce around 16 000 tonnes/year and this took nearly 3 years to design, construct and start up, the magnitude of the expansion plan becomes apparent.

The second plant, SL2, started up alongside SL1 at Mobile in the summer of 1995 with a capacity of around 20 000 tonnes/year, expandable to 30 000 tonnes/year by the construction of a third solution, spinning and fibre line. Approximately £90m was invested in this project.

The third plant was planned for Europe (Grimsby), for start up at the end of 1995 or early 1996. This was to be of a similar design to SL2 and with a capacity of around 20 000 tonnes/year also expandable to 30 000 tonnes/year by the construction of a third solution, spinning and fibre line. A fourth plant was initially planned for somewhere in the Far East.

*Courtaulds was taken over by Akzo-Nobel and the combined fibre operations renamed 'Acordis' during the year.

[In the event, the 1997 collapse of Far Eastern markets, and the reduction in demand for denim (see later) resulted in the postponement of the Far East plans and the construction of a bigger Grimsby plant which started in 1998. Ed]

4.3 Process description

This section provides a description of the process steps required for making lyocell. A block diagram of the process is shown in Fig. 4.1.

The principles of the process are simple. The pulp is first wetted out with dilute aqueous amine oxide to penetrate the pulp fibres fully. The subsequent removal of excess water is a very effective way of making a homogenous solution with a minimum of undissolved pulp particles and air bubbles. The solution is highly viscous at its operating temperature (90–120°C) and must be processed in similar high pressure equipment to that used in melt polymer systems. The fibres are formed by spinning into an air gap and then coagulating in a water/amine oxide bath. They are then washed and dried. The wash liquors are recovered, purified, concentrated then recycled.

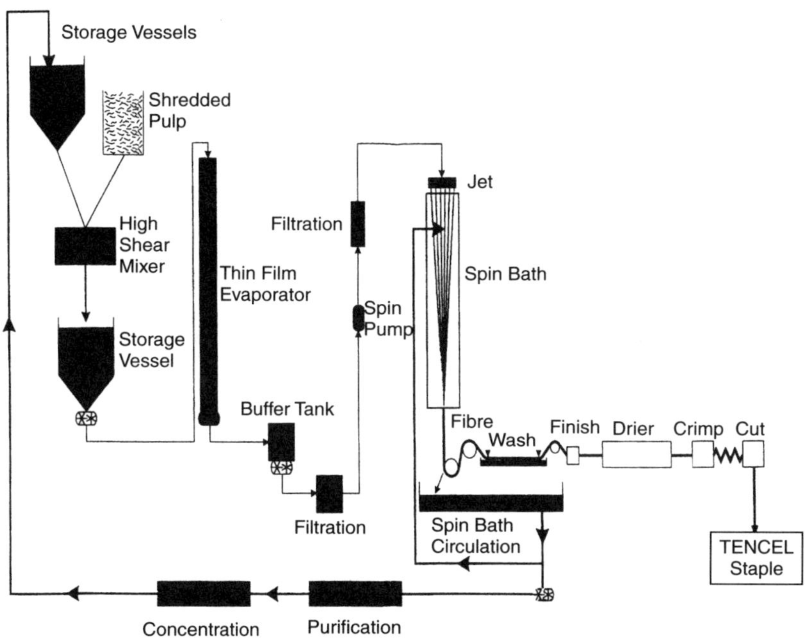

4.1 The Acordis Tencel® process. (Reproduced by courtesy of Acordis.) NMMO solution recovered from wash and spin-bath liquids is purified and concentrated and returned to the upper storage vessel for re-use in mixing. Unlike both viscose staple and the Lenzing Lyocell process, the fibre is washed, dried and crimped in tow form prior to cutting.

The process description below applies to the two commercial scale operations of Acordis and Lenzing. Variations in detail have been cited in patent applications and the literature but these are at a much smaller scale of operation.

4.3.1 Pulp and premix

Pulp is the principle raw material of the lyocell process in terms of cost and volume. In order to achieve the high quality properties of lyocell fibre, the pulp has to be of a good quality. Typically the degree of polymerisation (DP) of the pulp is in the range 400–1000 units – Tencel® fibres have a DP of 500 to 550.

The pulp is pulled from the reels into a shredder which cuts the pulp into small pieces for mixing with the amine oxide solvent. The amount of pulp fed to the mixer has to be measured accurately so that the cellulose content in solution is closely controlled.

The cut pulp is conveyed to vessels where it is mixed with a 76–78% amine oxide solution in water. A small quantity of a degradation inhibitor is also added to the mixer, other additives such as titanium dioxide (for producing matt fibre) can also be added. The mixing is achieved at 70–90°C in a ploughshare mixer that contains a number of high speed refiners to break the pulp down and aid solvent wetting. The resultant slurry consists of swollen pulp fibres and has the consistency of dough. This premix is dropped into an agitated storage hopper from which it is accurately metered to the next stage of the process.

4.3.2 Solution making

Premix is heated under vacuum to remove sufficient water to give a clear dark amber-coloured viscous solution of the cellulose. Typically the solutions contain 10–18% cellulose.

The evaporation of water from premix to make solution is achieved in a wiped thin film evaporator. This is a long vertical cylindrical vessel with steam heating in jackets around the vessel. A shaft down the centre of the vessel with blades attached to its circumference is rotated to smear the material around the heated surface to promote the evaporation process and to transport the solution down the vessel. The position, number, shape, size and type of these blades are chosen to maximise the output of the solution evaporator. The evaporator vessel is operated under vacuum to reduce the temperature (c. 90–120°C) at which the water evaporates, this is important because the amine oxide solvent in solution can undergo an exothermic degradation process if it is overheated. The specific evaporator used is a Filmtruder, shown in Fig. 4.2.

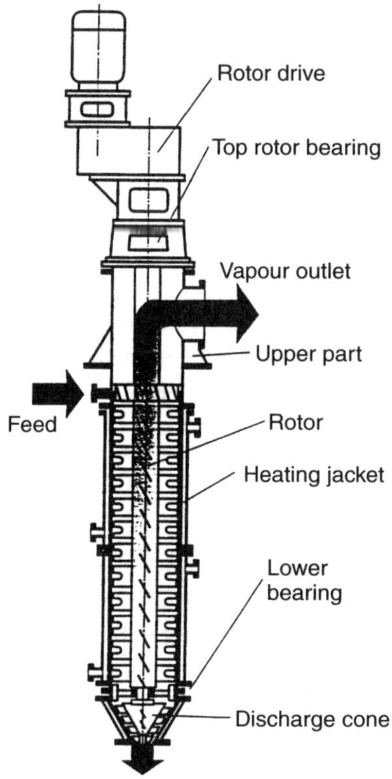

4.2 Thin-wall evaporator (Buss Filmtruder) where the cellulose dissolves as the solvent is concentrated by water evaporation. (Reproduced by courtesy of Acordis.)

4.3.3 Solution transport

The solution leaving the Filmtruder is pumped by a number of specialised pumps in series through the transport system. The transport system consists of a solution cooler and a hydraulic ram buffer tank which feed into the solution primary filters. Owing to the viscous nature of the solution the pressures involved in pumping the solution can be as high as 180 bar.

A complication of the process design is caused by the tendency of the amine oxide in solution to degrade exothermically. Exotherms can be caused by faulty operation of equipment or chemical contamination of the solution. If an exotherm occurs, the temperature of the solution increases rapidly and it decomposes to volatile amines and water. This causes very rapid increase in pressure that would be sufficient to rupture the high pressure equipment with very serious safety implications. To allow for this possibility, bursting discs are provided at strategic positions throughout the plant to give pressure relief in the event of an exotherm. The bursting discs are of a special design to prevent any flow dead spots. The discs vent into disentrainment pots which separate the solid degradation products and allow the gases to be vented to atmosphere. Understanding how and why exotherms occur and the development of a safe way of venting exotherms when they do occur were keys to the scale-up of the whole process.

4.3.4 Solution filtration

Prior to spinning it is necessary to filter various impurities out of the solution. Most of the impurities are introduced with the pulp feedstock, the principle ones being undissolved pulp fibres or inorganic compounds such as sand and ash.

The solution is passed through two stages of filtration. The primary filtration is centrally located and consists of sets of sintered stainless steel media candle filter elements. The secondary stage of filtration is achieved by candle filter elements associated with each spinning machine position.

Filters are washed for re-use by an off-line process involving rinsing with hot amine oxide, chemically decomposing the residual compounds and then ultrasonic washing.

4.3.5 Spinning

For spinning, the solution is split into substreams which serve a number of spinning positions. The solution is then supplied to each jet, via a filter, by a metering pump. It is then extruded and spun through an air gap into a spin bath containing dilute amine oxide solution.

Each jet consists of thousands of tiny holes through which the solution is extruded into fibres. Just below each jet face is a small air gap across which air is blown by the cross-draught system to condition the fibres. After passing through the air gap the fibres, or tow, are pulled down through the spin bath where the cellulose is regenerated in dilute solvent. The fibres are drawn, or stretched, in the air gap by the pull of traction units, or godets.

The design of the spinning assembly proved critical to the successful scale-up of the process and achievement of commercially attractive fibre properties and manufacturing costs.

Spinning the solution directly into an aqueous bath necessitates using very dilute cellulose solutions (producing high costs) and generates fibres with properties that are generally inferior to viscose. Extruding the solution into an air gap enables more economical, higher cellulose solutions to be spun. Furthermore, when the solution is extended or drawn in the air gap it is also oriented so that good strength and elongation properties can be imparted. A draw ratio of between 4 and 20 is typical – within this range fibre properties are similar. At lower ratios fibre tenacities are reduced and at higher ratios spinning stability deteriorates.

When the solution is drawn in the air gap it will readily break unless it is also cooled by means of a gas flow. This tendency to rupture is worse at higher draw ratios and spinning speeds but is reduced if highly viscous solutions are used. This necessitates the use of relatively small spinneret holes relative to the high polymer viscosity so that special jets needed to be designed.

The main limitation of air-gap spinning is the tendency of neighbouring filaments to touch and stick together. This limits the packing density of the spinneret and hence the productivity of the spinning machines. This is compounded by the need to cool all the filaments adequately using an air flow velocity sufficiently low that it does not disrupt the stable flow of the filaments.

The two commercial processes overcome these constraints in different ways. Acordis arranges the spinnerets into rectangular strips whereas Lenzing uses a circular array. Both processes use a controlled flow of gas across the filament arrays to stabilise and control the process.

4.3.6 Fibre washing

The fibre tows from each end are brought together into one large tow band for processing down the fibre line.

The first process on the fibre line is washing, the solvent is washed from the fibre with hot demineralised water in a series of wash baths. In the Tencel® process the fibre is washed as a single large continuous tow through a series of wash troughs each of which consists of a wide shallow bath containing a number of wedges. These wedges deflect the tow band alternately up and down as it is pulled along the trough, which serves to allow dilute solvent into the tow band where it is then squeezed out. Wash liquor leaving the wash line goes into the spin bath system. The washing water is fed countercurrent to the tow band at a rate to keep the spin bath liquor concentration at the required level (see Fig. 4.3).

4.3.7 Fibre treatments

After washing the fibre is treated in a number of ways:

- The fibre could be bleached if required.
- Soft finish is always applied to make processing easier.
- Antistat is applied.
- Other treatments to give specific fibre properties can also be performed.

4.3 Lyocell tows entering the tow-washing process. (Reproduced by courtesy of Acordis.)

4.3.8 Fibre drying

After washing and finishing the fibre is dried in conventional fibre drum dryers. These consist of a series of perforated drums that the fibre passes over. Steam heated air is sucked through the fibre as it passes over the drums and exhausted to atmosphere (see Fig. 4.4).

4.3.9 Crimping, cutting and baling

Dry fibre is crimped before being fed to a radial blade cutter for cutting into staple and then baled. The bales of staple fibre are then dispatched to customers (see Fig. 4.5).

4.4 Inspecting lyocell tow as it is condensed after drying and prior to crimping. (Reproduced by courtesy of Acordis.)

4.5 Bales of lyocell ready for despatch. (Reproduced by courtesy of Acordis.)

Lenzing use a process where the fibres are cut after spinning and it is then very similar to the wash/dry process used in viscose.

4.3.10 Solvent recovery

Diluted amine oxide solvent from the spin baths on the spinning machine and various other process sections is collected to recover the relatively expensive solvent. Over 99% of the solvent used is recycled and recovered by the process.

The amine oxide will slowly oxidise the cellulose during the process, particularly at the elevated temperatures often used. The reaction will reduce the DP of the cellulose (giving poorer fibre properties) as well as generating coloured compounds that would detract from the whiteness of the fibres. The amine oxide degrades to *N*-methyl morpholine plus other amines. The reaction is strongly catalysed by transition metals such as copper and iron. To control this it is essential that a stabiliser such as propyl gallate is incorporated – this appears to act both as an antioxidant and a chelating agent.

Solvent recovery consists of two main processes, ion exchange of the dilute solvent, then evaporation of the excess water to a concentration required in premixing.

The ion exchange process consists of cation and anion beds which remove various ions that would destabilise the solution and the colour contaminants that would otherwise build up in the solvent. The ion exchange resins and regeneration procedures have been developed especially for amine oxide.

The amine oxide is reconcentrated in a steam heated multiple effect falling thin film evaporator. The tendency of the solvent to degrade exothermically means that the process control of the operation has to be such that the amine oxide can never be overheated. The water overheads can be reused to wash the fibres so minimising the environmental impact of the process.

4.4 Lyocell conversion

Lyocell is similar in strength to polyester and stronger than cotton and all other man-made staple fibre cellulosics. It also has very high dry and wet moduli for a cellulosic fibre in both the dry and wet states. These properties (see Chapter 8 for details) allow customers great scope for making strong yarns in blend with virtually all the other commercially available staple fibres. They also lead to excellent efficiencies in converting these yarns to woven and knitted fabrics.

All man made cellulosics lose strength and modulus when wetted, but lyocell reduces by much less than the others. This is important in determining how the properties of the fabric are developed during dyeing and finishing. However the fibres do fibrillate during wet abrasion and thus specific techniques are required to achieve the best results – these are discussed later.

Once lyocell fibre has been produced, either as cut staple fibre or continuous tow it will be converted to yarns and fabrics by a range of conventional textile processes. The most common way of using lyocell fibre is as cut staple, with 1.4 and 1.7 dtex fibres being cut to 38 mm and converted into a spun yarn using machinery developed over many years for handling cotton fibres which are similar in linear density and length to lyocell.

The following comments apply to the processing of Tencel® fibres. The Lenzing Lyocell is made by a wet cut route and has very different processing characteristics.

4.4.1 Yarn manufacture

TENCEL® can be processed via established yarn manufacturing routes, using conventional machinery with few major changes to settings or procedures. Its processing performance is influenced by the following properties:

- it possesses a non-durable crimp;
- it has a high modulus;
- there is little fibre entanglement.

Thus TENCEL® will open easily with little nep. In sliver and roving, the fibres pack well together, give high cohesion and require high draft forces. It yields yarns with high tensile strength and few imperfections. It blends well with other fibres, especially other cellulosics. It adds strength to the final yarn and enhances the performance and aesthetic values of final fabrics.

The following comments refer to fibre processed by the cotton system.

4.4.1.1 Blending and carding

The open state of the fibre tufts facilitates ease of bale skimming, bales rise only a small amount on opening, are extremely stable and are consistent in openness. The production of TENCEL® staple fibres ensures that the filaments are not entangled and that removal of foreign matter is not required. Hence minimal opening is required to separate the fibre tufts ready for carding. It should be remembered that each additional machine in the opening line will add nep.

The vast majority of short staple TENCEL® fibre used in yarn manufacture is carded using chute feeds and revolving flat cards. Conventional settings are often employed, but particular attention is needed in chutes and card web take-off mechanisms. Few changes to conventional card settings are required to achieve even webs. Card wire should be suitable for 1.7 dtex and finer fibres.

4.4.2 Lyocell in nonwovens and papers

Although lyocell is primarily used as a textile fibre, it offers significant potential in both nonwoven fabrics and papers owing to its strength, biodegradability and potential for fibrillation. Only very small amounts of lyocell are presently used for nonwovens, but the importance of these end-uses will increase as lyocell capacity grows.

4.4.3 Nonwovens

A nonwoven is broadly defined as a 'textile structure made directly from fibre rather than yarn'. The fabric is usually made by producing a 'web' of fibres which is then strengthened by 'bonding' using various techniques, for example:

- Thermal bonding – bonding carried out by the application of heat
- Hydroentanglement – bonding achieved by entangling the fibres using very fine jets of high pressure water (see Fig. 4.6)
- Needle bonding – the fibres are entangled by a set of barbed needles which are punched through the web.

The key properties of lyocell which make it suitable for nonwovens are:

- high strength (dry and wet)
- biodegradability
- easy processing
- absorbency
- potential to fibrillate (in wet processes).

Preparation of the fibre web can be done in many different ways, however two main methods are carding and wet-laying.

Not all webs are suitable for bonding by all methods, however, both wet-laid and carded webs are suitable for hydroentanglement. This is the area in which work has concentrated on lyocell, owing to the fibre's ability to fibrillate in wet processes. Hydroentanglement produces bonding by entangling the fibres together. This is done by passing the web under rows of high pressure water jets (up to 200 bar). As this is a wet abrasive process lyocell can be made to fibrillate at high water pressures, producing submicron fibrils, which enhance filtration properties.

Hydroentanglement gives strong lyocell fabrics, which compare well to polyester and are stronger than viscose.

Lyocell can be bonded using a wide range of other techniques, which

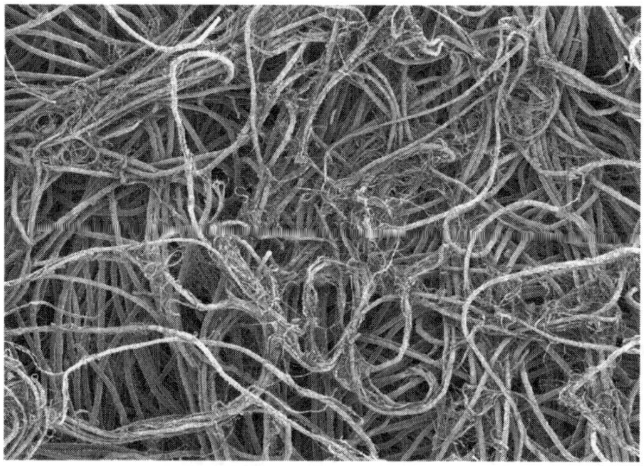

4.6 Surface of a durable wipe made by high pressure entanglement of 100% lyocell. The microfibrils give the surface a chamois-leather like texture.

will not be discussed in detail here. It is being trialled and developed in nonwovens for a wide range of end-uses, including:

- surgical swabs, drapes and gowns
- floppy disc liners
- filtration applications
- semi-disposable workwear
- lining materials.

4.4.4 Papers

The ability of lyocell to fibrillate means that it can be processed and made into a paper, like wood pulp (see Fig. 4.7). The main steps in paper making are:

1) Beating and/or refining – these two processes take a dilute suspension (stock) of short fibre (around 5 mm), and mechanically treat the fibre in order to fibrillate it. This is done either by passing it between a plate and a large grooved roller (beating) or pumping it through a pair of rotating grooved discs (refining). The forces exerted on the fibres cause the fibre to fibrillate, and can also cut the fibre (depending on the severity of the treatment).
2) Sheet formation – the refined stock is diluted further and then pumped to the paper machine where it is deposited onto a porous metal belt or 'wire'. The water is sucked through the belt, leaving the fibre on top as a paper sheet. The paper is then passed through a series of drying cylinders and presses before being wound on a roll.

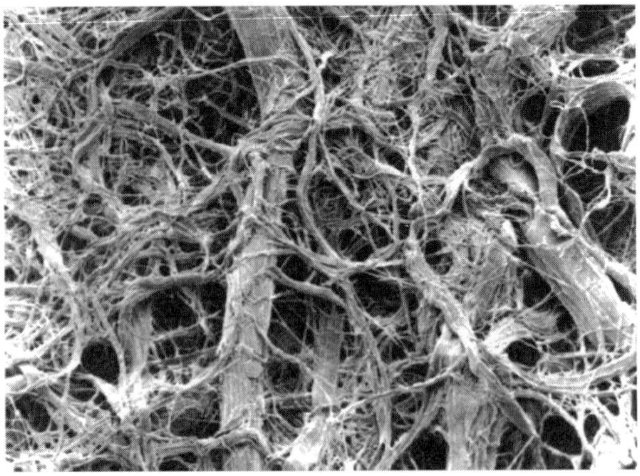

4.7 Surface of 100% lyocell paper held together purely by hydrogen bonding between the high surface area fibrils.

Lyocell can be processed into a strong paper. The properties depend upon the amount of fibrillation generated – increased fibrillation gives more paper strength, and also affects other properties. In general lyocell papers are strong, with good opacity and low air resistance owing to the circular nature of the fibres and fibrils. The end-uses being explored are therefore mainly filtration applications, and also as an additive to improve the properties (e.g. strength) of a standard paper made from wood pulp.

4.5 Dyeing and finishing of lyocell

The dyeing and finishing of lyocell fabrics is the key to their success. There are three characteristics of the fibres that can be manipulated to give fabrics with attractive and differentiated aesthetics – the ease of fibrillation, the high nodulus and the wet swelling characteristics. Fibrillation can yield the characteristic 'peach skin' surface touch of fabrics made from this fibre, but unwanted and uncontrolled fibrillation can also impair the fabric quality. Much of the dyeing and finishing development has been focused on this aspect. The high modulus or stiffness of lyocell contributes to the full handle of lyocell fabrics but allowances must be made in dyeing/finishing procedures to accommodate this property. The high level of wet swelling, particularly in caustic soda, allows fabric to process through dyeing and finishing more readily and yields a full and flexible handle. However, this swelling can cause crease damage to the fabrics if they are not processed correctly. Thus, although lyocell is broadly similar to cotton and viscose in preparation, dyeing and finishing, converters should refer to the manufacturers current technical information sheets for specific advice on the details.

4.5.1 Dyeing of lyocell

As lyocell is a cellulosic fibre, it can be dyed with colours normally used on cotton. Compared with unmercerised cotton, lyocell, except with a few reactive and vat dyes and a number of direct dyes (pale shades), dyes to a heavier depth by exhaust techniques and therefore many shades can be attained at a lower cost, particularly with reactive colours.

Vat, direct and reactive dyes are the principal classes of dye used on lyocell. The actual choice of the dye class depends, amongst other things, on the extent to which the dyes and dyeing methods are economical, brightness of shade, fastness requirement, machine availability, and processing reliability.

4.5.2 Dyeing mechanism

The dyeing mechanism for most classes of reactive dyes is very similar. First the reactive dye is exhausted on to the cellulose fibre using salt. In the second stage of dyeing, alkali is added to fix the dye.

Many of the modern reactive dyestuffs contain two or three reactive groups. A key discovery, made early in the development of lyocell, was that these multifunctional dyestuffs can crosslink the fibre and thereby prevent or inhibit the fibrillation of the fibre. Since manipulation of this fibrillation is critical for the development of the fabric aesthetics (see later), understanding how to control this crosslinking by the dyestuffs is also of critical importance.

This discovery also led to the patenting of 'colourless dyes' as a means of preventing fibrillation of lyocell fibres. The critical benefits of this approach are that the good physical properties of the fibre are maintained together with a good dye affinity. Most conventional cellulose crosslinkers, for example easy-care resins, severely inhibit both characteristics.

4.5.3 Control and use of fibrillation

Fibrillation can be defined as the longitudinal splitting of a single fibre filament into microfibres (see Fig. 4.8). The splitting occurs as a result of wet abrasion, particularly against metal. The fibrils formed can be so fine that they become virtually transparent and give a frosty appearance to the finished fabric.

The samples in Fig. 4.9 show an example of a non-fibrillated (a) and a highly fibrillated (b) lyocell fabric. The fibrillated fabric gives a frosty appearance. In cases of extreme fibrillation, the loose fibres on the surface of the fabric fibrillate and then tangle together to form very light coloured pills. The appearance of the fabric then becomes totally unacceptable.

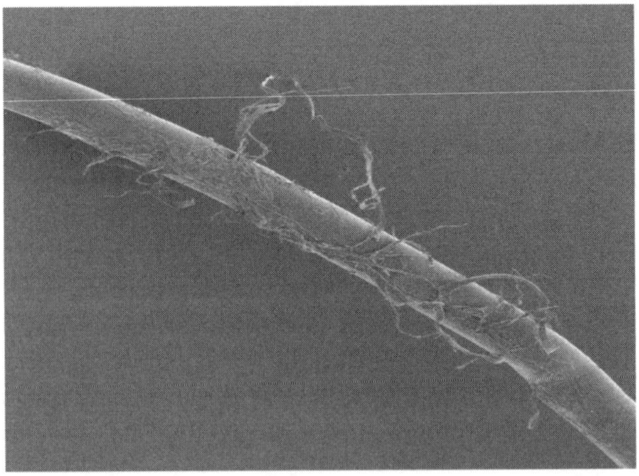

4.8 Micrograph showing fibrillation of a single fibre.

(a) 

(b) 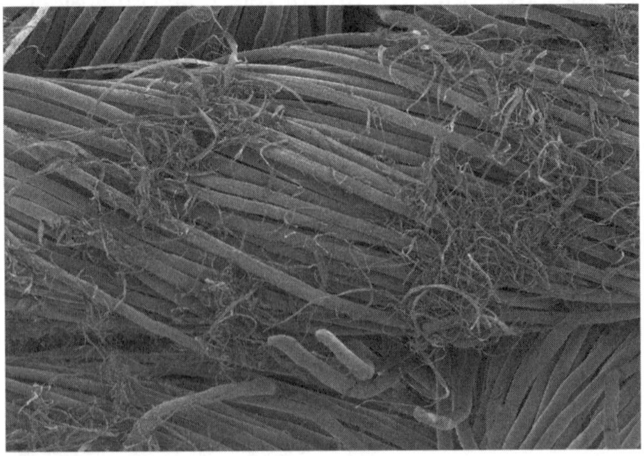

4.9 (a) Woven fabric surface before fibrillation. (b) Woven fabric surface after fibrillation.

4.5.4 What causes fibrillation?

Any process that abrades the fibre in a wet condition will generate some fibrillation. Processing of the fabric in rope dyeing equipment (jets, winches) where the fabric rubs against itself and metal can lead to fibrillation. A number of factors will accelerate fibrillation, such as high pH, high temperature, lack of lubrication, high machine loadings and vigorous machine action. Three methods are being employed to control fibrillation and details of these are given below.

4.5.4.1 Application of enzymes

Once fibrillation has occurred it can be removed by the use of specific cellulase enzymes. These need to be carefully controlled, but are very effective at polishing the fabric surface to remove any unacceptable fibrillation.

Enzymes will not prevent the recurrence of fibrillation of the fibres but, in conjunction with the optimum processing procedures – see later – they can give very attractive and durable fabrics.

4.5.4.2 Application of easy care resins

Resins which can crosslink with the fabric are frequently used after dyeing. This embrittles the fibrils and enables any fibrillation occurring during the dyeing process to be easily removed. This process is particularly suited to woven fabrics as these are prepared and dyed open width and so are free of fibrillation before dyeing.

Resin finishing can have a significant influence on the final aesthetic of lyocell fabrics. However, the amount of finish required to give the resistance to fibrillation and easy care performance is lower than for cotton so much of the natural softness of lyocell can be retained.

Lyocell fabrics produced by this route have clean bright colouration, a full, soft aesthetic and an excellent performance in use. They have proved very comfortable to wear, are durable and retain their 'as new' aesthetic. In blend with polyester they are proving to be excellent for industrial applications, workwear and careerwear.

4.5.4.3 Application of crosslinking chemicals

Lyocell's tendency to fibrillate can also be alleviated by the use of compounds which crosslink with the cellulose chains (see Fig. 4.10). Two chemicals have been marketed as fabric treatments, Axis (from Tencel Ltd) and Ciba 4425 (see Fig. 4.11). Both these compounds will crosslink with lyocell under alkaline conditions either before or after the fabic has been dyed. Treated lyocell is both rope dyeable and durable to laundering and there are no significant changes in fabric properties.

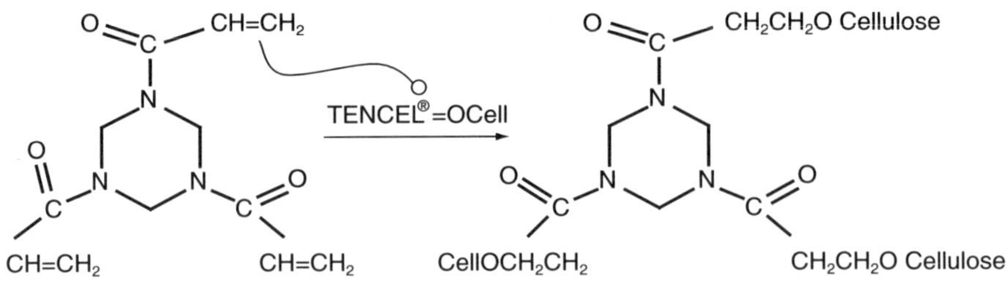

4.10 Crosslinking with triacrylamido-trihydrotrazine. (Reproduced by courtesy of Acordis.) This trifunctional colourless dye crosslinks the fibrillar structure of lyocell and allows fabrics to be processed with minimal fibrillation.

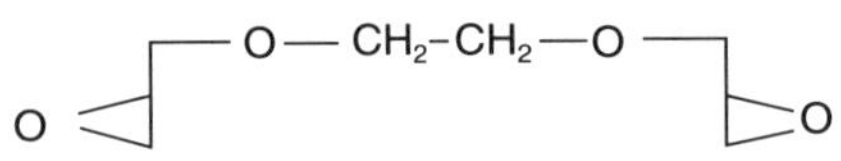

Ethylene glycol-diglycidylether

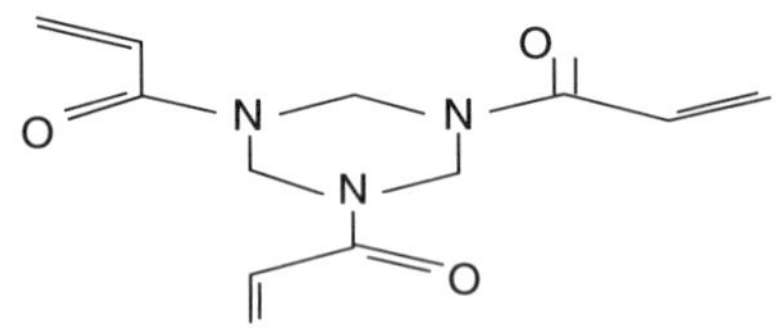

Triacrylamido-trihydrotriazine
AXIS

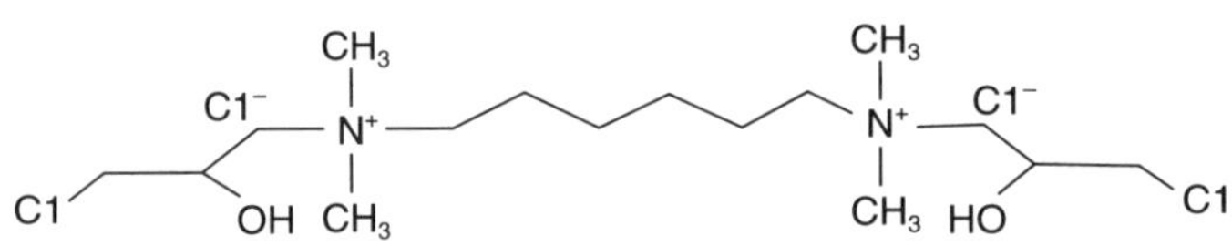

Bis-hydroxychloropropyl-*N*, *N*-tetramethyl-hexamethylenediamine

CATIONON

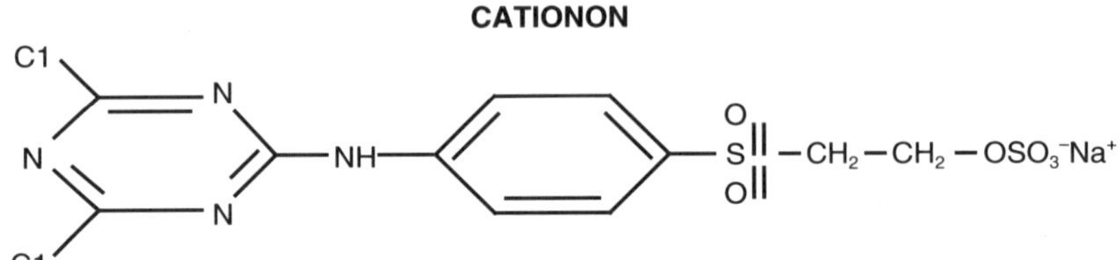

Cibatex AE 4425

4.11 Crosslinking: some of the compounds that can be used to control fibrillation in lyocell processing Axis and Cibatex AE 4425 are used commercially, the other two have been shown to work on a small scale (Reproduced by courtesy of Acordis.)

Figure 4.12 shows lyocell fibre SEMs.

Pre-treatment of the fibre during the manufacturing process has proved the most successful way of using these chemicals. In the A100 process developed by Tencel®, the AXIS chemical is applied to the washed tow in the fibre production line. It is then dried, crimped, cut and baled in'the normal

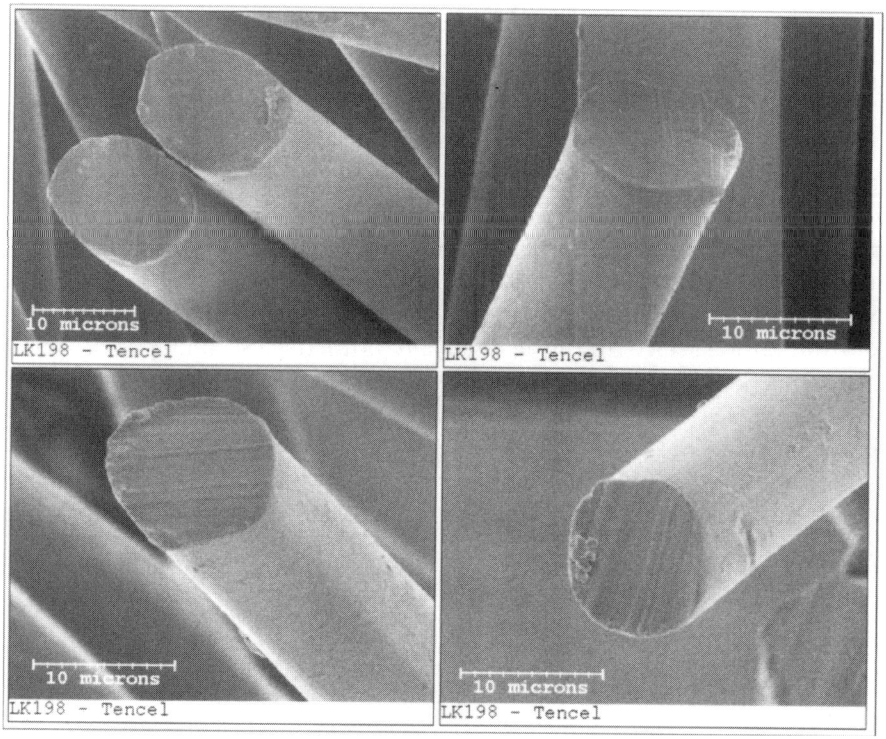

4.12 Lyocell fibre scanning electron microscope (SEM) micrographs.

way. The A100 fibre has similar strength, elongation and modulus properties to the standard fibre and has an enhanced dye uptake that gives more economical dyestuff costs and strong deep colouration. Its resistance to fibrillation means it can be processed on most dyeing machines and the fabrics produced have a good performance in subsequent washing.

Tencel® A100 was launched in 1998 and has proved to be very popular and effective at extending the range of applications for Tencel® fibres. It is particularly beneficial for jersey knit applications because of its excellent colouration, attractive aesthetics and good wash/wear performance. The warm touch of jersey fabrics made from A100 complement the cooler touch of 'peach skin' fabrics made by the enzyme route.

4.5.5 Using fibrillation to create novel hands

The characteristic handle of the first commercially successful lyocell fabrics came from the presence of many fine microfibrils on the surface of the fabric. This gave the so called 'peach skin' aesthetic. Much of the development of dyeing/finishing technology was aimed at achieving this touch without suffering the loss of fabric appearance caused by uncontrolled fibrillation. Additionally there was the challenge of achieving this aesthetic in a commercially economic way.

When a lyocell fabric is first processed in the wet state the loose fibres on the fabric surface fibrillate and form easily seen pills. This is called primary fibrillation and is aesthetically unacceptable.

Primary fibrillation can be removed by enzymes or other treatments so that a fabric with a clean surface can be generated. If this fabric is then agitated in the wet state, short and fine fibrils are formed on the apex of the yarns in the fabric. This secondary fibrillation is only visible as a slight frost on the fabric surface and is aesthetically attractive. Secondary fibrillation also gives the fabric its characteristic 'peach skin' touch, which can be modified to the required aesthetic by the use of a variety of proprietary softeners, eg silicones.

The most attractive fabrics made by this route are given a tumbling process after dyeing. This both generates bulk and softness in the fabric and optimises the surface texture.

4.5.6 Processing routes to making 'peach touch' fabrics

It is important that the fabric produced to give a 'peach touch' is designed to give a good performance in washing. There are a variety of measures that need to be taken to achieve this. Yarn and fabric constructions should be sufficiently tight to prevent the formation of loose fibres that can fibrillate and pill in use. This is more difficult to achieve in knitted fabrics, but singeing of the fabric prior to the primary fibrillation step is very beneficial.

Pre-treatment of the fabric in caustic soda can be effective in improving both fabric performance and aesthetic. In caustic soda, lyocell swells very significantly in diameter but very little in length. The maximum effect is at 10 to 12% soda but stronger mercerisation concentrations can also be used. The swelling of the fibres will increase the diameter of the yarns in the fabric and this will cause the fabric to shrink. This step needs to be carried out with the fabric in open width form and if the fabric is allowed to shrink during caustic treatment its bulk will be significantly enhanced. Furthermore the fibres become set into this new configuration when the fabric is subsequently washed. The fabric then has greater bulk and flexibility – in particular the wet stiffness of the fabric is much reduced. The latter is important for further processing since the fabric is less prone to crease damage marks during the dyeing. The caustic treatment also gives more rapid fibrillation removal in processing and a reduced tendency to fibrillation in domestic use.

When the fabric is subjected to mechanical action in the wet state in preparation and dyeing the surface hairs receive the majority of the abrasive action, therefore fibrillation will occur predominantly on these surface fibres. The fibrils formed are relatively long and are able to become entangled leading to an extremely matted appearance. It is important that the fabric is 'worked' until the surface fibres are fibrillated to their maximum extent. Failure to achieve this will result in incomplete hair removal and an unstable surface appearance in the finished fabric.

A cellulase enzyme is used to clean the fibrillated hairs from the surface of the fabric. Fibrillation of the surface hairs reduces the fibre diameter and this in turn enhances their vulnerability to enzymatic attack. Thus complete removal of the surface hairs can be effected, with minimal effect on the bulk of the fabric. The hydrolytic fibre degradation involved in this process results in a loss of weight, typically 4%, and is accompanied by some reduction in fabric strength. Preliminary local trials are therefore strongly advised. The use of an enzyme on greige fabric is not effective in complete hair removal unless excessive chemical treatment is used which leads to unacceptable loss of fabric strength.

In order to generate a stable finished fabric appearance, it is necessary to fibrillate the fabric a second time. The characteristics of this 'secondary' fibrillation are, however, very different to the 'primary' fibrillation. With the absence of surface hairs, fibrillation is confined to the yarn cross-over points and the high points of the fabric construction. Furthermore, the positioning of these fibrils means that they are physically unable to entangle and therefore pilling does not occur.

Secondary fibrillation produces two effects. A small pile is created on the surface of the fabric which gives the special touch and feel characteristics known as 'Mill Wash' or 'Peach Skin'. The fibrils are only a fraction of the size (diameter) of the fibres, therefore they appear much lighter in colour, even though they contain the same amount of dye. It is this optical effect which gives the dusted mill-wash appearance. Secondary fibrillation can be produced by either a simple washing treatment or by jet-dyeing the fabric and it should be generated to a level that will not increase on subsequent washing.

Dyeing can be carried out either before or after the fibrillation process, but certain dyes can affect the primary fibrillation characteristics and enzyme efficiency. For this reason application of dyes after the fibrillation and enzyme treatment stages is preferred. Softener can be applied to the fabric either by an exhaust or a pad technique, but the best aesthetics are obtained when the fabric is finally tumbled. This treatment maximises fabric bulk and removes any loose fibres and lint from the fabric surface.

The fabric can be finished with a low level of an easy-care resin but care needs to be exercised in selecting the system to be used since some resins can weaken the fibrils so that they fall off during subsequent washing. If this occurs the fabric can lose its 'peach touch'. In order to maximise the peach skin aesthetic, resinated fabrics should be tumble dried before resin curing.

4.5.7 Machinery selection and process methodology

The choice of processing route and the type of equipment used have been of vital importance to the commercialisation of lyocell. These two factors are very much interrelated and this has given a great source of variety (and early frustration!) to the users of lyocell. Lyocell fibre has unusual swelling characteristics: on immersion in water the high degree of swelling causes a pronounced increase in fabric stiffness. This makes the fabric susceptible to creasing in rope processes. The apex of the crease can fibrillate during the mechanical agitation that occurs during most wet processing and this shows up as white lines on the fabric after it is dried. Since this is generally unacceptable, it can be readily appreciated that careful machine selection is important, particularly for heavyweight fabrics. As described above, treatment of the fabric in caustic soda can be very effective in reducing the tendency of the fabric to crease damage.

Peach skin processes rely on surface abrasion and the machinery must be capable of giving evenly distributed abrasion to the fabric surface. Machines vary in severity of physical action, those with a gentle action require more severe chemical and temperature conditions to achieve the desired result.

Garment processing gives an ideal combination of the physical conditions for processing lyocell so this route processed most of the early commercial fabrics. In particular, garment processing of indigo denim was an established industry for tumbling and enzyme treating cellulosic garments. Most machine types designed for garment washing have been found to be suitable.

The early success of lyocell fabrics made by the garment route generated a demand for fabrics to be made by the more economical piece processing route. Simple processing of lyocell fabrics by open width/resination routes gave fabrics with a good technical performance but they did not have the 'peach touch' or the bulk of fabrics made by the garment route. Continuous processes offer advantages in terms of capacity and consistency. Preparation, hair removal, dyeing and tumbling can all be carried out continuously, but the machines used were not sufficiently aggressive to generate fibrillation. More recently, open-width routes have been developed that can generate the required aesthetics by using caustic soda pre-treatments (to generate bulk and softness) and sueding after-treatment (to generate the surface touch). These routes will become increasingly important because of their lower inherent cost.

The first 'large scale' piece-processed fabrics were developed in Japan using Nidom machines: i.e. large scale barrel machines with a similar mechanical action to the garment machines. They generated very attractive fabrics that had both the bulk and 'peach' surface touch of fabrics made in the smaller garment machines. Fabrics made using these Nidom machines were commercially successful in the early days of Tencel® when the products commanded a very high price in the Japanese market. However their relatively low productivity and high labour costs restricted the extent of their adoption.

The most successful jet machines for piece processing woven fabrics are now 'air-jets' working on lengths of fabric in rope form. These rope-processing machines must be able to re-orient the fabric frequently to ensure uniform fibrillation across the surface and to prevent crease damage marks appearing. The air jet rope-processors most easily meet these requirements for woven fabrics but knitted fabrics are less prone to damage marks and a wider range of soft flow jets can be used successfully.

In order to achieve the best aesthetics, secondary fibrillation should be enhanced by a final tumbling process. This can be carried out dry, or with water present. Tumbling under wet conditions gives additional bulk to the fabric. Wet tumbling machines can also be utilised for the fibrillation and enzyme processes.

In woven fabrics, the high degree of swelling in water allows the 'peach skin' surface, once developed, to be stable to domestic laundering. When the fabric is wetted the fibres swell and are locked into position. Therefore it is very difficult for unfibrillated hairs to be drawn from the yarns within the fabric onto the surface.

In knitted fabrics, the structure is generally more open so that it is more difficult to lock the fibres into the body of the fabric. The fibres can therefore more easily be drawn onto the fabric surface and fibrillate. Good results can be achieved by combinations of the factors normally used to minimise pilling: high quality yarns, high yarn twist levels, tighter stitch length and the use of easy care resins.

Even though peach-skin fabrics in lyocell are inherently stable, their aesthetics and performance can be enhanced by the use of a fabric conditioner during domestic laundering and by tumble drying.

4.5.8 Economical processing routes

Following the success in developing all the principles described above, the next challenge has been to reduce the costs of carrying these out.

Most of the early focus was on achieving high first quality yields in the dye houses since there were high levels of rejects due to white line damage marks. Progress here was rapid and once the key process parameters were recognised and dye houses became experienced, good first quality yields were routinely achieved.

The next steps have been to reduce chemical and process-time costs. Enzyme treatment/fibrillation removal was initially very expensive because it took significant machine time and the enzymes were costly. Combining process stages has significantly reduced the process times. For instance the generation of secondary fibrillation can occur during the dyeing process when an air jet is used. The price of enzymes has declined as the market has grown in scale and competitiveness and the quantities required have been reduced.

Most recently, Tencel Ltd. has developed process routes that combine preparation stages and use oxidative systems instead of enzymes. These novel routes have greatly reduced the dyeing/finish costs of lyocell to the point where it can begin to be cost competitive with the processing on cotton fabrics.

4.6 Lyocell marketing

Following its launch, the development of Tencel® has focused on yarn development, fabric development and latterly, garment development. Garment development is now significant.

4.6.1 Yarn development

Short staple yarns include 100% Tencel® as well as a wide variety of blends with cotton, polyester, silk, cashmere, wool, viscose, linen, and elastane.

Long staple yarns include 100% Tencel® and blends with wool, silk and cashmere.

4.6.2 Fabric development

The first commercially available fabrics in Tencel® emerged with an emphasis on denim, chambrays and piece-dyes. Developments in this field anticipated the so-called denim revival which was given a fillip by the revolutionary Tencel® denim/chambray type – dubbed the 'cashmere of denim' by the press for its extraordinary properties of lustre, drape and softness.

A wider portfolio of fabric types includes velvets, corduroys, colour woven, gabardines, twills, jerseys, prints, voiles, lace, coated fabrics and crêpes, most of which were first shown at Premiere Vision (October 93) by the 27-strong 'partnership' of Tencel® weavers from France, Italy, Germany, Spain, Austria and Switzerland. The commitment of these weavers to the systematic development of individual fabric types, worked, and continues to work, to a unique 'windows of opportunity' matrix which individual customers with a reputation for excellence in a given product area were invited to fill.

The early focus on the premium-branded casual wear market was supplemented by fabric innovation for formal blouse and dresswear, classic shirting and suiting, lingerie and nightwear, rainwear and evening wear and in active sportswear.

4.6.3 Expansion of products and markets

The marketing of Tencel® has followed a carefully planned and controlled strategy. Considering how small the volume of Tencel® production is relative to the whole fibres market, and the investments made by Courtaulds in developing Tencel®, it was essential that Tencel® was initially positioned at the top end of the fibres markets. The properties of Tencel® enable the finished textiles to be of the highest quality thus giving the differentiation required for its market positioning. However, in order to realise the potential for making quality products from Tencel® the textile processing has to be of the highest quality. To ensure this is the case and that no inferior products reach the marketplace with the Tencel® name Tencel® was sold to technically innovative and competent textile manufacturers. As the production of Tencel® increased, it was necessary to move into the middle sector of the textile market and to develop the new dyeing/finishing routes described above, and the new fibres like Tencel® A100. These built on the unique physical characteristics of Tencel® to generate a wide range of attractive and novel fabric aesthetics at economical cost, enabling the textile trade to broaden its range of markets and to successfully meet the price points required by the retail customers.

The combination of comfort and performance of Tencel® fabrics is now enabling them to grow in both household textiles and workwear applications. Attractive towelling fabrics are being marketed that exploit the excellent water absorbency, deep colouration and wet stiffness of Tencel®. The high dry strength of Tencel® woven fabrics makes them desirable as backing for abrasives while in nonwovens its high wet strength is ideal in a wide range of absorbent wipes. The combination of easy fibrillation and wet strength makes short-cut fibre attractive in premium paper and filtration applications. With increased manufacturing capacity this area is important in maintaining a relatively steady sales output. Tencel Ltd. is also looking to exploit other applications in cellulosic films and continuous filament yarns. These are being pursued by joint developments with companies specialising in those fields. (See also carboxymethylated lyocell in Chapter 6).

The very positive environmental characteristics of the Tencel® process underpins the marketing strategy in all applications.

5
Cupramonium processes

KENJI KAMIDE AND KAZUNARI NISHIYAMA

5.1 Short history

5.1.1 Who invented the cuprammonium process?

In 1857 a German chemist Matthias Eduard Schweizer (1818–1860; a chemistry lecturer at the Oberen Institute-Schule in Zürich who progressed in 1852 to professor) reported in his article[1] *Das Kupferoxyd-Ammoniak, ein Auflösungsmittel für die Pflanzenfaster*, that cotton, linen cellulose and silk dissolve at room (*gewöhnlicher*) temperature in cuprammonium solution (ammonial copper oxide solution). Schweizer previously observed that ammonia reacts with metal oxide to form a basic compound (*Kupferoxyd-Ammoniak*).[2] The colour of the solution is due to the complex cupric tetramine hydroxide $Cu(NH_3)_4(OH)_2$. Schweizer wrote that '*das Ammoniak mit gemmissen Metalloxyden Verbingen bilden Könne*'.

Schweizer is generally credited as the inventor of the solution. However, it is somewhat curious that Schweizer did not apply for a patent for his 'invention', although he described his new observation as '*eine höchst interessante Eigenschaft*'.[1] Note that in the article there was no description of the experimental conditions such as the composition of copper ammonium solution, the concentration of cellulose or silk, temperature, and so on.

There is a story that John Mercer, a famous calico printer in Lancashire, UK, who developed a process for improving the lustre and dyeability of calico cotton cloth by dipping it in sodium hydroxide solution under tension, made the actual discovery that cuprammonium solution was a suitable solvent for cellulose. However, he applied this solution to coating calico to make it waterproof. Wheeler wrote in his book,[3] citing Parnell's *Life of John Mercer*, that Mercer dissolved cellulose in cuprammonium solution, and then regenerated the cellulose with sulphuric acid some time before Schweizer's 'invention'.[4] This is the first mystery in the history of the cuprammonium process.

William Crooks (later, Sir William), who once cooperated with Joseph

Table 5.1 Prehistory of cuprammonium cellulose industry

Year	Inventor	Remarks
18??	John Mercer (UK)	Use of cuprammonium cellulose solution for treatment of cotton cloth
1857	M E Schweizer (Germany)	Dissolution of cotton cellulose into cuprammonium solution
1881	William Crooks (UK)	Fibres from cuprammonium cellulose solution for electric lamp, BP 2612 (1881)
1882	Edward Weston (USA)	Fibres from cuprammonium cellulose solution for electric lamp, BP 4458 (1882)

Wilson Swan (1828–1914) to develop an electric light bulb through his knowledge of vacuum technology, obtained a patent[5] on the carbonization of filaments of paper, cotton or linen treated with a solution of ammoniacal copper oxide to obtain a denser carbon filament. Edward Weston, a pioneer of artificial silk by the cellulose nitrate method, took out a similar patent[6] on carbon conductors for electric lamps. In 1885 Spiller also produced regenerated filaments from a cellulose–cuprammonium solution system. Table 5.1 summarizes the prehistory of the cuprammonium cellulose industry. In 1890 Louis Henri Despeissis, a French chemist, successfully produced (although on a laboratory scale) artificial fibre from cellulose dissolved in cuprammonium solution.[7] This was the most significant technological contribution to the regenerated cellulose industry, although his patent lost validity in 1892, because of non-payment of the patent maintenance fee. More sadly, soon afterwards he died. Despeissis was the first to attempt to produce pure cellulose fibre for clothing directly from cellulose, not from cellulose derivative fibres such as cellulose nitrate fibres (the collodion process, nitrate process, or Chardonnet process).

The principal process of the Despeissis patent is illustrated in Fig. 5.1.[8] In this process, cellulose (cotton or wood fibre) dissolved in aqueous cuprammonium hydroxide solution was extruded through a capillary to form fibres with a given cross-section area. Fatty and resinous materials attached to cellulose were removed in advance by washing in order to accelerate dissolution, and the solution was filtered through sand and asbestos to remove any undissolved matter. The solution was extruded into a bath of dilute acid (for example hydrochloric acid, formic acid, succinic acid, tartaric acid, or citric acid), alcohol and concentrated cresol solution. When cellulose in cuprammonium solution comes into contact with the above solution,[5] it precipitates immediately to give hard solid filaments, copper and ammonium being removed in the spinning bath. The filaments were

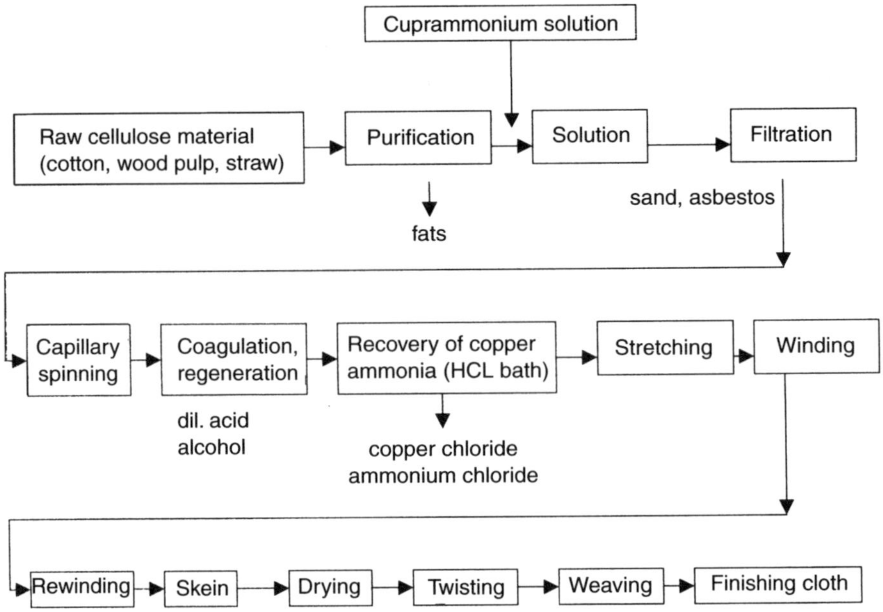

5.1 Production of artificial silk by the cuprammonium method (L H Despeissis, 1890).[7]

stretched in dilute hydrochloric acid using a winder, spool and drum. The remaining copper and ammonia in the filaments were reacted chemically to form the easily dissolved copper chloride and ammonium chloride. Dilute hydrochloride solution could be replaced with other solutions, which can precipitate cellulose. Washed filaments were dried and reeled. It was difficult to make a concentrated cellulose solution that could be employed on an industrial scale. The process described in Despeissis's patent was not commercialized in the original form.

In summary, the technology of cuprammonium solution was making progress as early as the 1850s–1880s in the following way:

1 Cuprammonium solution was found to be an adequate solution for cellulose.
2 Cellulose dissolved solution was first utilized as a coating reagent for cotton cloth and paper. (In the case of viscose, it was used to give waterproofing and rotproofing properties.)
3 Coarse short monofilaments were then produced for production of filament for electric light bulbs.
4 Endless thin filaments were finally produced on a laboratory scale as artificial silk for clothing.

In 1892, two years after Despeissis's invention, Rheinische Glühlampenfabrik, Dr. Max Fremery & Co. Kommerzial Gesellschaft at Aachen established a process of producing lamp filaments from cellulose in cuprammonium solution, based on their own unpatented 'invention'. The process was operated with economic success for several years and was the first commercial application of the cellulose-cuprammonium solution system. For example, in 1895 the company produced 2500 bulbs per day and eventually reacted production of over 4000.[6–9] Within five years, the company had developed a stretching process using rotating cylinders to make much finer filaments suitable for clothing.

Max Fremery (German chemist) (1859–1932) and John Urban (Austrian engineer) (1863–1940) applied for a patent in 1897 under the name of a German chemist Hermann Pauly,[10] who was a director of the Textilschule in Münchengladbach,[9] and was a Professor of chemistry at Wurzburg in 1924. It should be noted that Hermann Pauly was not the true inventor. The Pauly patent was applied for seven years after the Despeissis patent. The characteristics of the 'Pauly' (in fact, Fremery and Urban's) process is the dissolution of cellulose in aqueous cuprammonium solution at low temperature. For example, the specification of German Patent (Kaiserliches Patent) 98 642 (1897) indicates 'below 30°C' and the British Patent 28 631 (1897) claims 'as cool as possible'. Pauly insisted in the latter patent that 'The solubility of the cellulose in the ammoniacal solution of cupric oxide decreases as the temperature increases and on the other hand, the decomposition of the cellulose in the above solution increases as the temperature increases' and 'it is advantageous to keep the vessel, . . . , as cool as possible'. Pauly pointed out that the competitive cellulose nitrate fibre is 'extremely easily combustible' and that 'no (cellulose nitrate) process has, up to the present, been carried into practice'.* Pauly did not refer to the preceding technology and patents of the cuprammonium process in 'his' patent specification, although Fremery and Urban had already been engaged at least for six years in the business.

Figure 5.2 is a flow sheet of the manufacturing process of cuprammonium rayon by Hermann Pauly's method.[12] The figure is based on German Patent 98 642 (1897). Comparison of the two patents is made in detail in Table 5.2.[13] It is clear that the 'Pauly' patent lies within the range of the Despeissis patent. The former claims are much narrower and more restrictive than the latter. For example, Pauly referred to acetic acid as a coagulating liquid,

*It is noteworthy that at the time Pauly applied for 'his' patent, Chardonnet's company ('Société Anonyme Pour la Fabrication de la Soie de Chardonnet') was already producing (1894–1898) cellulose nitrate artificial silk, at the Besançon with a production capacity of between 100–1000 kg day^{-1} at a price of 21.75–30.0 francs kg^{-1}. The company made a profit of 20 200 French francs and paid a dividend of 6.25% in 1898.[11]

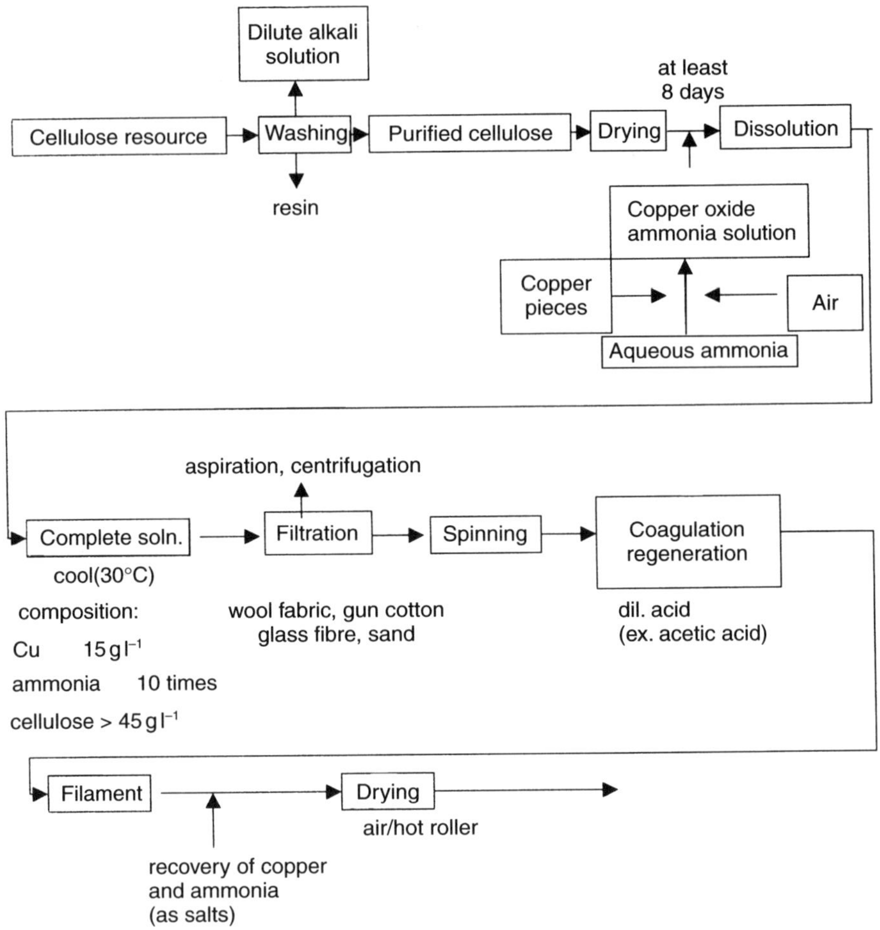

5.2 Herman Pauly's (actually, Fremery and Urban's) procedure for the production of cuprammonium artificial silk (1897).[10]

although this was omitted in the Despeissis patent. This fact strongly suggests that Fremery and Urban tried to ignore the existence of the preceding French patent, but as a matter of fact, were in contravention of that patent, which was already invalid and in effect only the published technical literature.

Why did Fremery and Urban apply for their patent under another person's name? VGF, which was established by them later stated in the book[14] 'from tactile economic considerations'. Why was the 'Pauly' patent finally accepted at Kaiserlichen Patentamt? The following remark in Wicht's 'Glanzstoff'[15] may help to gain some understanding. 'While

Table 5.2 Comparison of the Despeissis method and the 'Pauly' method

Items	Despeissis patent (1890) (FP203741)	'Pauly' patent (1897) (DP98642)
Cellulose resource	Cotton,woodpulp, straw	(Any cellulose)
Purification of cellulose	Washing	Treatment with dil.alkali soln.
Cuprammonium solution (preparation)	No description (based on Schweizer method)	Several methods
(composition)	No description	Cu; 15 gl^{-1} ammonia; >150 gl^{-1}
Spinning solution (purification)	Filtration	Filtration
(filtering materials)	Sand, asbestos	Woollen cloth,gun cotton, glass fibre, sand
(cellulose concentration)	No description	>45 gl^{-1}
(stock)	No description	Low temperature desirable (30°C)
Spinning method	Capillary	No description
(coagulation bath)	Dilute acid (hydrochloric acid, sulphuric acid, formic acid, tartaric acid, citric acid), alcohol, conc. phenol	Dilute acid (acetic acid)
Recovery of copper and ammonia	Dip into hydrochloric acid (in the form of salt)	As salt
Scope	Production of filaments from cuprammonium soln	Production of filaments from cuprammonium solution, prepared at low temperature

processing your patent application at the Kaiserliches Patentamt (German patent bureau) it has, for example, become evident that there has already been a French patent (F.P. 203 740-1890) which had expired already. In addition, this patent did not include a useable process for a consistent production of marketable threads.' This remains a mystery even today (i.e. the second mystery) in the history of the cuprammonium rayon industry.

An Alsatian textile engineer Joseph Foltzer, who published the first book[16] on regenerated cellulose technology (in France in 1903), called cupro rayon by the name 'Despeissis silk'. He had worked in Germany and Switzerland, and later took out some patents (for example, French Patent 345 687, German Patent 165 577 (1905), Swiss Patent 37 584 (with E. Weliss), German Patent 209 923 (1908) and German Patent 229 677 (1908)). After his book was published, Foltzer returned to Alsace and was arrested on suspicion of unlawful disclosure of German secrets. Arnold Hard wrote in his book[17] 'Germany considered Foltzer had committed a grave offence in publishing details of what they claimed was their own secret process'. Perhaps Foltzer was arrested for political reasons because he had revealed the 'secret' that there was no secret in German cuprammonium technology.

In 1899 an eminent Alsatian textile chemist Emil Bronnert (1868–1924), with Fremery and Urban, applied for a series of patents,* separately and jointly, before establishing their company (VGF): Vereinigte† Glanzstoff-Fabriken A G (VGF) started with capital of 2 million marks on September 19, 1899 financed by the Bergisch–Märkischen Bank.

After VGF was established they applied for two patents in the names Fremery, Urban and Bronnert (German Patent 115 989, date of application, Jan. 11, 1900: German Patent 119 230, Dec, 10, 1900). After that, all patents were obtained under the name of VGF until the end of World War II. The above patents, including, of course, the 'Pauly' patent, can be regarded as the fundamental VGF patents. The rights to these patents, owned by Fremery, Urban and Bronnert were estimated to be worth a million marks to VGF.

The first textile fibre factory using the cuprammonium process was opened at Oberbruch near Aachen, by expanding the experimental 1897 Rheinische Glühlampen factory to a capacity of 1.2 tonnes-filament/day.[18] Thereafter, a second factory was founded at Niedermorschweiler (now, Morschwiller le Bas) near Mulhouse. A third factory at Sachsen had been planned, but not realised: raw materials costs had to be reduced first.[19,20] This was to be achieved by:

1 moving away from using cotton wastes from carding, combing and other machines in the cotton industry and moving towards cotton linters;
2 moving away from copper sulphate (as a reaction product with aqueous sodium hydroxide solution) and moving towards copper hydroxide;

* Bronnert (German Patent 109 996, date of application; May 2, 1899), Fremery and Urban (German Patent 111 313, March 27, 1899), Fremery, Urban and Bronnert (German Patent 119 098, May 9, 1899: German Patent 119 099, May 13, 1899).

† Here, 'Vereinigte' means union of three groups which had a keen interest in the cuprammonium artificial silk business in Germany: the Aachen (lower Rhein) group (Fremery & Urban and the Bergisch Bank), the Elsass (middle Rhein) group (Bronnert and Th. Schlumberger), and the Glauchau (Sachsen) group (Arthur Lossow).

Table 5.3 Workforce and output of cuprammonium artificial silk in VGF in 1901 and 1911[22,23]

Year	Workforce (no. of persons)	Output (kg fibres) per day	Productivity kg/person · day
1901	376	168	0.48
1911	2517	2738	1.1
Ratio 1911/1901	6.7	16.3	2.3

3 increasing the recovery rate of ammonia to 85% and of copper to 95%;
4 changing from acetic acid and around 50% sulphuric acid coagulation to alkali solution coagulation.

Their achievements on recovery rate are not known but the above targets seem unrealistically high for the technology available at that time.

Cuprammonium rayon made by VGF ('Glanzstoff silk') appeared on the market in early 1900. VGF produced yarns of 100–200 denier, judging from their circulars and price lists.[21] This meant that the filament constituting yarn was much coarser than that of Chardonnet silk. An advertisement for cuprammonium rayon in July 1900 says:[17]

(1) The lustre is brighter than that of natural silk. (2) The lustre is exceedingly valuable for fancy thread for weft in silk, woollen and cotton goods. (3) The fibres are largely used for laces, galloons, fringes girdles, furniture stuffs, ribbons, tricots, embroidery etc. (4) The fibre dyes equal and uniform.

The productivity of VGF in 1901 and 1911 is illustrated in Table 5.3.[22,23] Over the 10 years they increased the labour force by a factor of 6.7 and production by a factor of 16.3. The productivity more than doubled. From the start of production, VGF recorded good profits until the outbreak of World War I (Table 5.4).[24–26] VGF's technology had spread from Germany to other European countries through licensed contracts, (e.g. in France, 1899), or in the form of a subsidiary company (Austria, 1904; England, 1907).

In July 1911, Hans Jordan, Vorsitzend (Chairman) of VGF, decided to buy all the rights to the German patents for the viscose rayon process and the Donnersmarcks Kunstseide und Acetatwerke plant near Stettin at a cost of 2 million marks.[23] He aimed to convert their fibre business from the cuprammonium rayon (process), which was even at that time yielding sufficient profit (see Table 5.4), to viscose rayon. VGF explained this transition of business in terms of future prospects.[27] The production of cuprammonium rayon reached a maximum (820 tonnes) in 1912, but it stopped in 1916 because of a shortage of raw materials.[28] However, Thiele's 'stretch-

Table 5.4 Profits of VGF in 1900–1913

Year	Capital (10^4 marks)	Total profit (10^4 marks)	Net profit (10^4 marks)	Dividend (%)
1900	250	—	6.5	—
1901	250	—	—	(9)
1904	250	—	170	—
1906	250	241.1	203.1	35
1907	250	252.8	214.1	40 (40)
1908	250	287.6	209.2	40
1909	500	444.2	309.9	30
1910–12	500	—	335.9 (average)	—
1913	750	—	—	38

Reproduced from K Kamide,[8] p. 245, Table 7.8, Y Kami[24], *Recent Artificial Silk*, Meibundo, 1927, p. 141 and W E Wicht,[9] p. 35.

spinning process' developed for J P Bemberg, a powerful domestic competitor, had succeeded in commercial production of a cuprammonium rayon that had filaments as fine (1.0–1.5 den) as Chardonnet silk, and with superior physical properties. Against this, Glanzstoff silk was uncompetitive.

The years 1910–1911 constituted a remarkable turning point in the artificial silk industry. Shortly after succeeding against the nitrate process, in 1907 the cuprammonium process was confronted with almost insurmountable competition from the viscose process. It became clear in 1909–1910 that the viscose process was the best. World production of cuprammonium rayon began to decay, passing through a maximum in 1909. The conventional cuprammonium process was clearly uncompetitive economically with the new viscose process and was inferior in quality to cellulose nitrate silk.

In converting to viscose, VGF was attracted by its cheaper raw materials (e.g. wood pulp) and simpler spinning process.

5.1.2 Invention of Thiele's 'stretch-spinning' and to the present day

Edmund Thiele, an exceptional German chemist, joined J P Bemberg A G (JPB; Barmen, Germany) in 1900, to study new spinning processes. From 1904 he worked together with Emil Elsässer on this project. Thiele invented the 'stretch-spinning' method[29–32] in 1902.

Compared with cellulose derivative solutions, cuprammonium solution is not easily stretched using the apparatus employed in the viscose process and so filaments had to be spun through smaller holes. Thiele considered the use of very small diameter capillary tubes disadvantageous.[29,30]

1 Cellulose solution must be carefully filtered.
2 Higher pressures are required to force the dope through small holes.
3 Small holes block more easily.

The above disadvantages, Thiele argued, would be partially obviated if a comparatively thick filament issuing from a larger opening (such as 1/4 or 1/8 th millimetre) could be drawn down by stretching. This was originally done by Lehner in nitrate silk production but such a process had never before been employed with aqueous solutions of cellulose. (The formation of the filament from cellulose nitrate solution is effected by gradual evaporation of the solvent, but precipitates immediately when cuprammonium solution makes contact with non-solvent.)

The main claims of Thiele's stretch spinning patents[29–32] are:

1 A process for the manufacture of filaments from cellulose solutions wherein an aqueous cellulose solution issues through comparatively large orifices into a precipitating medium that acts only slowly upon aqueous cellulose solution, the said issuing solution being substantially drawn out into fine filaments in such medium.
2 A process referred to in claim 1, using the drawing out of the cellulose solution into fine filaments to be effected by the action of its own gravity.
3 After having drawn out into fine filaments in a slowly acting precipitating medium, the filaments are subjected to the action of a more rapidly acting precipitating medium, both media coming into contact in such manner that filaments can pass directly from the slowly acting medium into the rapidly acting one.

Figure 5.3 shows Thiele's apparatus in its original form.[30,31] Thiele worked for J P Bemberg for only a few years and then moved from Barmen, where J P Bemberg's plant was situated, to Brussels, where he collaborated with R Linkmeyer, who was an entrepreneur, and filed another patent.[33] After Thiele left J P Bemberg, the engineering group, originally headed by Emil Elsaesser, continued to develop stretch-spinning technology.[34–39] Figure 5.4 illustrates the apparatus.

Surprisingly, Thiele and his successor's patents (most probably, German Patent 220 051, 1907),[34] were judged by a German court on 4th May 1907 to be anticipated by Pauly. The judgment is not easy to justify from a modern view point, because 'Pauly' was considered only to mimic the Despeissis patent (see Table 5.2) and the lack of novelty in the 'Pauly' patent was indirectly acknowledged by VGF at that time.[15] In other words, VGF won their patent defence war against J P Bemberg. This constitutes the third mystery in the history of the cuprammonium process. In this way, J P Bemberg A G had been forced to associate with VGF in the cuprammonium rayon business, although details of the above circumstances are not known.

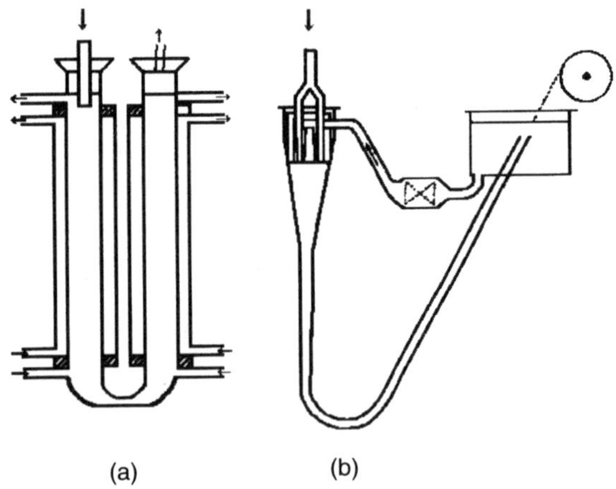

5.3 Apparatus for 'stretch-spinning' from cellulose solution (Thiele's[31] original method). (a), (b): Thiele[31] and British Patent No 8 083 (1902).[30]

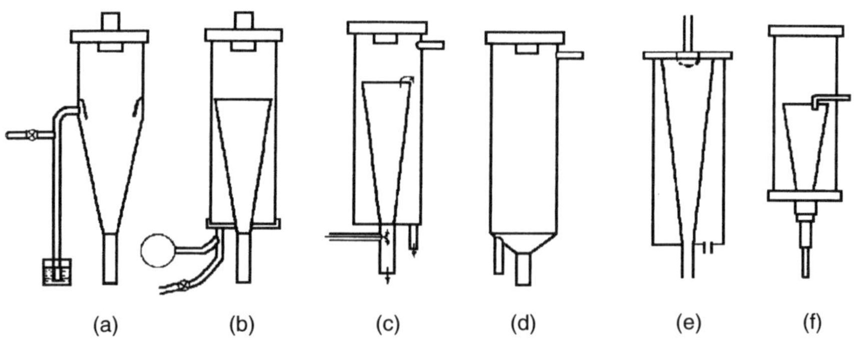

5.4 Apparatus for 'stretch-spinning' from cellulose solution (J P Bemberg Akt.-Ges.) (a), (b): German Patent 413 791 (1923),[35] (c): German Patent 413 790 (1923),[36] (d): German Patent 415 798 (1923),[37] (e): British Patent No 229 638 (1925),[38] (f): British Patent No 235 853 (1925).[39]

J P Bemberg A G founded a factory with a capacity of 500–600 kg per day of cuprammonium rayon using Thiele's procedure at Ölde near Barmen-Rittershausen. In 1916, in order to amalgamate the factories at Elberfeld (VGF) and Barmen (JPB), a mutual technical exchange was carried out and since then, cuprammonium rayon production was practically commissioned to JPB, which supplied the fibre products under the trade name of 'Bemberg'®. Glanzstoff did not reopen the production after World War I. In other words, a horizontal business division was created between VGF and JPB. Cuprammonium rayon was produced by JPB and viscose rayon by VGF. In 1925 J P Bemberg came under the control of VGF and in 1948 VGF took over 35% of the capital of JPB.

Thiele's technology was transferred to the USA (American Bemberg Corp., Elizabethton, Tennessee, 1925), Italy (Bemberg S A; Gozzano 1924), France (Le Cupro textile; Rennes), Japan (Asahi Bemberg; Nobeoka, 1928) and UK (British Bemberg; Doncaster, 1928). In 1926 Interessen Gemeinschaft (IG) Farbenindustrie A G (established in 1925) built the Dormagen factory. Beaunit Mills took over American Bemberg in 1948 as a division (Beaunit Fibers).

Table 5.5 summarizes the development and commercialization of the cuprammonium process from 1890–1930. It clearly shows that:

1 VGF and JPB started independently and almost simultaneously. VGF commercialized in 1900, but J P Bemberg succeeded in large scale production barely from 1916.
2 During the short time when Thiele worked at JPB, JPB applied for a patent (German Patent 162 866, 1900[40]). This suggests that JPB had its own engineers as well as Thiele.
3 Until 1907 JPB had obtained three patents (German Patents 162 866, 1900,[40] 174 508, 1904[41] and 220 051, 1907[34]), the last being the most technologically important, because it claims 'stretch-spinning', by which JPB could produce fibres on a large scale. Therefore, it seems probable that German Patent 220 051[34] was the main target of the legal struggle by VGF against JPB.
4 Funnels suitable for industrial use had been patented, after commencement of large-scale production and after the time of expiration of Thiele's original patents, during 1923–1925 (German Patents 408 889, 1923,[42] 413 791, 1923,[35] 413 790, 1923,[36] 415 798, 1923,[37] and British Patent 235 853, 1925[39]).

As the denier of single filaments in a yarn is decreased, the yarn exhibits superior handling quality (aesthetics) and, accordingly, can be sold at relatively high price. The denier of single filaments is also a measure of the sophistication of the production technology. Table 5.6[43] records the single filament denier of yarns that were available in several European countries in 1922. Nitrate processes had enjoyed a monopolistic position in supplying fine filaments before the commercialization of Thiele's stretch-spinning method in 1916. Viscose rayon belonged to another category competing on price rather than quality. Bemberg® competed with nitrate artificial silk. In 1922 Germany alone produced Bemberg®. The technology did not spread to other countries which continued the production of the coarser Glanzstoff-type rayon until 1924–1928. The number of single filaments in a Bemberg® thread was, in most cases, three quarters of the denier

Table 5.5 Cuprammonium rayon process in 1890–1930

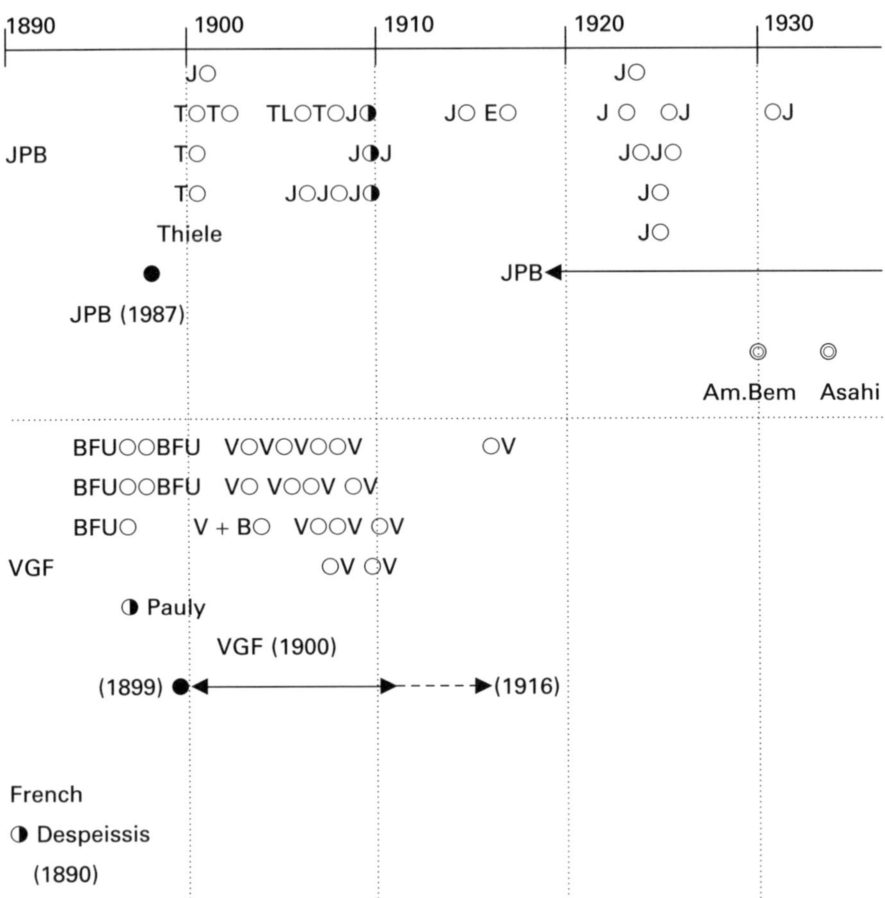

●: Establishment of company; ◄- - - -► Commercial production; ◑: Basic patent; ◎: Technology transfer; ○: Patent; T: Thiele; TL: Thiele-Linkmeyer; J: JPB (J P Bemberg); E: Elsaesser; BFU: Bronnert-Fremery-Urban; B: Bronnert; V: VGF (Vereinigte Glanzstoff Fabriken A G).

number: 100 denier consisted of 75 single filaments with a fineness of 1.3 denier.[45]

During the 1910s–1920s the names of the following companies were known in Germany as cuprammonium rayon producers, as well as VGF and JPB. The year in parentheses is the date established and marks indicated is the amount of capital at the time of foundation:[46] Hanauer Kunstseide Fabrik GmbH (1910; 2.5 million marks), Rheinische Kunstseide Fabrik A G (1907; 2.0 million marks), Glanzfüden A G (1910; 1.0 million marks),

Table 5.6 Single filament denier and number of filaments constituting a yarn (1922)

Process	Single filament denier			
	Germany	France	Switzerland	Belgium
Nitrate	1.4–1.8 (14 C2)	–	–	–
Cuprammonium	1.2–1.7 (57–227)	6.1–8.4 (20–21)	–	–
Viscose	2.4–7.7 (31–113)	6.6–10.0 (16–23)	8.2–10.0 (11–16)	8.1, 8.4 (17, 18)

The original table (Table 8.9) of ref. 43 was constructed using data in ref. 44.
( ): Number of filaments constituting a yarn.

Sachsischen Kunstseiden Werke A G (1915; 1.5 million marks). Hanauer Kunstseide was dissolved in 1921 after defeat by VGF in a patent war and Rheinische Kunstseide was also dissolved.

In Table 5.7 the cost of production and the price of the regenerated cellulose yarns produced by three processes during 1899–1914 are collected. The table was compiled by revising Kamide's table 8.5[47] and the detailed references of the original table are omitted for simplicity. Even if the scarce and scattered data available are considered, we can presume that: (1) the prices of Chardonnet (nitrate) silk were always higher than those of VGF (cupro) silk, probably owing to the difference in fibre quality. The price of yarns produced on a commercial scale decreased in the following order: nitrate process >> cuprammonium process > viscose process. (2) The cost of fibre production decreased in the following order: nitrate process > cuprammonium process > viscose process.

Cuprammonium rayon produced in 1900–1920 had the following drawbacks, which severely limited its applications: (1) poor mechanical properties, including low tensile strength (TS), in particular in the wet state (0.3–0.5 $g\,den^{-1}$); wet–dry ratio 0.1–0.25, (2) uneven dyeing and (3) high price, compared with the natural fibres. Because of (1), the products were utilized for embroidery, ribbons and strings. In the 1920s–1930s, TS(dry) exceeded 1.5 $g\,den^{-1}$ and TS(wet) increased to above 0.8 $g\,den^{-1}$, allowing the fibres to expand their usefulness into washable apparel applications.

The rise and fall of the leading companies producing cuprammonium rayon since the 1950s can be summarized as follows. In 1955 British Bemberg closed a cuprammonium rayon plant at Doncaster. During the 1960s five companies (except in Eastern Europe) were in operation for production of cuprammonium rayon: J P Bemberg A G (Barmen and

Table 5.7 Production cost and price of regenerated cellulose fibres for 1899–1914

Year	Nitrate process		Cuprammonium process		Viscose process	
	cost	price	cost	price	cost	price
1894	20 f/kg	25–27 f/kg				
1897		26.5 f/kg				
1899	17 f/kg	50 f/kg			2–3 s/lb (VSS)	
1900	11. 3 s/lb	12 s/lb (Tubize)				
1901						
1902				956 d/lb (120 d)		
			9.56 m/kg	14 m/kg		
1903		40 f/kg				
1904	7 s/lb	~12 s/lb		2.5 s/lb		
1906				3 s/lb		5.6 s/lb (150 d)
1908		8 s3 d/lb (Tubize)		$2\frac{1}{3}$ s/lb		
1909						
						6.5 s/lb (150 d)
1911		20 f/kg			2 s4 d/lb (150 d)	
					24.9 d/lb	
1912	12–13 m/kg	20 f/kg	9–10 m/kg		35.9 d/lb	63.3 d/lb
					8–9 m/kg (average)	
1913			12 f/kg		2 s2 d/lb 78.4 d/lb	
1914						
1915					28.1 d/lb (180 d)	

The table was constructed on the basis of Table 8.5 of ref 47.
f: franc. (France), s: shilling (UK), d: pence (UK), m: mark (Germany).
Tubize: Davenport & Co Ltd, Chardonnet Silk Mill, Tubize, Belgium.
VSS: Viscose Spinning Syndicate, UK.
Denier of fibre is shown in brackets.

Wuppertal, Germany), Farbenfabriken Bayer (Dormagen, Germany), Asahi Chemical Industries Co (Nobeoka, Japan), Beaunit Fibers (Elizabethton, Tennessee, USA), and Bemberg S A (Gozzano, Italy). Note that all of them were originally based on the technology of J P Bemberg A G. In 1970 there were four plants producing cuprammonium rayon in the world, again excepting Eastern Europe: (1) J P Bemberg A G (Wuppertal), 27 tonnes/day; (2) Beaunit Fibers (Elizabethton), 25 tonnes/day; (3) Bemberg SpA (Gozzano), 14 tonnes/day; (4) Asahi Chemical Industries Co (Nobeoka), 80 tonnes/day (→80 tonnes/day in the 1990s). Now, only Asahi survives, although the economic environment is not very favourable, as a monopoly supplier of cuprammonium rayon.

5.1.3 Characteristic features of research and development of the regenerated cellulose fibre industry

The regenerated cellulose fibre industry, formed in the late 19th to early 20th centuries, is different from the traditional cotton and wool fibre industry which originated between the late 18th and early 19th centuries and also differs from the modern and synthetic fibre industry that developed in the mid-20th century. Regenerated cellulose is called the second generation fibre industry, cotton and wool is called the first generation, and the synthetic fibre industry is called the third generation.[48] The second generation fibre industry has common characteristics that feature in various processes.[48] Table 5.8 shows the dates that the basic patents were filed or literature was published and of the first commercial production of three kinds of regenerated cellulose fibres, all belonging to the same generation.[49] It is interesting to note that the processes of synthesis of cellulose derivatives, dissolution of the cellulose derivatives or cellulose into solvents and spinning of the solution into fibres were studied and developed at different locations. Nobody succeeded in developing the complete process in one location. If Despeissis's patent is regarded as a basic patent, it took 9 years to attain its first commercial use. The cuprammonium rayon process got off the ground 3–4 years after the first commercialization although 'true take-up' was only realized when J P Bemberg A G started large scale production by Thiele's method. In this sense, approximately 20 years elapsed before the first small-sized factory was scaled-up.

The professions of the main inventors of the regenerated cellulose fibres are summarized in Table 5.9. It is obvious that the majority are professional chemists with academic backgrounds. Fremery, Bronnert, Lehner and Müller (the last two are not shown in the table) had doctoral degrees. This is a common feature of the regenerated cellulose rayon industry.

The industrial revolution, including the textile (cotton and wool) industry, had been achieved by numerous amateurs.[50] But the operation,

Table 5.8 Basic patents and first commercial production of three kinds of regenerated cellulose fibres

Process	Basic patents			First commercial production (year)	The time the first commercialization was scaled up	Place of first commercial production
	derivatives	spinning dope	spinning			
Nitrate	1846 (Schoenbein)	1847 (Maynard)	1884 (Chardonnet)	1884	1891–1895 50 kg/day	Besançon (France)
Cuprammonium		1857 (Schweizer)[a]	1890 (Despeissis) 1897 (Pauly)	1899 1200 kg/day (1900)	1902–1903	Oberbruch (Germany)
Viscose	1892 C-B-B[c]	1892 C-B-B[c]	1898 (Stearn)	1901 600 kg/day (1903)	1910[b]	Sydowsaue (Germany)
				1904 453 kg/day	1908	Coventry (UK)

[a]in academic journal, [b]sold to VFG, [c]Cross–Bevan–Beadle.

Table 5.9 Profession of main inventors of cellulose chemical fibres

Process	Name (age when first patent obtained)	Profession
Nitrate	J W Swan (55)	Chemist, electric engineer (President of Society of Chemical Industry and Institute of Electrical Engineers)
	H B Chardonnet (45)	Scientist (Ecole Polytechnique), Count
Cuprammonium	M Fremery (38)	Chemist (Dr.)
	J Urban (34)	Engineer (Diploma-Ingenieur)
	E Bronnert (29)	Chemist (Strasbourg, Dr.)
	E Thiele (?)	Chemist (Dr.)
Viscose	C F Cross (ca.40)	Chemist (Owen's College)
	E J Bevan (ca.40)	Chemist (Owen's College)
	C H Stearn (?)	Bank clerk (self-educated chemist)
	C F Topham (?)	Technician (glass blower), inventor of genius

management and improvements to heavily mechanized processes required large groups of engineers who had been educated in science subjects at various levels, instead of traditional craftsmen trained by apprenticeship. In technical education, Germany and France were most advanced at that time. Nevertheless, in a relatively limited area of Lancashire, UK, many inventors who participated in the process of industrial development of natural fibres (late 18th to mid-19th centuries) worked independently, developing machines in isolation. In contrast to this, communication between developers was extremely comprehensive and complex during the development of regenerated cellulose fibres.

Figure 5.5 shows such person-to-person relationships.[51] The numbers in Fig. 5.5 indicate the order of contact between people from longest ago to most recent. For example, Bronnert once worked with Chardonnet, then with Lehner and after that joined the group of Fremery and Urban, who worked together all their lives in a kind of *Gemeinschaft*. The VGF factory at Niedermorschweiler was burnt down during World War I. After the war Bronnert left VGF to form his own company. The *gemeinschaftlich* relationship observed between Fremery and Urban was also found between Stearn and Topham. Bronnert's splendid business career is easily understood if we recall that he was an Alsatian. Processes were no longer commercialized by individuals. The groups of Fremery–Urban–Bronnert at VGF and Thiele–Elsässer at J P Bemberg contributed significantly to the cuprammonium rayon process. However, these two groups were relatively small in size, greatly different from the large research groups of gigantic contemporary multinational companies.

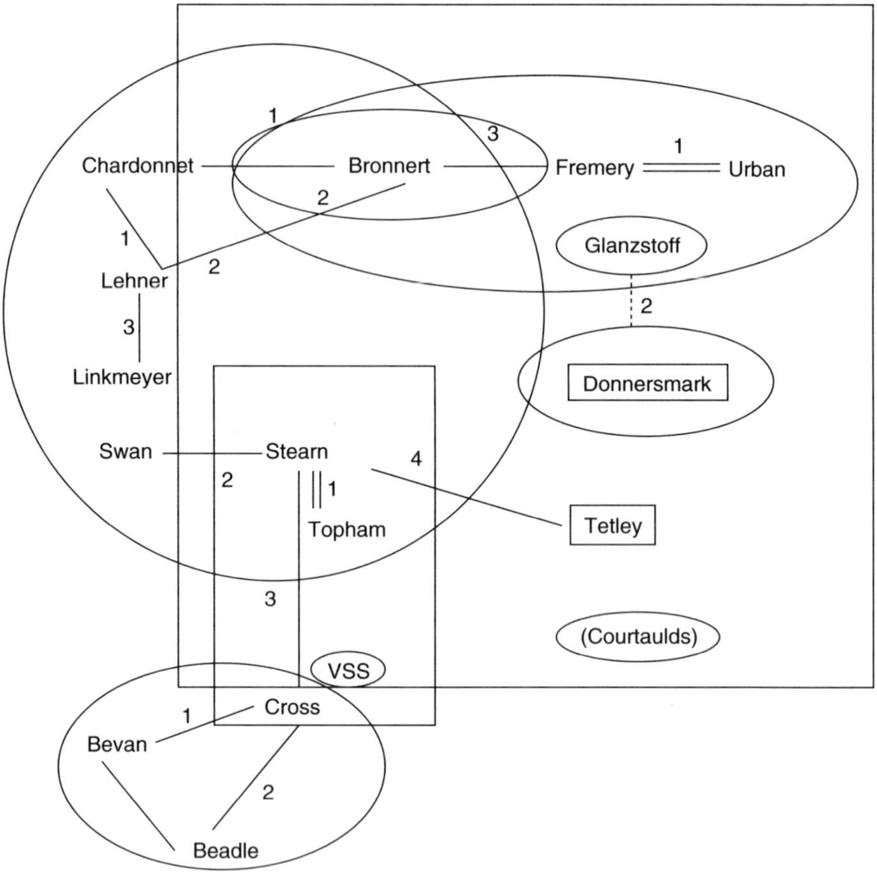

5.5 Person-to-person relationships between developers and entrepreneurs working on regenerated cellulose fibres. ═══ : gemeinschaftlich (comrade), ——— : gesellschaftlich (employer–employee), - - - - - : business relationships. A development group is shown by a large circle or rectangle.

The first commercial application of any rayon process was not directly concerned with textile fibres, but with electricity or coatings. Figure 5.6 shows the technological and personal relationships between the first commercial applications and the area of textile fibres.[52] In the cuprammonium rayon process the manufacturing technology of light bulb filaments was transferred, with further development, to the textile fibres. Chardonnet and Despeissis, who had intended from the first to produce textile fibres (artificial silk) directly were the exception, since France was a centre of the silk industry in Europe at that time. But Sir Joseph Wilson Swan (1828–1914) was actually the first man to commercialize rayon silk in 1883 via the nitrate

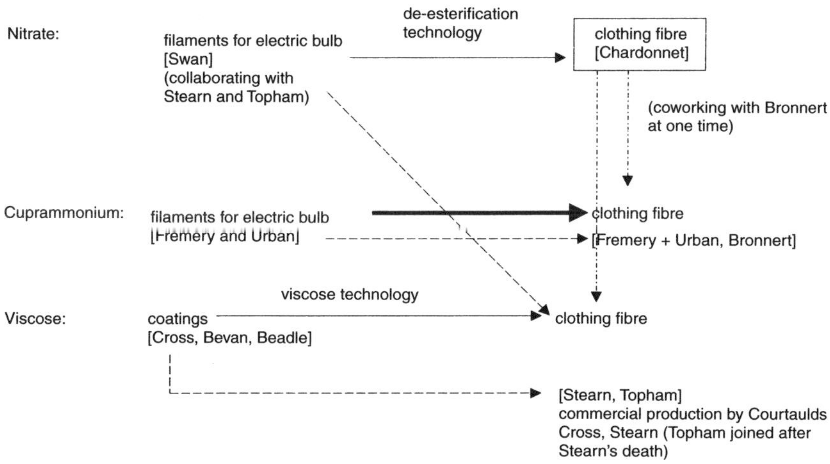

5.6 Personal and technological relationships between the commercial applications targeted initially and clothing fibres. [ ], Main developer; ( ), collaborator; ----▶, personal connection; ⟶, strong technological connection; ▭ , clothing fibre first commercial; ------▶ : technological stimulus (motivation for research); ⟶ : technological connection.

route, one year earlier than Chardonnet. Despeissis's first process could not be made commercial.

5.2 Science and technology of manufacturing processes

5.2.1 Cellulose

Cotton linter, the short fibres remaining on the surface of cotton (*Gossypium*) seed after ginning, is used exclusively for this process. It is a by-product of cotton seed oil mill production. Purified wood pulp (in particular, high α-pulp (α content > 95%) was once tried in place of cotton on a pilot scale at Asahi Chemical Industries during and immediately after World War II because of a shortage of cotton linters from overseas. It was not adopted commercially because of drawbacks in processability and the poor fibre quality,[53] both caused by the contamination of gel-like particles in the spinning solution. Cotton linter is now imported not only from the USA (Texas), but also from Mexico, Turkey, Pakistan, China, and the former Soviet Union. Raw cotton linter (with a cellulose content of about 70–75 wt% and a viscosity-average degree of polymerization (DP) = 2000–3000 (rough estimate)), is contaminated with fat, wax, proteins, pectins, inorganic materials, colouring materials, pepper trash, seeds and weeds. The composition differs markedly depending on the climate, soil and cultivating conditions under which the cotton grew. Cotton linter is purified by treating with 3–5% aqueous sodium hydroxide solution at 130–160°C for 2–6h, and then is bleached with aqueous sodium hypochloride solution, reduced with sodium sulphite and dehydrated to a water content of ~ 50 wt%. Strong correlations are shown between the breaking strength of the filaments in the wet state and the α-cellulose content of cellulose resources.[53]

5.2.2 Ammonia

At an early stage of commercial application, ammonia produced by decomposition of ammonium sulphate as a by-product of coke production, was utilized. Nowadays, ammonia is synthesized from natural gas. Very pure liquid ammonia is transferred from the ammonia factory and diluted with water to give a 25% ammonia solution.[54]

5.2.3 Cuprammonium solution

Cuprammonium hydroxide is formed by dissolving copper hydroxide in a concentrated ammonia solution. This so-called cuprammonium solution (*cupram; Kupfer-oxyd ammoniak Lösung; Kuoxam Lösung*) is a deep blue solution in which a kind of strong bivalent base, cupric tetramine hydroxide (CTH), $Cu(NH_3)_4\ (OH)_2$, is assumed to exist,[55] but has never been isolated. The concentration of CTH increases with temperature. On the other hand, the concentration of the ammonia dimer, NH_3–NH_3 which does not react with copper hydroxide to give CTH, increases with ammonium concentration so that the concentration of CTH attains a maximum in 6N NH_4OH.[56] CTH is extremely sensitive to light and undergoes rapid photodecomposition to copper oxide, having to be stored in a cold dark place.

In practice, copper sulphate $CuSO4{\cdot}5H_2O$ (purity > 99.2wt%, Fe_2O_3 < 0.08wt%, Al_2O_3 < 0.05wt%, ZnO < 0.04wt%) is extensively employed as a copper source. Metallic oxide contamination of copper sulphate changes to insoluble hydroxide in cuprammonium solution. Copper hydroxide is produced from copper sulphate via the following routes.

5.2.3.1 Direct method for production of copper hydroxide from copper sulphate

Addition of aqueous sodium hydroxide solution to aqueous copper sulphate solution at pH > 8–9 yields copper hydroxide. However, the copper hydroxide obtained by this reaction is readily oxidized by air and has not been adopted for commercial fibre production.

5.2.3.2 Indirect method for production of copper hydroxide from copper sulphate via basic copper sulphate

Hot (~55°C) dilute (~20 wt%) aqueous sodium carbonate solution is added to hot (~90°C) dilute (2–3 wt%) aqueous copper sulphate solution at a specific pH to give stable basic copper sulphate (BCS), $CuSO_4 \cdot 3Cu(OH)_2$ (or $CuSO_4 \cdot 2.5Cu(OH)_2$), (Equation [5.1])

$$4CuSO_4 + 3Na_2CO_3 + 3H_2O = CuSO_4 \cdot 3Cu(OH)_2 + 3Na_2SO_4 + 3CO_2 \qquad [5.1]$$

BCS appears as crystals, suspended in the mixture. The supernatant of the mixture is then decanted, and the precipitate (slurry) is dissolved in aqueous ammonium solution, in which any copper sulphate that did not react in the first step is transformed to $Cu(OH)_2$ by reaction with sodium hydroxide, (Equation [5.2])[57]

$$CuSO_4 \cdot 3Cu(OH)_2 + 2NaOH = 4Cu(OH)_2 + Na_2SO_4 \qquad [5.2]$$

In the solution $Cu(OH)_2$ reacts with ammonium ions to give cupric tetramine hydroxide (CTH), (Equation [5.3])

$$Cu(OH)_2 + 4NH_4OH = Cu(NH_3)_4(OH)_2 + 4H_2O \qquad [5.3]$$

In this way a cuprammonium solution is prepared. In addition to the $Cu(NH_3)_4$ ion, various copper complex ions, including ions of $Cu(NH_3)$, $Cu(NH_3)_2$, $Cu(NH_3)_3$, and $Cu(NH_3)_5$, as identified by a spectroscopic method, coexist with excess NH_4^+ in a $CuSO_4$–NH_3–H_2O system.[58] Hence, the cuprammonium solution should not be oversimplified as being aqueous CTH solution (i.e. the binary solution). The solubility of $Cu(OH)_2$ in aqueous ammonia is at most ~10 g l^{-1}.[57] The sulphate group, which coexists in the cuprammonium solution, lowers the solubility of cellulose in the cuprammonium solution, making the complete dissolution of cellulose difficult. Therefore, the sulphate content of BCS should be kept as small as possible.[59] Ammonia, which used to be a cheaper reagent than sodium hydroxide, was employed in place of sodium hydroxide.

In an alternative method, the rapid reaction of aqueous ammonium solution with copper sulphate solution at controlled pH (e.g. at pH = 8) with strong agitation was utilized for preparation of basic copper sulphate containing large amounts of $Cu(OH)_2$[60], (Equation [5.1′]), which is filtered and reacted with sodium hydroxide to yield pure hydroxide, (Equation [5.2])

$$CuSO_4 \cdot nCu(OH)_2 + NH_4OH \rightarrow Cu(NH_3)_{4\,\text{or}\,6}SO_4 + (n - m)Cu(NH_3)_{4\,\text{or}\,6}(OH)_2 + mCu(OH)_2 \qquad [5.1']$$

where n = 6–9.

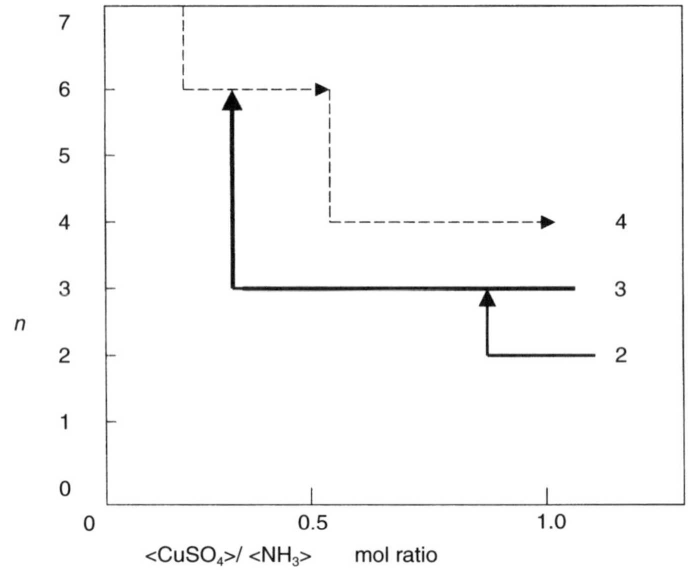

5.7 Number of molecules *n* of the precipitate (nCu(OH)$_2$). $CuSO_4$ as a function of molar ratio ⟨$CuSO_4$⟩/⟨NH_3⟩ for the mixture $CuSO_4$–NH_3–H_2O at 25 °C.[58] ———— : case when NH_3 is added to aqueous $CuSO_4$ solution; ------------ : case when $CuSO_4$ is added to aqueous NH_3 solution.

In Equation [5.1′], the precipitates appear if the molar ratio of copper to ammonium, designated as ⟨Cu⟩/⟨NH_3⟩, is larger than 0.2. The precipitate is a basic salt, composition nCu(OH)$_2$·$CuSO_4$.[58,61–66] 3Cu(OH)$_2$·$CuSO_4$ is formed in a nearly quantitative yield over a wide range of ⟨Cu⟩/⟨NH_3⟩ (see Fig. 5.7). To avoid precipitation, excess ammonium ion $(NH_4)^+$ is added to the mixture. The maximum amounts of copper hydroxide and of ammonia that are attainable, which can be exhausted in forming CTH in the solution, are unambiguously determined by the solubility limit of CTH in the solution. Then, 'excess' copper hydroxide and 'excess' ammonia are defined as those existing in other forms than CTH in the solution when CTH reaches saturation.

First, $Cu(NH_3)_4\ SO_4$ is formed by reaction of $CuSO_4$ with NH_4OH and this reacts with the remaining $CuSO_4$ by Equation [5.4]

$$3Cu(NH_3)_4SO_4 + 4CuSO_4 + 12H_2O = (6Cu(OH)_2)CuSO_4 + 6(NH_4)_2SO_4 \qquad [5.4]$$

The composition of BCS is governed by the molar ratio ⟨Cu⟩/⟨NH_3⟩. Relationships between the composition of the reaction mixture $CuSO_4$–NH_3–H_2O, as expressed by the molar ratio of $CuSO_4$ to NH_3, ⟨$CuSO_4$⟩/⟨NH_3⟩, and n moles of the precipitate(nCu(OH)$_2$)·$CuSO_4$, formed from the mixture at 25°C, are shown in Fig. 5.7. The relationship n–⟨$CuSO_4$⟩/⟨NH_3⟩ differs depending on the order of mixing of the three components: whether NH_3 is added to aqueous $CuSO_4$ solution or aqueous $CuSO_4$ is added to aqueous NH_3 solution.[58] At pH 7, $n = 4$–6 is obtained.

In commercial operations, the technological and economical effects should be carefully considered in order to choose the optimum conditions. For example, in the above case basic research indicated the use of $(6Cu(OH)_2)\ CuSO_4$, but it became clear that the new process had several problems in practical operation;[67] the concentration of Cu remaining in the supernatant phase increases (the yield of BCS decreases), compared with the case where $n = 3$ and where the value of n is >5 but varies depending on the delicate operating conditions which are not easy to control. Sedimentation of BCS ($n = 4$–6) suspended in the solution is extremely slow, washing out the ammonium sulphate that adheres to BCS is difficult and as a result, the concentration of SO_4^- in the spinning solution does not decrease.

In summary the cuprammonium solution is prepared via the route shown in Equation [5.5]:

$$CuSO_4 \xrightarrow{[5.1]\text{ or }[5.1']} \underset{\text{BCS}}{nCu(OH)_2 \cdot CuSO_4} \xrightarrow{[5.2]} Cu(OH)_2 \xrightarrow{[5.3]} \underset{\text{CTH}}{Cu(NH_3)_4(OH)_2} \qquad [5.5]$$

5.2.4 Dissolution of cellulose into cuprammonium solution

Cellulose is dissolved in cuprammonium solution, prepared in advance and stocked in a closed dark vessel. Cellulose is usually supplied wet to avoid possible oxidation.[68] Reducing agents such as salts of tartaric acid or sodium sulphate have been utilized to keep the quality of cellulose in solution high, but if oxygen is completely excluded, these reagents are not necessary.

In this case, the solution contains excess copper hydroxide in suspension and also excess ammonium. The dissolution of cellulose can be explained as follows:[69]

1 First, cellulose dissolves by reacting with cupric tetramine hydroxide ion CTH to form a complex, releasing ammonium ions. In the dissolving process, it is obvious that free ammonia is generated.
2 Then, excess, undissolved copper hydroxide suspended in the solution dissolves, to compensate for the amount of CTH consumed to form the cellulose complex, by reacting with (excess) ammonia to form CTH.
3 Newly formed CTH reacts with cellulose to form a complex dissolvable in the solution.

The existence of excess copper hydroxide and ammonium in solution allows us to keep the maximum concentration of CTH and, as result, enables us finally to prepare the highly viscous concentrated solution (cellulose concentration C_c is 10–12 wt%, viscosity is 1000–3000 poise at 30°C, specific gravity is 1.1 at 30°C). As the dispersion of cellulose into cuprammonium solution approaches the molecular dispersion, the viscosity of the dispersed system decreases somewhat. As dissolution of cellulose proceeds, excess copper hydroxide and excess ammonia are completely exhausted. Cellulose dissolves forming a cellulose–cuprammonium complex in cuprammonium solution.

The scientific study of the cellulose–cuprammonium solution system can be traced back to that of the alcohol–alkali–copper system in 1898 by Fr. Billnheimer.[70,71] In Table 5.10, the brief history of the research on the above systems is summarized. The cellulose complex is now considered to be cuprammonium complex coordinated preferentially to the hydroxyl groups at C2 and C3 giving a δ chelate.[78,79] In these studies, elementary analysis and spectroscopic methods such as optical rotation, optical rotary dispersion and circular dichroism were used together with UV.

Fast mixing of the appropriate amounts of copper hydroxide (or basic copper sulphate), aqueous ammonium and cellulose is now a highly effective direct method of preparing a concentrated cellulose solution for spinning. The solution is refined by filtration using a cloth of fine nickel filaments and degassed under reduced pressure to exclude air bubbles and excess ammonia from the solution. A small amount of aqueous sodium hydroxide is added simultaneously or after making the above mixture. It is assumed that with addition of sodium hydroxide the concentration of the cellulose complex increases. Dissolution of cellulose in cuprammonium solution is slightly exothermic and the temperature of dissolution is usually controlled to about 15°C. Any rise in temperature of the solution makes the dissolution rate slower. Mechanical agitation during the dissolution step is extremely important. The composition of the spinning solution is, for example, as follows: cellulose, 8–12 wt%; ammonia, 6–8 wt%; copper, 3–5 wt%.

Miyamoto *et al.*[79] indicated that in the dissolved state cellulose does not always coordinate with the cuprammonium complex. This means, that when about a half of the glucopyranose units are substituted as the diol anion salt of the cuprammonium complex, this cellulose derivative is soluble in the aqueous cuprammonium solution. If we can make the cellulose derivative described above at least copper-coordinated in a separate process, it might lead to a new way of producing the cellulose–cuprammonium solution from lower concentrations of copper and ammonia. Miyamoto *et al.* succeeded experimentally in preparing cellulose solutions using half the copper and one-third the ammonia used in the conventional procedure.

Table 5.10 Brief history of scientific research on the cellulose–cuprammonium hydroxide system

System	Complex	
(I) Polyhydric alcohol +alkali- (sodium hydroxide) copper	Various crystalline materals	Billnheimer (1898)[70,71]
(II) Copper hydroxide + aq.ammonia	Copper–alcohol complex soluble in aq.alkali solution $Cu(NH_3)_4(CO)_2$	Bonsdorf (1903)[55]
(III) Glycerol + copper ethyleneamine hydroxide (copper oxide + ethylene diamine) (En)	$[(C_3H_4O_3)_2Cu]\ [Cu(En)_2]$	Traube (1911)[72]
(IV) Cellulose + copper ethyleneamine hydroxide	$[(C_6H_8O_5)_2Cu]\ [Cu(En)_2]$ (isolated as precipitate by addition of ethanol to the system) glucose : copper = 2 : 1 (mol/mol)	Traube (1921)[73]
(V) Cellulose + cuprammonium	$[(C_6H_8O_2)_2]^{2-}[Cu(NH_3)_4]^{2+}$ $[(C_6H_7O_5)Cu]^{2-}[Cu(NH_3)_4]^{2+}$	Traube (1921)[73] Hess–Messmer (1922)[74]
	$(C_6H_{10}O_5)\ (C_6H_7O_2)$ {OH, OH → $Cu(OH)_2$; OH $[Cu(NH_3)_4(OH)_2]_{1/2}$}	Lieser–Ebert (1925)[75]
	Copper coordinates to C2,C3 hydroxyl groups of glucose residue	Reeves (1944)[76]
	1 : 1 salt of CTH cation + diol anion at C2 and C3 positions of pyranose ring	
	Chelate coordination of copper with two hydroxyl groups	Bukhari (1968–1976)[77]
	Cuprammonium complex, coordinates preferentially to hydroxyl groups at C2 and C3, giving δ chelate form	Miyamoto *et al.* (1995)[78,79]

By what mechanism does cellulose solid dissolve into aqueous cuprammonium solution? Even in the case where cellulose or cellulose derivative solid is not easily soluble in solvents, once it is dissolved, the solution is usually stable. Cellulose–aq.cuprammonium solution is typical, so the dissolved state as an equilibrium state should be rigorously distinguished from the dissolution process. These two processes had not hitherto been distinguished for the cellulose–copper hydroxide–ammonia–water system. When cellulose solid is mixed with aqueous cuprammonium solution, cupric tetramine ion $(Cu(NH_3)_4)^{2+}$ reacts with cellulose to form the 'complex'. In this case, the cupric tetramine ion first penetrates, in particular in excess ammonia, the space between two neighbouring glucopyranose rings causing breakdown of the intermolecular hydrogen bonds. Second, the cupric tetramine ion builds the cellulose–cuparammonium complex by breaking down the intramolecular hydrogen bonds of the glucopyranose units. A cellulose molecule, in which some glucopyranose rings form a 'complex' with cupric tetramine ions, disperses molecularly into the remaining solution. The underlying mechanism of dissolution of cellulose solid into solvent is very common in cuprammonium solution and dilute alkali solutions. Cupric tetramine ion acts to break the intramolecular hydrogen bonds. Miyamoto *et al.*[79] showed that the apparent degree of breakdown in the intramolecular hydrogen bond at O_3 H-O_5' (Here, O_3H means OH group at C_3 position on a glucopyranose ring and O_5' is ring oxygen on a neighbouring glucopyranose ring) of the undissolved part of cellulose solid, treated with cuprammonium solution and regenerated with dilute acid, increased with the concentration of the hydroxyl group C_{OH}, where cuprammonium complex is ionized. Their results suggest that cuprammonium ion breaks the intramolecular hydrogen bonds. It is now clear that the dissolution of cellulose does not depend on the formation of a 'cellulose-complex with a definite composition'. The cellulose-complex can be considered to play in dissolution as a kind of strong solvated solute, interfering effectively with the reassociation of dispersed cellulose molecules.

So, the key factors governing dissolution of cellulose are (1) breakdown of supermolecular (especially, intramolecular hydrogen) bonds in the solid and (2) formation of strong solvation (or interaction). Figure 5.8 shows schema of speculative concept of dissolution of cellulose into aqueous cuprammonium solution.

5.2.5 Fundamentals of wet-spinning

The fundamental technological concept behind the wet-spinning of cuprammonium rayon is the complete separation of the following three steps, stretching (deformation), coagulation (solidification) and regeneration. This is essential to produce better quality fibres with high productivity.

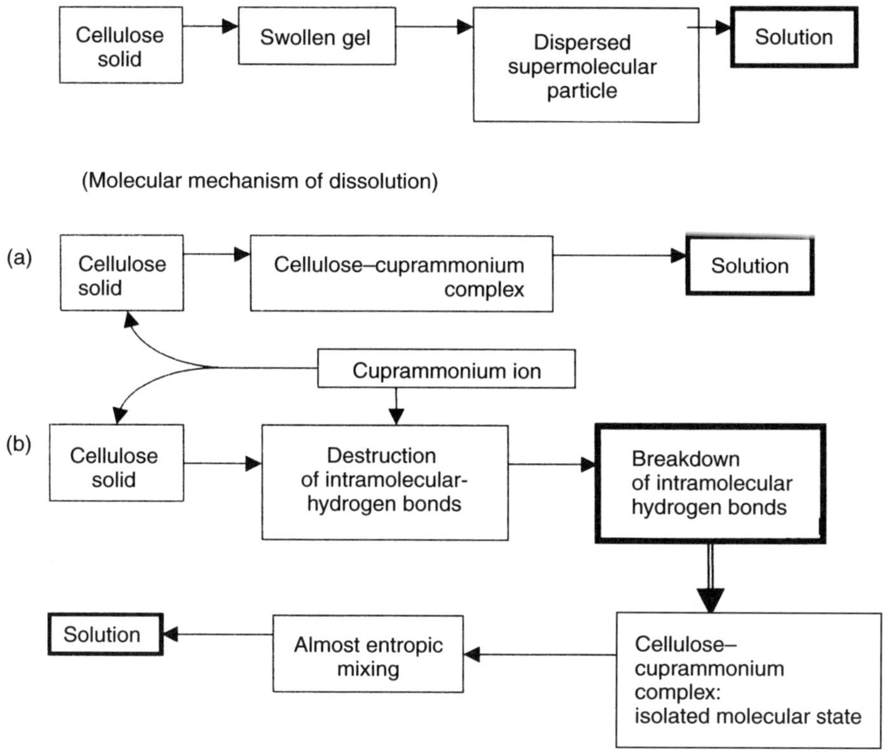

5.8 Dissolution of cellulose into aqueous cuprammonium solution: (a) classical concept, (b) new concept.

In the Despeissis[7] and VGF[10] patents, the last two steps are completely mixed together and indistinct and the first step is practically absent; Thiele intended to separate the first two steps from the latter, although the degree remained far from the ideal state. The Asahi Chemical Industries Co have developed an advanced and highly sophisticated process along the lines of the above concept (see Table 5.11).

5.2.6 Dynamics of stretch-spinning

Stretch-spinning, first used commercially by J P Bemberg A G (Germany) in the 1910s has been extensively developed in Germany, the USA and Japan. Viscous dope is extruded under pressure through a die (nickel or stainless steel), that has holes 0.6–1.6 mm in diameter, into a glass or plastic funnel flowing with degassed and de-ionized, warm spinning water (*Fall wasser*). The dope coagulates very gradually in the funnel and is stretched hydrodynamically up to some hundreds times, forming fibres which emerge

Table 5.11 Revolution in wet-spinning technology

Stage	Relation between three steps	Apparatus	Inventor(s) (period)
I	Coagulation + regeneration (no stretching)	Single bath	Despeissis, VGF (1890–1910)
II	Stretching + coagulation–regeneration	Funnel + bath	Thiele (JPB) (1900–1940)
III	Stretching–coagulation–regeneration	Double or triple funnel system + bath	Asahi (1950–1990)

+, simultaneously; –, separately.

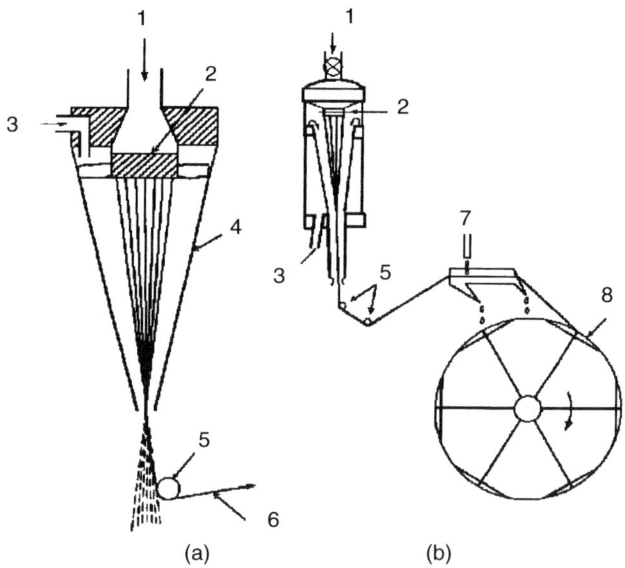

5.9 Stretch-spinning showing funnel (a), and hank-type spinning machine (b). 1: spinning solution; 2: spinneret; 3: spinning water; 4: funnel; 5: guide; 6: blue yarn; 7: dil. acid; 8: hank.

from the lower end of the funnel (Fig. 5.9(a)). The filaments are drawn round a rod to separate them from the entrained spinning water that they are carrying. The spinning funnel design has advanced remarkably in the last 70 years (see Fig. 5.17) and a more complete separation of drawing (deformation) and coagulation processes, which Thiele dreamed of a hundred years ago, has now been successfully realized in practice by Asahi (Nobeoka, Japan). Note that use of a single funnel does not allow effective separation.

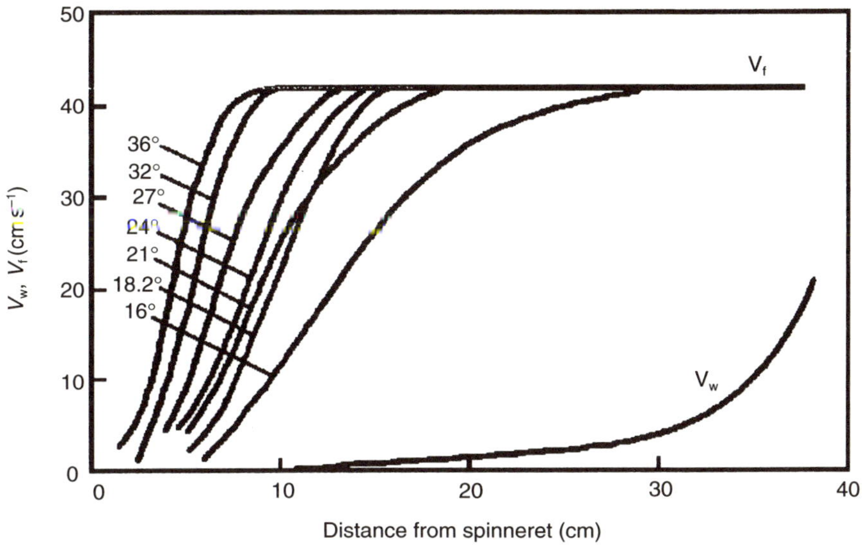

5.10 Line velocity of running liquid thread V_f and of the spinbath V_w plotted against distance from spinneret. The numbers indicate the temperature (°C) of the spinbath.[81]

The cellulose–cuprammonium complex is supposed to be decomposed during drawing in the funnel by hydrolysis removing part of the copper and about 70–80% of the ammonia from the complex, giving the copper–cellulose complex compound known as 'blue yarn'.

Five forces act on the running thread[80] (1) gravity (density of the threads), (2) buoyancy (density of the spin bath), (3) friction between the threads and the spinning water owing to their velocity difference, (4) surface tension on the boundary of thread and water, and (5) tension, generated from the take-up roll and applied on the filament. These forces are balanced in steady state operations. The density difference between thread and water is quite small, in particular at the inlet region of the funnel where the direction of gravitational force is opposite to that of buoyancy, so that forces (1) and (2) are negligible. The frictional force (3) is a predominant factor in controlling the drawing of filament in the funnel. Force (3) varies depending on the shape and length of funnel, the spinning velocity and the denier(diameter) of single filaments. Except very near the inlet region, the velocity of thread is greater than that of spinning water (Fig. 5.10).[81] The surface tension is also minimal. The region between the inlet to the middle part of the funnel plays an important role in controlling the quality of fibres. Provided that all other operating conditions are kept the same, the coagulating rate increases with an increase in cellulose concentration in the

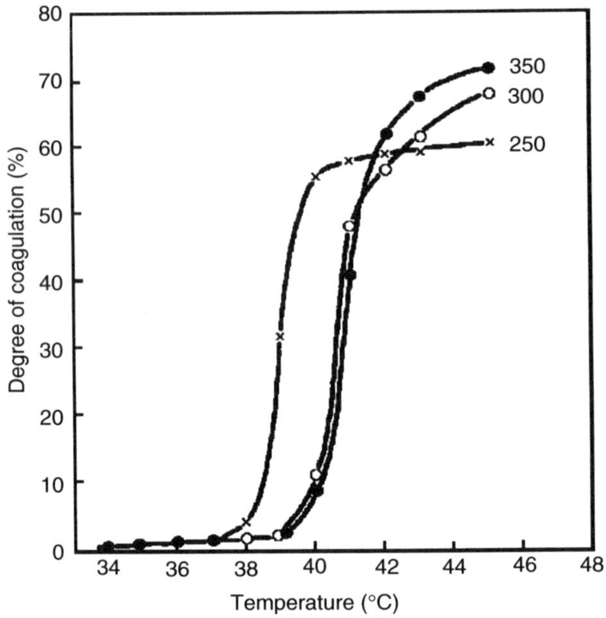

5.11 Degree of coagulation of threads, as measured by their ammonium content, as a function of temperature of the spinbath. The numbers show the flow rate (ml/min) in the spinbath.[84]

spinning solution and of the temperature of the spin bath[82,83] (Fig. 5.11) and with a decrease in the ammonium concentration. A low coagulation rate results in occasionally insufficient coagulation of the running thread within a funnel. A high coagulation rate indicates that the coagulation completes relatively high up in the funnel (Fig. 5.9) and the spinning tension increases, giving fibres with an irregular cross-section, which have poor yarn qualities such as low tensile strength and low tensile elongation. Viscosity-spinning solution too low is difficult to stretch in a funnel and solution too high up is not suitable for in-pipe transportation and filtration (in other words, processability is poor).

The characteristic of the 'flow stretching' process can be summarized as:

1 Use of a larger pore (~0.8 mm diameter) die becomes possible.
2 Use of a highly viscous spinning liquid becomes necessary.
3 Yarns can be made from high molecular weight cellulose and have better physical properties.
4 Fibres can be made with relatively uniform cross-section, with no microscopically visible skin unlike viscose rayon fibres (see Fig 5.24(b)).

5 No evolution of gas during coagulation. (In the viscose rayon process the decomposition of cellulose xanthate is accompanied by unavoidable evolution of toxic gas.) No microscopic voids in the final fibres.
6 Very fine fibres are achieved by high stretching.

Thiele anticipated points 1, 2 and 6 in his patent specifications.[29]

The thread, after leaving the lower end of the funnel, runs into a small bath containing sulphuric acid (5–10 wt%), where the cellulose is completely regenerated. Very pure sulphuric acid is supplied for this purpose.

5.2.7 Commercial stretch-spinning process

Hank spinning (or reel spinning) was first used commercially before the 1940s. In this process the untwisted yarn is wound up in a hank as illustrated in Fig. 5.9(b). The yarns are twisted in the form of skein before weaving. These additional operations are very labour intensive and expensive. The yarns made by the hank process have relatively lower tensile elongation. The centrifugal (or Topham box) collection method, in which the thread is wound in twisted form (the early viscose rayon process) gives yarns with better mechanical properties. For cuprammonium yarns this replaced the Hank spinning process during the 1930s and it was suspended (by Asahi) in 1943. The yarns were supplied to the weavers in the form of cake, which demanded careful handling, otherwise the fine filaments broke. It was discovered in the mid-1940s that a careful choice of the spinning conditions produces slightly self-bonding fibres (i.e. yarns in which the constituent filaments adhere to each other slightly owing to their self-adhesive properties, but after weaving the adhesion of filaments can be removed in a finishing step on the woven cloth). With this technique new 'Hank process' yarns can be woven without any further twisting. The Hank process therefore was revived during the 1950s, overtaking the centrifugal process, which was actually abolished in 1957 (by Asahi), being replaced by a continuous process, in which untwisted yarns are collected on 'cheeses'. The Hank process still survives although its importance has diminished considerably.

5.2.8 Thermodynamics of wet-spinning

Coagulation that occurs during the spinning process is a kind of phase separation. From the 1970s–1990s the thermodynamics of phase separation in polymer solutions was extensively studied for multicomponent polymer–solvent or polymer–solvent–non solvent systems.[85] Kamide and co-workers showed that when the initial polymer concentration (i.e. the solution concentration at the instance of phase separation) V_p^0 is smaller than the critical concentration V_p^c, the polymer-rich phase separates as small particles suspended in a medium (i.e. a polymer-lean phase) and these particles grow by amalgamation. Phase separation is a sequential occurrence of the following elementary steps (Fig. 5.12).[86] Figure 5.12 shows (a) generation of critical nuclei,[87] (b) growth of particle (diffusion),[85,87] (c) attainment of phase equilibrium (primary particle),[85,87] (d) growth of primary particles by amalgamation,[85,88] (e) formation of secondary particles,[85,88] (f) coagulation by contact between secondary particles,[85,88] (g) the coagulated membrane,[85,89,90] (h) regeneration and drying.[85,89] Detailed molecular mechanisms of the above steps were disclosed by Kamide, Iijima and their co-workers.[87–93]

In wet-spinning, the shape of the secondary particles changes dramatically to microfibrils. Stretching a non-coagulated thread of cellulose cuprammonium solution may be regarded as simple flow of the spinning solution, mixed with a few of the initial steps in Fig. 5.12. When tension is absent, the interstitial space between particles gives a porous structure, that is, pores are formed by contaction of the secondary particles. The delicate control of the morphology of the fibre was achieved successfully in the case of simple polymer solutions such as the acrylonitrile/methylacrylate copolymer–aqueous nitric acid system.[94] It is easily assumed that the wet-spinning process of cellulose cuprammonium solution may obey the above-mentioned thermodynamic mechanism, but the cellulose–cuprammonium solution system is tremendously complicated to study. For example, the spinning solution is composed of a cellulose–cuprammonium complex, cupric tetramine hydroxide (CTH), ammonium ion, copper ion, hydroxide ion, non-ionized ammonia, water, and so on, and the composition of each component, which is not yet fully understood, varies depending on the nature of starting materials, and on the temperature of the coagulating liquid. If some specific solvents other than water and acid are chosen as coagulant, macroscopical liquid–liquid phase separation occurs even for cellulose–cuprammonium solution.[95] Acetone provides an example and the phase-diagram of this system was investigated by approximating the cellulose–cuprammonium solution + acetone system as a quasi-quaternary system ((cellulose + copper) + ammonia + water + acetone).[95]

Figure 5.13 shows electron micrographs of particles in cellulose–cuprammonium solution coagulated by an aqueous ammonia–acetone mixture.[96] Phase separation of the above cellulose solutions obeys the particle growth mechanism shown in Fig. 5.12.[97]

If we stop the process of coagulation half way, the secondary particles can be separated.[95] Using the cellulose–cuprammonium–coagulant system, microporous hollow fibre membranes were developed[98] and are commercially produced by Asahi at Nobeoka. Iijima *et al.* studied the phenomenological effects of solvent-casting conditions on the pore

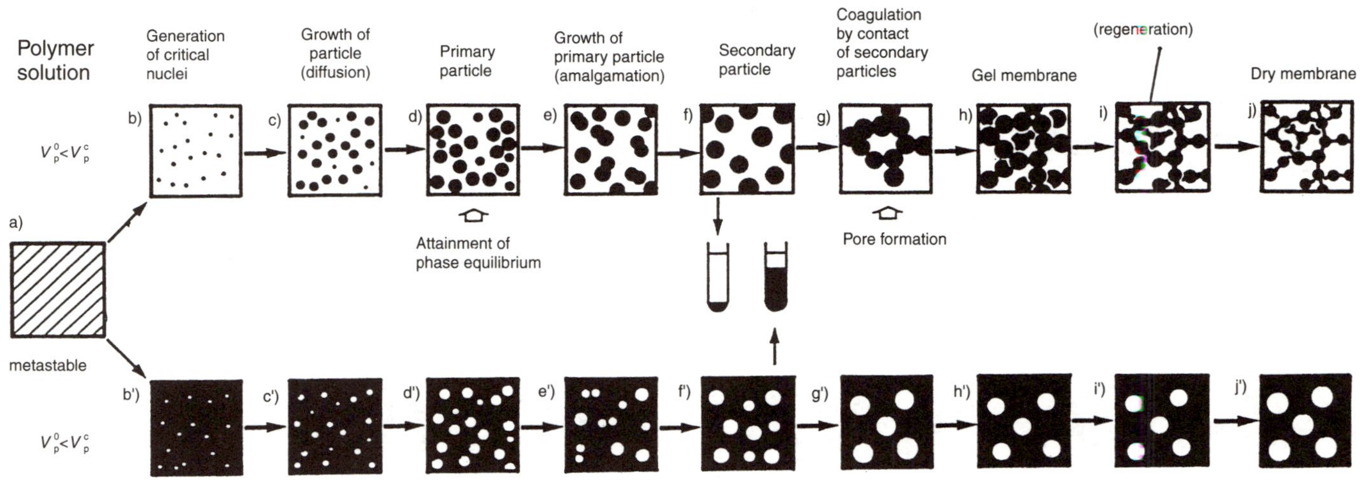

5.12 Elementary steps in porous membrane formation by the microphase separation method. V_p^o, polymer volume fraction of the solution when the phase separation occurs; V_p^c, polymer volume fraction of critical solution point.[89]

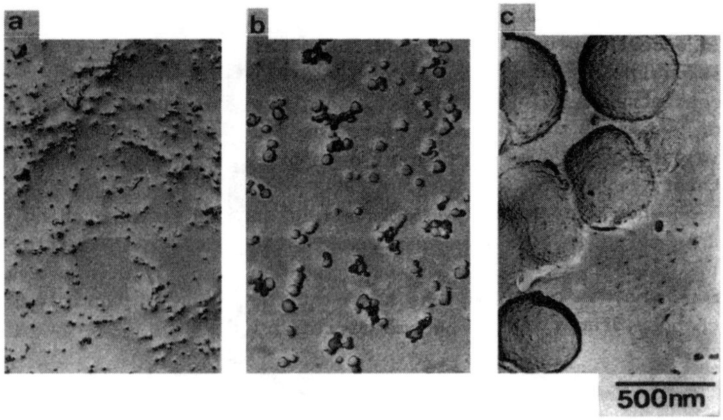

5.13 Electron micrographs of particles in cellulose cuprammonium solution coagulated by aqueous ammonia–acetone mixture (steps d, e, and g in Fig. 5.12). (a) primary particles (step d); (b) intermediate stage of the conversion of primary particles to secondary particles (step e); (c) the secondary particles in contact with each other (step f); starting solution (the weight fractions of cellulose, $W_{cell}{}^{S}$; copper, $W_{cu}{}^{S}$; ammonia, $W_{NH3}{}^{S}$ and $W_{H2O}{}^{S}$ are 0.08, 0.316, 0.1122 and 0.7762, respectively). Composition of coagulant, acetone/ammonia/water (30.00/0.56/69.44, wt/wt/wt).[94]

characteristics of cellulose membranes regenerated from cuprammonium solution. The acetone concentration $W_{acetone}$ of the system is one of the most important factors determining phase separation characteristics like the two-phase volume ratio (i.e. the volume ratio of the polymer-lean phase to the polymer-rich phase) and the composition at the point of phase separation. The ammonium concentration, W_{NH3}, in the cellulose/cuprammonium/acetone/water system influences the size of the secondary cellulose particles. Membranes consisting of large secondary particles have a larger mean pore size, and this fact agrees well with the prediction of Kamide's lattice theory[90] on pore size distribution. Iijima *et al.* also studied the supermolecular structure and virus separability of regenerated porous cellulose membranes.[92] According to them, a porous cellulose membrane has a 'cavity-vein' structure (Fig. 5.14). Viruses were captured by two different mechanisms: accumulation in cavities and blocking in narrow veins. Viruses moved downstream in the membrane, leaving in the filtrate, however, without exceeding a 'critical' filtration volume V_c, the porous cellulose membrane achieved complete rejection of viruses from the filtrate (i.e. aqueous serum protein mixture such as blood) (Fig. 5.15).[92] V_c depends on filtration pressure and virus concentration in the filtrand (i.e. supplying liquid).

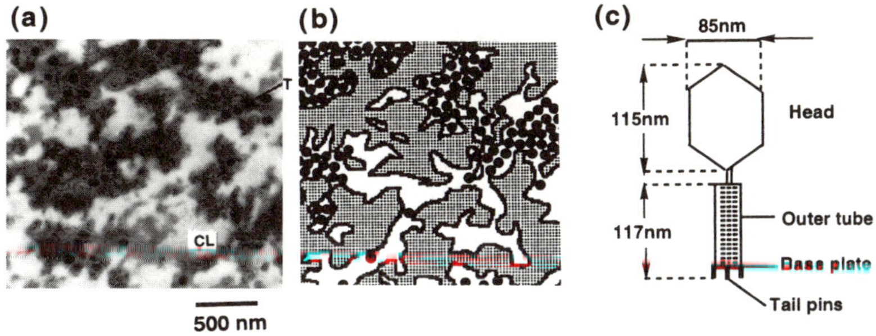

5.14 Cross-section of a porous regenerated cellulose membrane and captured viruses in the membrane. (a) transmission electron micrograph of a cross-section of porous regenerated cellulose hollow fibre membrane: pore diameter by flow rate method 2 r_f = 41 nm; a T4 phage suspension was filtered through the membrane by perpendicular filtration at a pressure difference ΔP = 200 mmHg; brackish particle (T) T4 phage; white area (CL) cellulose; grey area, pore; (b) schematic representation of cavity-vein structure: shadowed area, cellulose; black particle, virus particle; (c) sketch of T4 phage.[92]

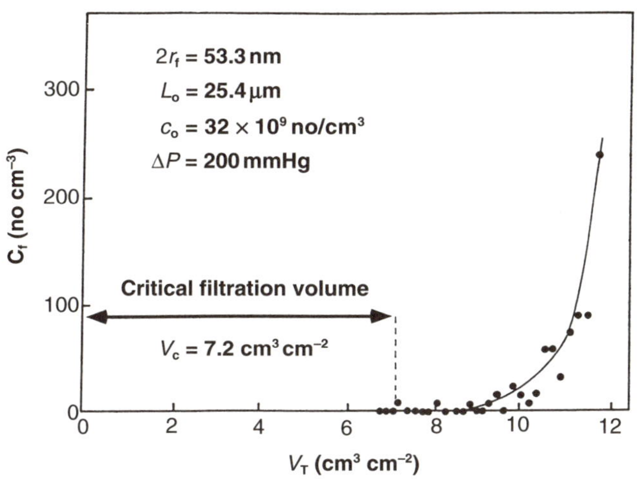

5.15 Change in virus concentration C_f during filtration.[92] The volume of each fraction was ca. 0.2 ml, V_c was 27.2 ml cm^{-2}, $2r_f$ was 53.3 nm, C_o is concentration of virus, L_0 is thickness of membrane, V_T is total volume of rinsed saline water per unit area of membrane during rinsing, ΔP is pressure difference.

5.2.9 Molecular characterization of cellulose and its derivative solutions

Until the late 1970s, very few comprehensive and reliable studies on the molecular characteristics of cellulose derivatives (CD) had been reported,

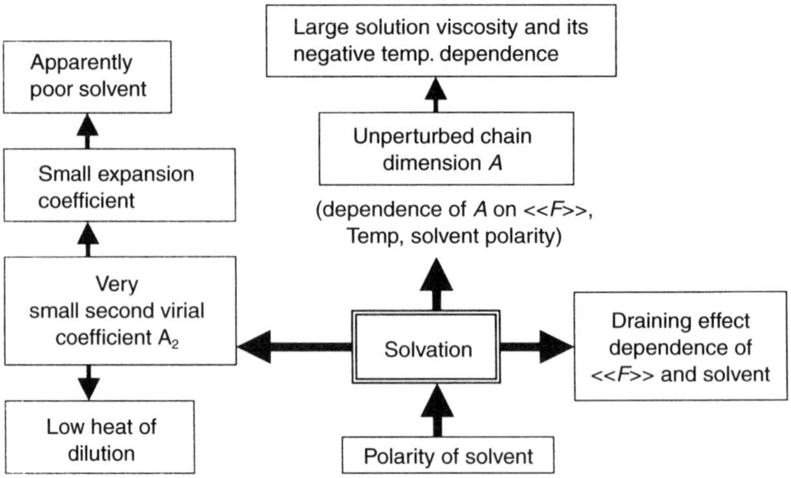

5.16 Relationship between solvation and some characteristic features of cellulose derivatives solutions[99] (<<*F*>> is the total degree of substitution).

because of experimental difficulties encountered in the determination of the substituent groups, fractionation, and the removal of gel-like materials from the solution. Moreover, the experimental data, that had been obtained unsystematically for some cellulose and CD solutions, could not be explained reasonably in terms of the known and established solution theories applicable to many synthetic flexible polymers. For this reason, the molecular characterization of cellulose and CD remained unsolved even in the 1970s. During the 1980s–1990s, the molecular properties of cellulose and its derivatives (in particular cellulose acetates) solutions were synthetically studied.[99–101] Needless to say, these studies were very valuable providing basic knowledge for planning in an industrial fibre-forming process. Unfortunately, cellulose–cuprammonium solution had not until then been a target of this kind of study except at a rather primitive stage of research in the 1930s–1940s.

Common phenomenological features of cellulose and CD solutions are: (1) a small second virial coefficient, (2) a low heat of dilution, (3) a large solution viscosity and (4) its negative temperature dependence. These can be reasonably and consistently explained by solvation (Fig. 5.16), which had been quantitatively evaluated by adiabatic compressibility and nuclear magnetic resonance measurements.[99]

Table 5.12 collects some molecular parameters of cellulose solutions. The table was constructed from tables 11 and 13 from Kamide and Saito[99] and table 7 from Kamide and Saito.[102] From Table 5.12 we can conclude: (1) cellulose dissolves molecularly in cadoxen, FeTNa and 6 wt% aqueous LiOH,

Table 5.12 Some typical molecular parameters of cellulose solutions

Molecular properties	Cellulose		
	cadoxen	FeTNa	6 wt% (LiOH)
$K_m \times 10^2$ ($cm^3 g^{-1}$)[a]	3.85	5.31	2.78
a[b]	0.77	0.78	0.79
X[c]	0.2–2.2	0.3–2.9	—
$A \times 10^8$ (cm)[d]			
method 2B[e]	1.53	1.96	2.18
method 2D[f]	—	2.27	—
method 2G[g]	1.57	2.17	2.13
method 2K[h]	1.56	2.07	2.13
most probable	1.55	2.12	2.15

[a,b] Parameters K_m and a in the Mark–Houwink–Sakurada equation; between the limiting viscosity number[η] and the molecular weight M: $[\eta] = K_m M^a$.
[c] Draining parameter, X defined by $X = (3/2\pi)^{1/2}(d/a')N^{1/2}$ (where a' is the length of a link, d is the hydrodynamic diameter of segment in pearl-necklace model, N is the number of segments). If $X \gtrsim$ 10–20, the polymer chain can be regarded as a solvent-impermeable sphere.
[d] Unperturbed chain dimension.
[e] From the linear expansion factor calculated from the second virial coefficient, M and radius of gyration $<S^2>^{1/2}$.
[f] From $<S^2>^{1/2}$, M and the non-Gaussian unperturbed chain a_2.
[g] From [η], a_2 and the molecular dependence of parameter $\Phi(a_\Phi)$.
[h] From K_m, a, a_2, and a_Φ.

(2) cellulose molecule behaves as Gaussian chain in the above solvents in the unperturbed state, (3) the excluded volume is small, but the draining effect can never be ignored. Not only the physical properties shown in Table 5.12, but also many other properties of the above solutions show similarities to those of CD solutions (see Table 5.10[99], Table 5.11[99], Table 5.13[99], and figs. 6.3.8[99] and 6.3.9 from Kamide and Saito[99]). In other words, cellulose complex solutions like cellulose–cadoxen and cellulose–FeNa solutions have the same features as those of CD solutions, which have been investigated more systematically. Judging from the similarity between parameters K_m and a in the Mark–Houwink–Sakurada equations (see Table 5.12, footnotes a and b), solutions of cellulose in aqueous NaOH, cuoxam (cuprammonium hydroxide), cuen (cupriethylenediamine hydroxide) and EWNN (iron–tartaric acid–sodium complex solution) may behave similarly.

5.2.10 Continuous spinning process

The continuous wet-spinning process is defined as a process in which the extrusion of the spinning dope, and the final winding of the thread on to a normal textile package, take place continuously on one machine or on an integrated set of machines which, to all intents and purposes, form one machine.[103] From 1905 the improvement of spinning apparatus including solidification, drying and winding was carried out at VGF under the guidance of Eduard Boss (1877–1935; he joined VGF in 1905 and married Fremery's daughter).[104–106]

According to Barker and Alleston,[103] the continuous process can be classified as follows:

5.2.10.1 Simple continuous method

This consists of guiding yarn over and under rollers. The original idea was proposed by the Société Française de la Viscose (SFG)[107] in 1906 and by J Foltzer[108] in 1908. Both remained just ideas until commercialized by J P Bemberg before the end of the First World War: in the 1940s J P Bemberg operated their machines for spinning direct to warping beams and produced about 1000 tonnes of warps by the end of the war.[109] This was called the Dureta method.[110,111] Afterwards J P Bemberg developed the process with the collaboration of I G Farbenindustrie (later Farbenfabriken Bayer).[103] Their process consisted of a spinning machine (400 ends, 12.8 m long), a washing machine (8.5 m long) and a drying machine (4.4 m long) (overall length 25.4 m). The threads were formed into a sheet, passing through acid treatment and into a wash bath in this form. After sizing and drying they were wound on to a beam.[101] The Farbenfabriken Bayer operation at Dormagen used this type of continuous spinning machine, selling the product under the trade name of Cupresal®.[112,113]

5.2.10.2 Continuous belt method

The yarn is dropped on a very slowly travelling belt in a loose spiral form and transferred through various treatment baths on the belt. Boss proposed this method in 1906,[103,114,115] but it was not commercialized. Forty years later, difficulty in controlling the tension and shrinkage in the above method was pointed out.[101] In 1972, Miyazaki *et al.* at Asahi invented[116] this technology and in the 1980s–1990s Asahi succeeded in applying it commercially, which will be described in more detail in Section 5.2.12.

5.2.10.3 Advancing reel or helical reel method

An early suggestion for the use of a reel was made in VGF's patent literature in 1910.[117,118] In 1932 William Henry Furness invented an advancing reel method with a specially designed reel suitable for the cuprammonium rayon process (United States Patent No 1770750 (1930) and Japanese Patent No 101 379 (1933)).[119,120] The Furness method was taken over by the Industrial Rayon Corp (USA)[103,121,122] and commercialized in 1938 for high-tenacity rayon production.[122] The success of the commercial application of this method is mainly due to use of a chemically resistant small plastic reel and the isolation of each treatment to one or more small advancing reels.[103]

5.2.10.4 Simple roller method

In 1952 Hugo Hofmann invented a continuous machine,[123] which was purchased by Beaunit Mills and used commercially at Elizabethton, Tennessee.[124]

Table 5.13 summarizes the progress of continuous spinning technology. Note that in the 1950s the spinning speed of this technology was, without exception, in the range of $< 70\,m\,min^{-1}$.

Some properties of the end products of regenerated cellulose fibres depend greatly on the residual internal strain, which was fixed in the fibre in the spinning and finishing process, and in particular, in the tension-drying step, and which when released brings about significant shrinkage in the after-treatment process (e.g. the drying step of scoured cloth). Therefore, the properties of the products produced by the continuous process are usually different from those produced by hank or cake processes. The operating conditions of continuous spinning and after-treatment processes and their formulae for the filaments produced by the continuous method should be optimized.

Asahi's continuous process originated in the USA.[125] The process that was licensed in 1953 from Beaunit fibres gave untwisted, hard yarns wound on to a beam, unsuitable for rewinding and mainly used only for knitting. This continuous process had first to be improved by coating the yarns with carefully chosen softening oil before drying in order to produce soft fibres. In this way, continuously spun yarns successfully widened the end-use application in woven fabrics, resulting in closure of the cake-yarn process. Both hank and cake processes were batch processes, and were extremely limited in the face of the dramatic improvements in the processability and yarn quality that were urgently demanded in the 1950s–1960s in order to compete economically with newly developing synthetic fibres. That the production of untwisted cuprammonium fibres in the factory was not to be a fatal disadvantage was due to significant technological advances in winding and weaving machines during the 1950s, and due to a change of fashion to bright yarns. The continuous spinning process conformed to the 'faster, cheaper, better' philosophy quite well.

Table 5.13 Development and commercialization of the continuous wet-spinning process of cuprammonium rayon 1900–present

Class	Method	Fundamental idea or patent(s)	Commercial process
I	Simple continuous method	(1) Société' Française de la Viscose; DP Nr 192 406 (1906) (2) Foltzer J; DP Nr 209 923 (1908) (3) DP Nr 763 735 (1940) ⟹	JPB (afterwards jointly with IG) 1940s–1950s Barmen (Dureta method) JPB (Dureta method) Bayer (ex IG) Dormagen
II	Continuous belt method	(1) Boss E (VGF), DP Nr 235 134 (1906) (2) Miyazaki, *et al.*, USP No 3 689 620 (1972) ⟹	Asahi, Nobeoka 1980s–1990s
III	Advancing reel or helical reel method	(1) VGF, DP Nr 239 822 (1910) (2) Furness W, USP No. 180 0828 (1931); JP No 101 379 (1931) ⟹	Industrial Rayon Corp (USA) (Viscose)
IV	Simple roller	(3) Hofmann H, US Patent 2 587 619 (1952) ⟹	Beaunit Mills Elizabethton ↓ 1953 Asahi Chem. Nobeoka

⟹ Commercialization, ⟶ transfer of technology.

5.2.11 Technology innovation in the conventional spinning process

The spinning velocity is a good measure of the sophistication of the production technology. On continuous machines, high velocity cannot be attained by improving the spinning process alone, but requires innovation in many other areas, closely correlated with spinning speed, such as purification of cellulose and preparation of a highly uniform spinning solution.

During the 1920s–1940s, the spinning velocity in the plant operation was far below 50 m min^{-1}. From 1940–1965, numerous energetic attempts were made to improve productivity in the manufacture of fibres, based on the fundamental concept of wet-spinning already described.

The following are key criteria for improving the conventional process:

1 A small frictional resistance between thread and coagulation bath (i.e. bath resistance).
2 Uniform drawing (stretching) of liquid thread with little or no coagulation.
3 Sufficient coagulation of the running thread under the low tension.

Figure 5.17 shows 'stretch-spinning' apparatus invented at Asahi during the 1950s–1970s for increasing the spinning velocity. In 1950 Munakata and Maeda proposed the introduction of a long narrow guiding pipe as a second coagulating bath positioned under the spinning funnel.[126] In this patent, the pipe was expected to accelerate rather than coagulate the thread. In the patents that followed, this idea of a 'two-step funnel method' was advanced.

In 1954, Suwa, Mihara and Nagao invented a second type of funnel coagulation bath, to which fresh coagulating liquid is constantly supplied and which is positioned at a distance from the first upper funnel, adjustable according to the spinning speed.[127] They controlled stretching in the free dropping zone before the coagulation bath.

In Nagao's patent in 1956,[128] the second bath is a 'liquid brake', slowing down the half-coagulated thread. This idea developed into a liquid injector. In order to control the tension applied to a running thread, selecting the coagulating bath length (ℓ_1 and ℓ_2 in Fig. 5.17(c)) and free dropping length (L_1 and L_2 in Fig. 5.17(c)) was shown to be effective.[128]

In 1962 Aizawa and Ishida proposed a stretching apparatus consisting of a conical tube (ℓ_2), long straight tube (ℓ_2) and a coagulation zone (ℓ_3) (Fig. 5.17(d)), wherein the flow rate of the coagulating liquid is suddenly diminished, thereby permitting completion of the coagulation in a relaxed state.[129] They showed how this method could produce cuprammonium rayon, which is superior in both tenacity and elongation (particularly the latter), having a homogeneous cross-section and easy resin finish. In 1969 Naniwa invented a method, where a liquid retarding the coagulation of the liquid thread is

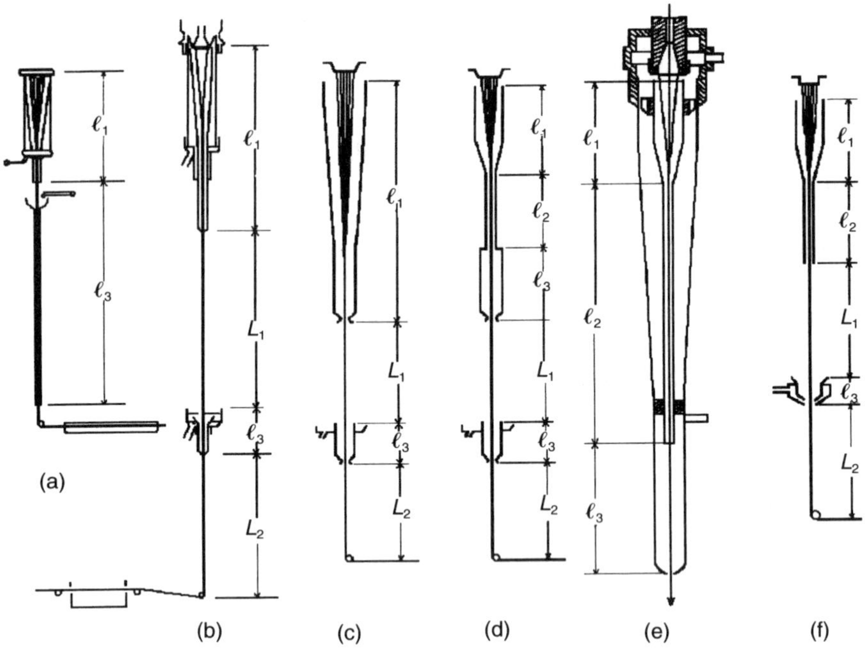

5.17 Stretch-spinning apparatus invented at Asahi during the 1950s–1970s. (a) Munakata-Maeda (1950);[126] (b) Suwa-Mihara-Nagao (1954);[127] (c) Nagao (1956);[128] (d) Aizawa-Ishida (1962);[129] (e) Naniwa (1969);[130] (f) Makita *et al.* (1972).[131]

supplied to a first bath and a coagulating liquid is supplied to a second bath. For this purpose two kinds of spinning water, differing in temperature, are supplied to each bath[130] (a double spinning water method).

Makita *et al.* invented in 1972 an injector to decrease the speed of the running thread below the funnel-type stretching tube midway down the free-dropping zone[131] (Fig. 5.17(f)). Using this method, the maximum spinning velocity attainable technically was estimated to be 700–800 m min^{-1}, and the maximum operation velocity, by which fibres can attain the quality approved for commercial products, was about 500 m min^{-1}. In 1972 Miyazaki *et al.* proposed a double-funnel type stretching tube (see Fig. 5.19),[116] in their new continuous spinning process (Net process), which will be described later. The distance L_1, measured from the lower end of the minor funnel and the upper part of the spray unit, together with the conditions (feed and temperature) at the primary liquid bath substantially affect the spinnability of the filament. The length L_2, measured between the lowest part of the spray unit and the level of the movable deflector pin, as well as the conditions (feed rate and temperature) at the secondary coagulating bath substantially influence the coagulation properties of the filaments.[116,131]

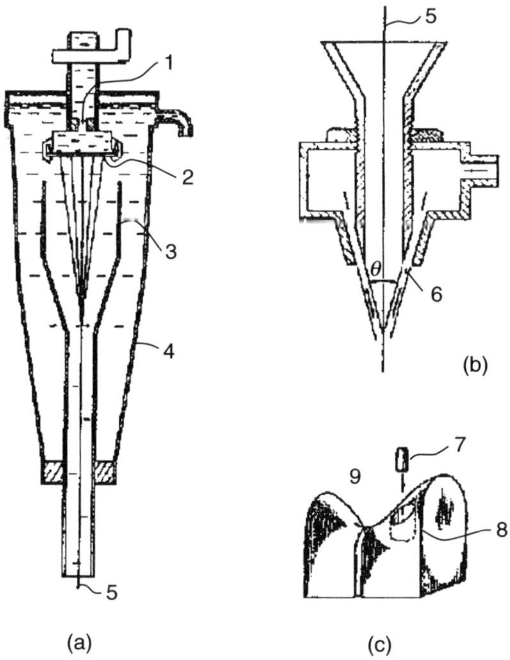

5.18 Enlarged view of a spinneret-funnel assembly (a), liquid injector (b) and trough of a saddle guide (c). 1: spinneret; 2: extrusion orifice; 3: inner funnel; 4: outer funnel; 5: thread; 6: ring nozzle; 7: supply nozzle; 8: liquid-receiving blind hole; 9: thin filament passage groove.

In 1986 Moroe and Iwase invented an apparatus consisting of a stretching nozzle, which enables the running liquid to stretch up to several thousand times instantly at its outlet, a liquid injector and a gas injector to separate the thread from the coagulating liquid.[132] Using this invention the production of fibre became possible at 1000–1500 m min^{-1}.

The most advanced process at present consists of a spinneret-funnel assembly (double-funnel type), a liquid injector (as a brake), and a gas injector (as a separator) (Fig. 5.18).

(1) Spinneret-funnel assembly (double-funnel type)

The spinneret is provided with a number of fine orifices, each of which has a bore diameter of 0.8 mm (Fig. 5.18(a)). The spinning solution is supplied by a metering pump through a filter, and a feed pipe from reservoir, to inlet pipe, fitted with an on-off control cock, and a reduced inlet opening. (2) Hydraulic braking and secondary bath liquid injection unit (liquid injector). Figure 5.18(b) shows the unit that provides a liquid braking effect on the liquid entrained by the filaments, which emerge from the spinneret-funnel assembly, shown by a single line in Fig. 5.18(a). This unit comprises an inside guide funnel and outer box-shaped casing member, concentrically arranged to inject a thin liquid ring jet of coagulating liquid towards the running bundle of continuous filaments.

5.2.12 Innovation of continuous process: development of the Net process

When Asahi introduced Beaunit-type machines in 1954, the take-up velocity was $73\,\mathrm{m\,min^{-1}}$. When the two-step funnel method, developed by Asahi, was combined with the original continuous process, the spinning velocity increased up to $81\,\mathrm{m\,min^{-1}}$ (1961). Two years later the continuous finishing section was improved by Asahi and also introduced to the plant, and the spinning velocity attained to $95\,\mathrm{m\,min^{-1}}$. In 1966 the two-stage spinning water method was invented at Asahi and the spinning velocity increased up to $110\,\mathrm{m\,min^{-1}}$. In 1971 Asahi developed the vacuum-free drop-spinning method and the spinning velocity reached $130–150\,\mathrm{m\,min^{-1}}$. In 1974 the epoch-making continuous process (Net process) was brought into commercial operation at a velocity of $380–500\,\mathrm{m\,min^{-1}}$ (see next section).

When the whole operational speed of the yarn manufacturing plant is above $400\,\mathrm{m\,min^{-1}}$ the following three major problems occur:

1 First there is an unacceptable number of broken filaments and considerable generation of 'fly' in the spinning and after-processing stages. This is because the entrained bath liquid impinges on the secondary bath pool, which leads to disturbance of the pool and poor filament quality. A process and method for eliminating such liquid disturbance in the secondary bath pool is therefore necessary.

2 Second, it is difficult to provide tension-free yarn for the after-treatment. Then, a process and apparatus for reducing tension is needed.

3 The third problem is that filament breakage is substantial and unacceptable. It is caused by contact between the extruded and coagulated filaments and the stationary filament guides (pins or combs), which are subjected to a gradually increasing deposit of copper hydroxide that will become the harder copper oxide. This hard oxide deposit, when accumulated on the contacting surface of the guide, will cause filament breaks, whose frequency increases as the spinning velocity increases.

To overcome these problem, Miyazaki *et al.* proposed the new 'Net' process[116] (Fig. 5.19).

According to their invention, the filaments leaving the spinneret (1) in a vertically downward direction pass through a first coagulating bath (double-funnel type) (2) and the fluid entrained by the filaments increases by means of a concentric angular-directed jet (liquid brake) (3) of fluid prior to passage through a second coagulation bath (lower funnel) (4). The filament may then pass over a bar which changes the direction of the filament and removes

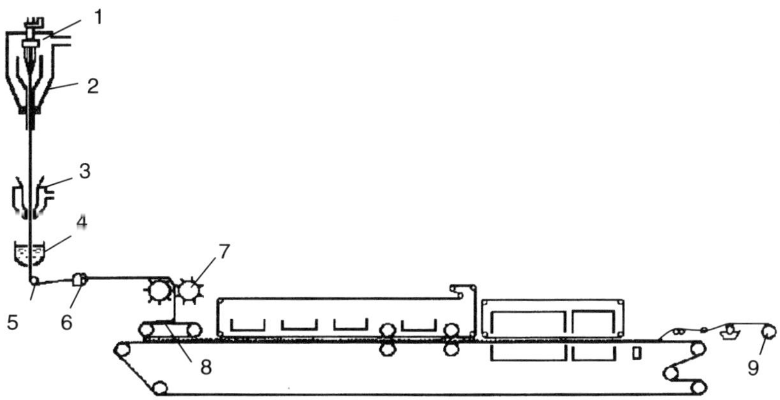

5.19 Continuous wet-stretch-spinning process (Net process).[116]
1: spinneret; 2: double-funnel type coagulation bath; 3: liquid jet brake; 4: second coagulation bath; 5: guide bar; 6: saddle guide (see, Fig. 5.18(c)); 7: roll; 8: travelling web; 9: cheese.

entrained fluid. Further coagulation bath treatment may be provided by passage through the saddle guide (6) (see also Fig. 5.18(c)) prior to passage between a pair of vane-type rolls (7) that shake the remaining bath fluid from the filaments. The filaments are then loosely deposited on a travelling web (8) for further treatment. Specific apparatus for carrying out the process includes elastic-covered vanes on the rolls and several decelerating fluid jet arrangements. Here the cross-section of a lower funnel (3) is shaped like a cup opening out at the bottom. The stationary arranged guide bar (5) serves to guide the filaments so as to deflect their passage from the vertical to a substantially horizontal position leading them to a saddle stationary filament guide (6). Regeneration is by a dip treatment for the filaments in aqueous sulphuric acid solution on the saddle guide. The tensioned state of filaments can be transformed by a conveyer (8) into a perfectly loose state, which can be maintained through the after-treatment stages. The yarns are finally wound up into cheeses. The process can be operated at the rate of 500–1000 m min^{-1} (the maximum operation velocity for a long run may be 500 mm min^{-1}). The processability of the process was recently improved by Moroe and Iwase[132] to allow 1500 m min^{-1} operation. This process is now in extensive commercial use at the Nobeoka Plant of Asahi (Japan). Even now, the continuous spinning process continues to be developed.

5.2.13 Staple fibre and other products

At present, the factory production of cuprammonium rayon is carried out overwhelmingly by the ordinary continuous yarn spinning process and the new continuous yarn spinning process (Net process). However, Asahi started Bemberg staple fibre production in 1960, spun-bond nonwoven fabric in 1974, hollow fibre for artificial kidneys (i.e. blood dialysis) in 1975, and hollow fibre membrane for blood filtration (Planova®) in 1985. The first two are fruitful new applications of the Net process for textile filaments. The last two result from fundamental research on the phase separation thermodynamics of the cellulose–cuprammonium solution system (see Section 5.2.8).

5.2.14 Spinning velocity

At the early stage of development of rayons, the spinning velocity V_s had not attracted the attention of development engineers, and V_s was in the range of several to 10 m min^{-1}. In fact, spinning velocity was mentioned in very few patents. For example, among about 300 Japanese patents, registered between 1905 and 1940 on regenerated cellulose fibres, only nine patents described the spinning velocity. This situation did not change until the 1950s. We could find only six patents that disclosed V_s, between 1941–1949. In contrast to this, 60 patents described the spinning velocity between 1950 and 1965.

Figure 5.20 shows the change in the spinning velocity, described in the patent literature, in the production of cuprammonium and viscose rayons,

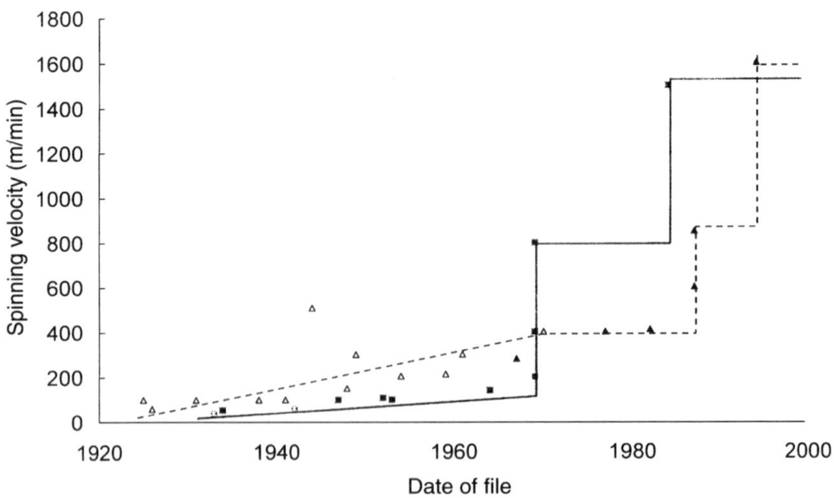

5.20 Improvement in the spinning velocity of cuprammonium and viscose rayons. ———: cuprammonium rayon; ------------: viscose rayon; ■: cuprammonium rayon (Asahi); □: cuprammonium rayon (others); ▲: viscose rayon (Asahi); △: viscose rayon (others).

plotted against their file date. V_s of cuprammonium rayons remained below 100 m min^{-1} until 1945–1950, reached c. 100 m min^{-1} in the 1950s and c. 150 m min^{-1} in the 1960s, and then started to increase markedly in the late 1960s to 800 m min^{-1}, and 1500 m min^{-1} was recorded in 1984. A similar change was also observed for viscose rayon. It is clear that a great technological advance has occurred since the late 1960s and is still in progress. The trends observed in Fig. 5.20 correlate well with the change in the operation velocity at the plant: 73 m min^{-1} in 1954, 81 m min^{-1} in 1961, 95 m min^{-1} in 1963, 110 m min^{-1} in 1966, and 139–150 m min^{-1} in 1971.[133] The above change also corresponds to commercial introduction of new technology like the two stage-funnel method, continuous finishing, the double- (or multiple-) spinning water method, and the dropping method.[133]

5.2.15 Recovery of waste materials

The economic significance of recovery of the solvents used in the process was well recognized even at the beginning of its commercial use. For example, VGF described in the complete specification of British Patent No 1283 (1905) that 'The economical value of the process above described resides in the fact that both the ammonia and copper can be recovered from the acid solution'.[134] The acid solution, in this case, is the regenerating bath liquid and VGF considered the recovery of waste materials only from the regeneration bath at that time. In November 1945 Hegan and Ingham reported that Bayer's Wuppertul plant recovered copper from the bath by precipitation, but did not recover ammonia.[135]

The recovery rate of copper by Asahi improved from c. 70% in 1930 to 99.9% in the 1980s (Table 5.14). The following developments were made in this period:

Table 5.14 Copper recovery rate at Asahi

Year	Recovery rate (%)
~1930	70–75
1940	75–80
1945	80–85
1948	85
1951	96
1961	99
~1970	99.7
~1980	99.9

Data from Asahi.[133]

1 quick sedimentation of small amounts of copper in suspension
2 filtration by a sand filter, then partial replacement of sand with active charcoal
3 positive use of CTH formed by adding ammonium in the recovery process
4 use of ion-exchange resin, whose function has been upgraded over 40 years.

Figure 5.21 illustrates the copper recovery system of Asahi. Note that at the Nobeoka plant the waste materials from all the production processes (filaments, staple fibres, nonwoven fabrics and hollow fibre membranes) are recovered totally and the optimum operating conditions of the recovery plant are determined considering the production plan of various cuprammonium cellulose products and the seasonal factors. The recovery of waste materials is now not done for economical reasons, but for environmental protection. Cuprammonium cellulose fibres are now manufactured almost entirely without new copper input.

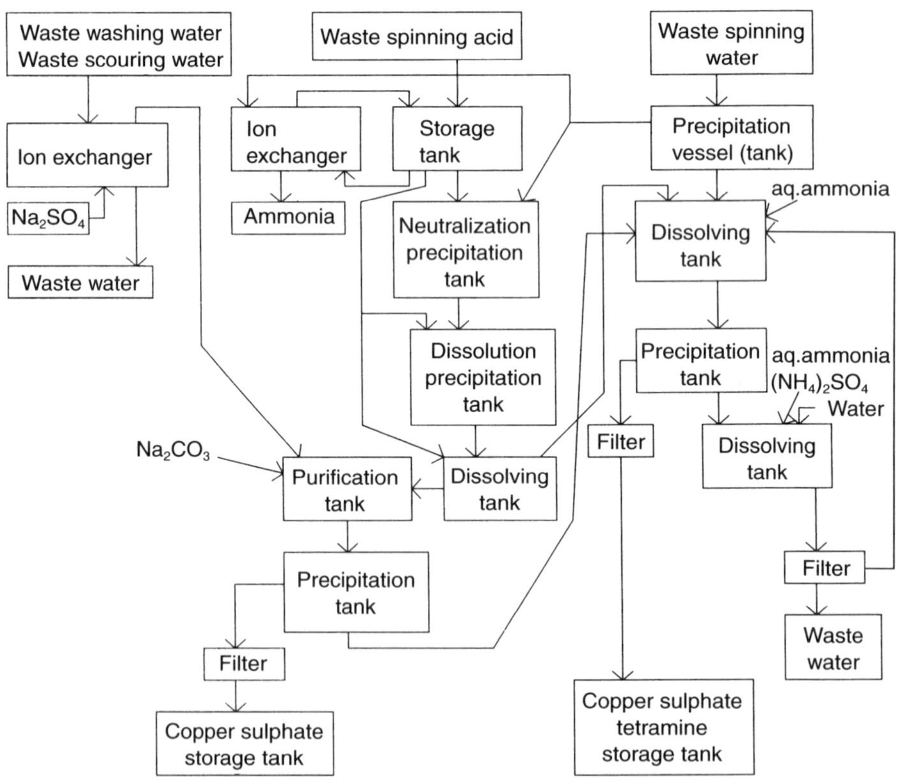

5.21 Copper recovery system (Asahi Chem. Industries).

In contrast, the recovery of ammonia is difficult, although enthusiastic attempts were made for many years. For example, Asahi established as early as 1935 the first commercial plant in the world for the recovery of ammonia, where ammonia was effectively evaporated from the waste from spinning water and then absorbed by sulphuric acid in the form of ammonium sulphate.[136] Then, at Bayer's Dormagen plant, distillation of ammonia from waste spinning water under reduced pressure was developed[137–139] (see also German Patent 738 115). Now, ammonium is recovered by ion-exchange, if necessary in combination with the above distillation method, and is used directly for preparation of the spinning solution and partly used in plant for miscellaneous neutralizing agents.

5.2.16 Labour productivity and advances in technology

Figure 5.22 shows the plots of the labour productivity, plotted as kilograms of fibre per day-person, for cuprammonium and viscose yarns versus year. Starting from 0.48 kg-fibre/day-person in 1901 (see Table 5.3), the productivity of the cuprammonium rayon process has changed over 100 years, passing through three phases: phase I (before World War II) where there was a very gradual improvement through choice of optimum operating conditions; phase II (during World War II when the factory system was virtually destroyed; and phase III (after the war) which has seen explosive advances caused by the development of novel processes. The cuprammonium process made progress in parallel with the viscose process, now attaining approximately 100 kg-fibre/day-person, which is about 200 times the amount at the beginning of commercial production and about 20 times that of the pre-war zenith.

Harrop made a rather stereotypical comment on the Japanese rayon industry between the wars saying that 'Japanese engineers displayed a remarkable ability both to imitate engineers in other countries and also simplify and improve the ideas of others'.[143] His comment may be properly applicable to any countries whose industrial conditions are behind the most advanced and who have an eager desire to level up their native industries. For example, from Roman times to immediately before the industrial revolution England operated a variety of textile technologies such as fulling (from Rome, 2nd–3rd century), weaving (wool from Flanders, 11th–15th century); worsted from the low countries, 13th and 16th century; silk from France, 17th–18th century; and cotton (in the form of fustian) from the low countries, 17th century, and dyeing and finishing (wool from the Netherlands, 17th century).[144] Harrop commented only on the conditions essential to 'catch-up' with modern technology, but not satisfactory conditions. Other than this, nobody has explained reasonably why any one country (or one company) has been prosperous up to the present time, inventing and

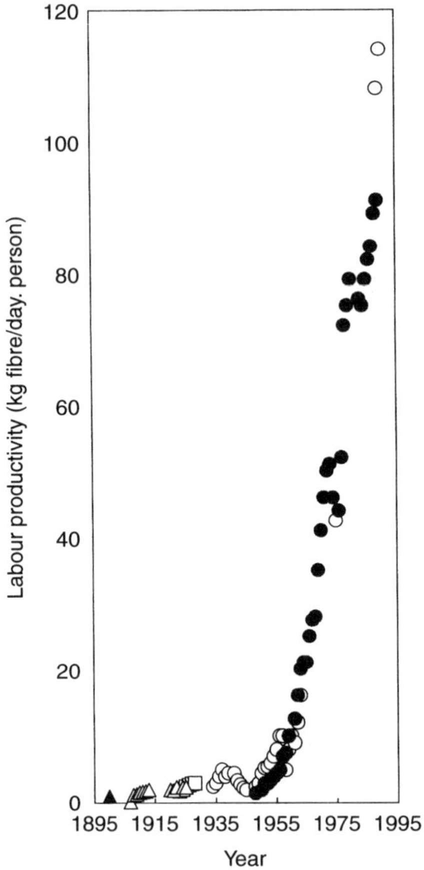

5.22 Labour productivity of cuprammonium and viscose rayons from 1901–1990: ▲: VGF (Germany; cupra);[140] △: Courtaulds (UK: viscose);[140] □: Teikoku Jinken (Teijin) (Japan: viscose);[141] ○: Asahi (Japan: viscose);[142] ●: Asahi (Japan: cupra).

developing wide expertise, which even Despeissis or Fremery, Urban or Thiele could not have imagined. This success in revolutionary innovation of wet-spinning technology for cuprammonium rayon production owes much to the continuous and very long-term investment in fundamental research and technology. It is not due to cheap labour and a genius for imitation!

Figure 5.23 demonstrates the hierarchy of development from basic research to process development in the cuprammonium cellulose fibre industry. Research and development (R&D) covers several different disciplines, ranging from the molecular to the technological level. After World War II, the three lower levels flourished remarkably. The achievements gained in the fundamental disciplines were transferred to higher ones (as

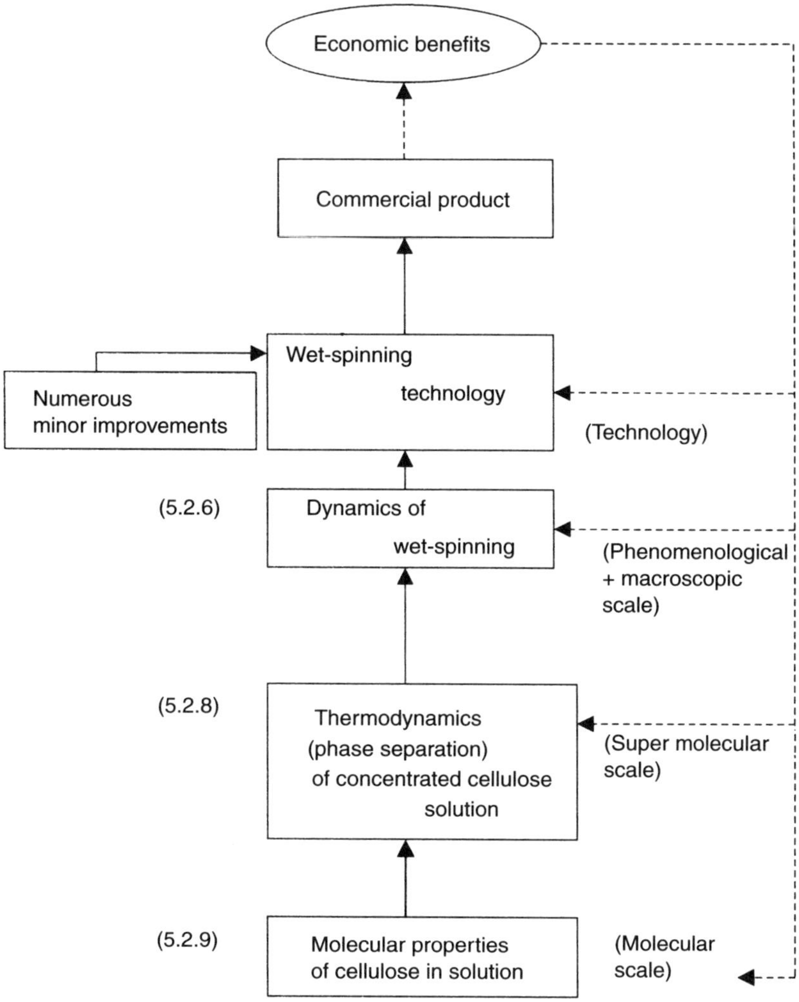

5.23 Science and technology of wet-spinning of cuprammonium solution. Numbers in parentheses indicate section numbers.

shown by the solid lines in the figure), leading directly to innovation in process or products, although transfer and communication were often difficult, owing to the lack of common language among the groups belonging to various disciplines. Broken lines in the figure represent capital investment in R&D, which motivates and/or rewards achievement. Provided that the R&D machine works well, a continuous cyclic flow will occur (i.e. positive circulation) and in addition, if the R&D machine can be quickly modified to respond to external change, enduring circulation can be anticipated with some confidence. At present, without the basic research underpinning the hierarchy, no revolutional progress can be anticipated. Cuprammonium processes are no exception. As early as 1907–1911 many people engaging in the business of cuprammonium artificial silk, judged that the cuprammonium process had no future. Now, it is widely considered that the regenerated cellulose fibre industry, including the cuprammonium process is out-of-date and has no interest for young scientists or engineers.

This industry is, however, very artistic and the process involves numerous minor improvements as well as the remarkable innovations described in this chapter. The former are not usually published or patented, but their contribution to the industry should never be underestimated. In this sense, the motivation of the workforce operating the plant is just as important to the success of the plant. The plant is in fact an accumulation of tiny, but invaluable pieces of 'know-how', gained by a large number of 'nameless' workers. However, it should not be assumed that any entrepreneur can readily enter this business nowadays through simply purchasing the patents in question or reading the relevant scientific and technological documents.

5.2.17 Production output

Table 5.15 collects the production output data for cuprammonium rayon from 1920–1940. The last column denotes the world output of all kinds of rayon. J P Bemberg and Asahi produced more than half of world output of cuprammonium rayon in the 1930s. Output of cuprammonium rayon was 3.46% of total rayon production in 1930, decreasing to 2.62% in 1935 and 1.85% in 1939/1940. At present (1990s), only a few percent is estimated for cuprammonium rayon (filaments and staple fibres) within world rayon production. The official record of annual production of cuprammonium rayon in Japan is tabulated in Table 5.16. The last column of the table is the world production of rayons. Unfortunately, the records of countries other than Japan are not separated from those of viscose rayon. Now, Asahi produces approximately 90% of the world's cuprammonium fibres. Its total world production can be roughly estimated at c. 20000–25000 tonnes per year.

5.3 Morphology and properties

5.3.1 Morphology

Figure 5.24(a) is a scanning electron micrograph of cuprammonium rayon. The denier of a filament is about 1.4, the filament surface is very smooth and the denier is uniform. Figure 5.24(b) is an electron micrograph of the cross-section of a filament, which was sampled immediately after regeneration and not dried after spinning. Cuprammonuim rayon has a

Table 5.15 Production output of cuprammonium rayon (1920–1940)

Year	Output (1000 tonnes)			
	J P Bemberg[a]	Asahi Chemical	World[d]	World[e] (all rayons)
1920	—		—	15
1922	0.440	—	—	
1923	0.500	—	—	
1924	0.520	—	—	
1925	1.000	—	—	
1926	1.000	—	—	
1927	1.960	—	—	
1928	2.291	—	8.2	
1929	2.507	—	8.1	
1930	1.836	—	7.2	208
1931	1.371	0.390[b] (0.345)[c]	5.9	
1932	1.605	1.900 (1.893)	5.8	
1933	2.504	2.670 (2.659)	9.6	
1934	3.488	3.440 (3.425)	12.5	489
1935	3.468	4.460 (4.443)	12.8	
1936	—	5.220 (5.112)	16.7	
1937	—	5.120 (5.088)	16.3	
1938	—	(4.979)	18.0	
1939	—	(4.047)	20.8	
1940	—	(3.558)	—	1127

[a] Yamazaki;[145] [b] Yamazaki Part 1, Chapter 3, 4.2;[146] [c] Ishii *et al.*;[147] [d] Lieser p. 59;[148] [e] Ishii (from Textile Organon).[149]

(a) (b)

5.24 (a) Scanning electron micrograph of cuprammonium rayon (Bemberg®); (b) electron micrograph of cross-section of cuprammonium rayon (Bemberg®) sampled immediately after the regeneration stage and not dried.

Table 5.16 Production output of cuprammonium rayon in Japan (1947–1998)

Year	Output (10^3 tonnes/year)		
	Filament[a]	Staple fibre[a]	World (rayons)
1947	0.270	—	949[b]
1950	3.523	—	1612 (664)[c]
1955	6.391	—	2278 —
1960	14.587	1.044	2608 (937)
1965	18.429	3.020	3338 —
1970	27.124	3.360	3431 (991)
1975	22.508	1.474	— (821)
1980	22.080	3.480	— (828)
1985	21.970	1.590	— (690)
1990	24.925	0.831	— —
1995	20.626	1.536	— —
1998	19.218	1.678	— —

[a] *Year Book of Textile Statistics*;[150] [b] *History of Japanese Chemical Fiber Industry* (cited from *Textile Organon*);[151] [c] World Rayon Filaments Production.[152]

Table 5.17 Mechanical properties of cuprammonium rayon filaments (75 denier/45 filaments)

Mechanical properties	Bemberg	
	Continuous	Continuous (NP)
Tensile strength (g/d) (dry)	2.5	2.7
Tensile elongation (%) (dry)	12	13
Tensile strength (g/d) (wet)	1.6	1.6
Tensile elongation (%) (wet)	26	26
Wet shrinkage (%)	−0.5	2.4
Boil shrinkage (%)	5–6	1–2

circular cross-section and no visible skin, but does have a microscopically thin outer skin, which is more porous, when undried, than the compact inner part. A cuprammonium rayon filament can be regarded approximately as a fine flexible uniform cylinder. This morphological character is apparently responsible for the superior feel and brightness of the fibres.

5.3.2 Mechanical properties

Figure 5.25 illustrates the stress–strain curves of cuprammonium and viscose rayons. Some typical physical properties are collected in Table 5.17.

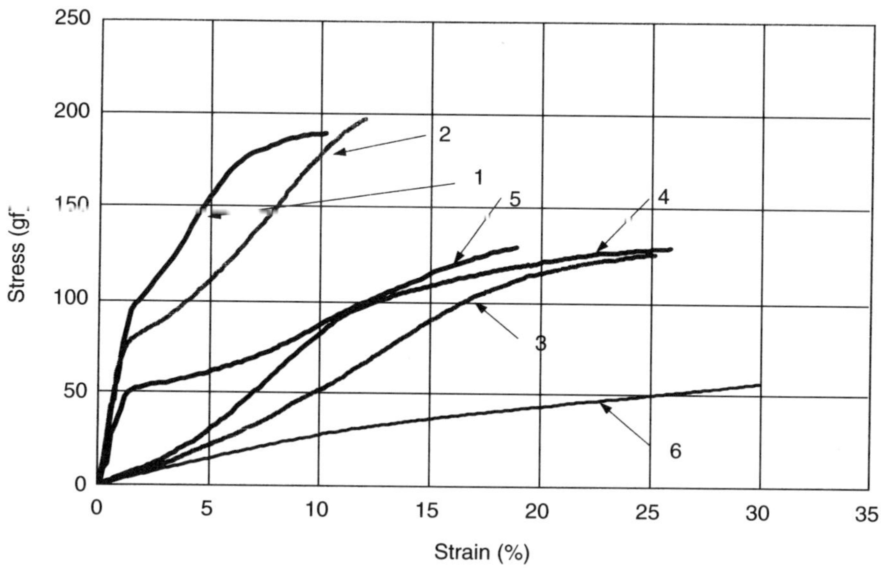

5.25 Stress–strain curves of cuprammonium and viscose rayons at 20 °C (relative humidity 65%, 84 dtex). Curve 1, Bemberg (continuous process) dry; curve 2, Bemberg (net process) dry; curve 3, Bemberg (net process) wet; curve 4, Bemberg (continuous process) wet; curve 5, viscose (cake) dry; curve 6, viscose (cake) wet.

Note that mechanical properties depend, to some extent on the spinning and after-treatment conditions, and the mechanical properties of the commercial products are designed to fit the end-use. Cuprammonium rayon has higher tensile strength (TS) and lower tensile elongation (TE) than viscose rayon, whose TS ranges from 1.7–2.3 g denier^{-1} and TE from 18–24%. Table 5.18 shows TS in wet and in dry cuprammonium rayon from the 1950–1990s. The ratio of TS in the wet state to that in the dry state (W-D ratio) is now 0.70–0.73, compared with 0.45–0.55 for viscose rayon. It is to be noted that TS (wet) exceeded 1.5 g in the 1970s.

5.3.3 Hydrophilic properties

Figure 5.26 is a contour map of equilibrium water content (W_c) plotted against environmental temperature and humidity.[154] In the lower humidity and lower temperature region ($T < 40$°C and <60%), W_c is temperature-dependent. In the high humidity region (>80%), W_c is humidity dependent.

Table 5.18 Wet–dry ratio of TS of Asahi Bemberg fibres (75 denier)

Year	Method	Tensile Strength (g/d)		Ratio
		dry	wet	
1951	Hank	2.03	1.22	0.60
	Centrifuge	2.14	1.31	0.61
	Continuous[a]	2.10	0.93	0.44
1961	Continuous	2.17	1.03	0.47
	Continuous[b]	2.14	1.32	0.62
1976	Hank	2.18	1.31	0.60
1977	Continuous	2.54	1.64	0.65
1978	NP	2.54	1.55	0.63
1988	Continuous	2.46	1.73	0.70
1998	Continuous	2.56	1.88	0.73

Data from Munekata in *Chemical Fibers*, 1956, p. 281 (data were obtained in 1951);[153] *30 years History of Asahi Bemberg Plant*, 1962, and data from Asahi Bemberg plant.
[a] America Bemberg made; [b] Beaunit Mills made.

5.4 Products and application

5.4.1 Filaments and staple fibres

Table 5.19 collects correlations between process and products of filaments, staple fibres, and non-woven filament cloth. Table 5.20 summarizes end-uses of fabric and knitted materials made from filaments and staple fibres and their characteristics. Cuprammonium rayon fabrics have the following features when they are worn: (1) high water content, (2) low tribo-electric voltage, (3) moisture absorbing and releasing ability, (4) no cling, (5) less ravelling, (6) no stickiness, (7) smooth wearing through sleeve, (8) low friction coefficient, (9) low noise at rubbing with lining, (10) good pliability, (11) excellent dynamic drape behaviour, (12) good iron setting ability, and (13) excellent wearing comfort. These are closely correlated to the fineness, circular cross-section and hydrophilic properties of fibres, and these are more manifest when wool is used as a face side fabric, to which cuprammonium rayon cloth is added as lining.

5.4.2 Nonwoven fabric

Cuprammonium filaments constituting nonwoven fabric are almost perfectly round, adhering to each other at some places. Self adhesion is formed during the sheet forming and refining process. Bemliese,® a trade

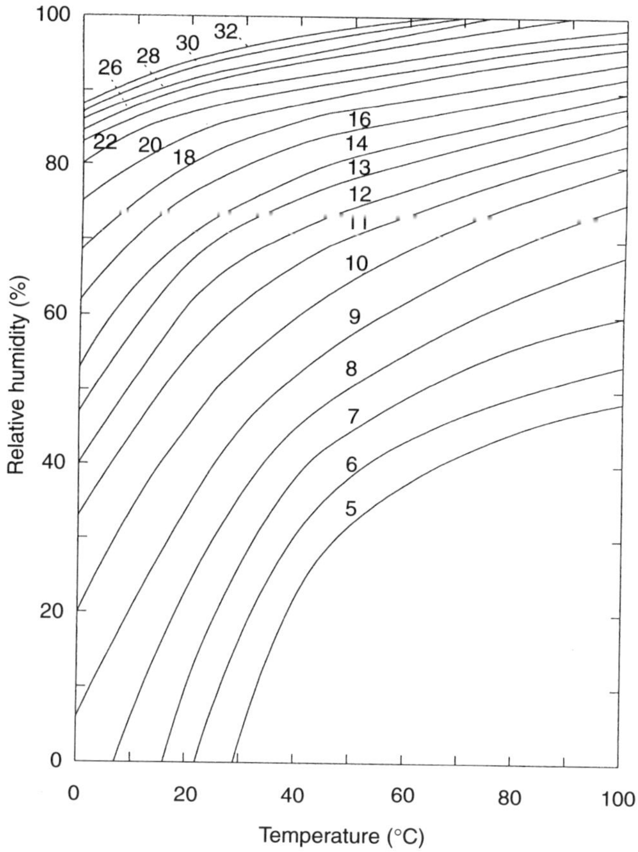

5.26 Relationship between the equilibrium water content of cuprammonium rayon filaments (75 denier/45 filaments)[154], relative humidity and temperature. The numbers denote the equilibrium water content (%).

Table 5.19 Commercial products and their manufacturing methods (Asahi Chemical Ind. Co.)

Type	Method	Registered name	Spinning velocity (m min^{-1})	Products (dtex)
Filament	Hank	Bemberg	50–70	22–330
	Continuous		90–170	17–165
	Continuous (not process)		400–1000	84–165
Staple fibre	BF		100–150	1.4–3.3
Nonwoven filament cloth	Wet-spunbond	Bemliese	30–50	15–60 (g m^{-2})

Table 5.20 End-uses and characteristics

	End-uses	Characteristics
Bemberg® fabric, knitted	a. Lining b. Lingerie c. Dress fabric (velvet, outer, inner) d. Native costume (duppatta, saree) e. Interior (curtain)	a. Superior hydroscopicity and anti-static b. Extremely soft touch and fine texture c. Silk-like brilliance and dyeability
Bemliese® nonwoven fabric	a. Domestic uses (cosmetic puff, wet wiper, tea bag) b. Medical uses (hospital swab, towel) c. Industrial uses (clean room wipe, industrial wipe) d. Agricultural uses (seeder tape, seeder net)	a. Binderless (pure cellulose) b. Sanitary and soft c. Dust- and lint-free d. Highly liquid absorbent and retentive e. Anti-static f. Ecologically safe
Hollow fibre	a. Artificial kidney	a. High strength b. Easy to produce thin-walled membrane c. Little protein and blood corpuscle due to high hydrophilicity d. Heat resistant and γ-ray resistant e. Safety and long-term actual results

mark of Asahi, is distinguished from other cellulosic nonwoven fabrics, because it is made of continuous filaments of pure cellulose and has no binder. The following characteristics are due to the sheet forming component and water-jet treatment: (1) residue-free, (2) lint free, (3) non-toxic, (4) high absorbency and high liquid retention, (5) anti-static electricity, (6) biodegradable, (7) bulky and soft, (8) uniform thickness and density.

Being lint-free, soft, flexible and hygienic with a high capacity, Bemliese® has various applications as follows: (1) consumer uses: cosmetic puff, wet wiper, tea bag, various kinds of disposable uses, (2) medical uses: hospital swab, towel, gauze, adhesive bandage, (3) industrial uses: clean room wiper, industrial wiper, (4) agricultural uses: seed tape, covering net for sowing.

Characteristics and end-uses of Bemliese® are also summarized in Table 5.20.

5.4.3 Cuprammonium cellulose hollow fibre membrane

As most old cellulose industrial products were replaced by synthetic polymers, cellulose and its derivatives were continuously forced to attain new markets for their survival. For this reason, a broad open-minded approach was helpful. One attractive market for cellulose is undoubtedly the membrane business. Cellulose membranes have the longest history of use in the membrane industry[155] and even now are keeping their top position.[156] Unlike regenerated cellulose fibre, cellulose membranes are superior because of the following properties: (1) chemical stability, (2) safety (biological compatibility), (3) low price, (4) high tenacity in the wet state, enabling the preparation of thin membranes, (5) ease of control of pore size, ranging from 1–100 mm in diameter, and (6) porosity.[155,156] In comparison with synthetic polymer membranes, cellulose membranes have: (1) large removability of low-molecular-weight nitrogen metabolites such as urea, (2) better balance of material permeability and water ultrafiltration rate, and (3) high dimensional stability and good processability in module manufacturing.[157] Now, most cellulose membranes are hollow-fibre membranes. The traditional technology of regenerated cellulose fibres for textiles combines well with membrane science and technology.

The cuprammonium process is the most lucrative because a tremendously wide range of variation in the sequential steps of dissolution–deformation–coagulation–regeneration is possible. However, the target markets have changed, for example, from packing material (i.e. sausage skin) and moisture-proof cellophane to separation media such as haemodialysis and ultrafiltration (UF).

There are two types of cellulose membranes with respect to their morphology: symmetrical and asymmetrical membranes – cellophane (viscose rayon) and cuprammonium cellulose membranes for haemodialysis are typically symmetrical membranes. Asymmetrical cuprammonium cellulose hollow fibre membrane developed in the 1980s–1990s is now used as a virus separation medium.

5.4.4 Artificial kidney

Cuprammonium cellulose hollow fibre membranes are utilized as one of the main elements in a blood purification device, the hollow fibre artificial kidney (AK), produced and sold by Asahi Medical Co Ltd (Tokyo, Japan), a subsidiary of Asahi, which was first in the world to produce a dry-type hollow-fibre artificial kidney commercially in 1974. During the 25 years or so that have followed, regenerated cellulose hollow fibre membranes have been considerably improved in the following areas:[158,159] (1) pin hole, (2) break-down during dialysis, (3) membrane thickness (100 μm → 7 μm), (4) purity (low molecular weight non-cellulosic component), (5) sterilization method, and (6) biological compatibility. There are two types of artificial kidney that use cuprammonium hollow fibres: the AM-SD series (conventional hollow fibres) and the AD-BIO series (biocompatible AK with standard and middle-flux ranges). In the latter series, the biocompatibility of the cellulose membrane is improved by polyethylene glycol grafting technology. Specifications for AK are as follows: surface membrane, 0.8–2.0 m^2; inner diameter, 180 µm; membrane thickness membrane, 15 µm. Ordinary haemodialysis membranes are believed to have a mean pore diameter of 2–3 nm. Biocompatible membranes have a larger mean pore size (4–10 nm). The sieving coefficient S_c, is defined as the ratio of substrate in the filtrate to that in the supplying solution of β_2-microglobulin (molecular weight 11 800; molecular size c. $4.5 \times 2.5 \times 2.0\,nm^3$) = 0.6 and S_c of albumin = 0.02.

The world market for artificial haemodialysis apparatus was roughly estimated to be 7.2×10^7 units/year in 1999. Cuprammonium cellulose membrane occupies a 35% share, and cellulose acetate membrane has a 24% share.

5.4.5 Virus removal filter

In 1987 Asahi Chemicals developed the BMM (Bemberg Microporous Membrane) process as a new technology for effective removal of viruses. Asahi marketed its filter under the trade name Planova®. Planova® is a validated virus removal filter used in the purification of various plasma-derived and biopharmaceutical products such as globulin, blood coagulation factors and interferon. Planova® 'has been filed with the Biologic Master File of the US Food and Drug Administration.

Planova®'s features are: (1) the effective removal of known and unknown viruses and other contaminants by a size exclusion mechanism, (2) a high recovery rate of proteins owing to the hydrophilic nature of the membrane, (3) a high filtration rate and a large filtration capacity, (4) the availability of two filtration methods: dead-end and tangential (i.e. parallel flow) filtrations,[160] (5) various filter sizes.

Three types of membrane are on the market: Planova 15 N (mean pore size, measured by the water flow rate method, 15 ± 2 nm), Planova 35 N (pore size 35 ± 2 nm; virus removal filter) and Planova 75 N (mean pore size, 72 ± 4 nm) used as a prefilter.

5.5 Conclusion and future prospects: does the cuprammonium rayon industry have a future?

Table 5.21 summarizes three major drawbacks of the cuprammonium rayon industry in the 1910s and their solutions in the 1990s. During these 90 years

Table 5.21 Drawbacks of the cuprammonium rayon process encountered in the 1910s and at present

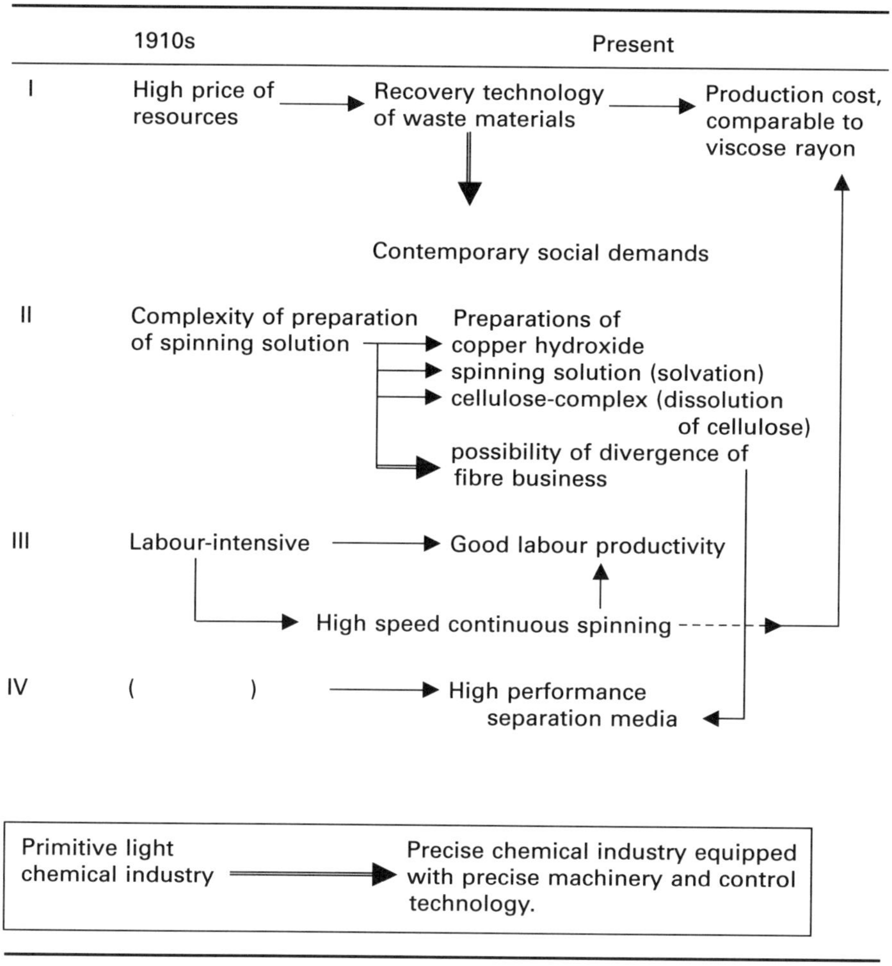

the cuprammonium process made a slow metamorphosis from a primitive light chemical industry to a precise chemical industry.

Cellulose is synthesized from carbon dioxide and water and is biodegradable, being the most abundant naturally occurring organic compound. Therefore, there is no doubt about the very long term future of the cellulose industry after complete exhaustion of fossil fuels, although we cannot tell exactly when this will happen. The problem is whether the existing cellulose industry, including the cuprammonium rayon process, which at present is under fierce competition from synthetic polymers, can survive until that time. Cellulose is an earth-friendly material, but the regenerated cellulose fibre industry is heavily handicapped by the huge areas of land and amounts of water required. These are the negative legacy of the 19th century.

Now is a time of global free competition (laissez-faire; this does not mean fair competition) with people trying to make immediate profit, or make a fortune overnight. We may expect some specific improvements to arise from basic research and technology, which is now being accelerated. The next hundred years' history of the cuprammonium process may differ greatly from that of the last hundred years.

References

1 E Schweizer, 'Das Kupferoxyd-Ammoniak, ein Auflösungsmittel für die Pflanzen faser', *J Prakt Chem*, 1857 **72** 109.
2 E Schweizer, 'Über des Unterschwelsäure Kupferoxyd-Ammoniak und die Ammoniak basischen', *J Prakt Chem*, 1856 **67** 430.
3 E Wheeler, *The Manufacture of Artificial Silk*, (cited by, Hard A, in *The Story of Rayon*, United Trade Press, 1939, p 38).
4 C G Schwalbe, 'Chemie der Cellulose', 1st ed, p 145 cited in *Chemical Fibers*, ed. A. Munekata 1956, Chap. II (2), p 238 (as ref 6) Maruzen.
5 W Crooks, *Electric Lamps & c*, British Patent 2612, June 1881.
6 E Weston, *Carbon Conductors for Electric Lamp*, British Patent 4458, Sep. 1882.
7 L H Despeissis, France Patent 203 741, May 1890.
8 K Kamide, *History of Textile Industry*, Society of Textile Machinery of Japan, 1993, p 233, Fig. 7.4.
9 W E Wicht, *Glanzstoff zür Geschichte der Chemifaser, eines Unternehmens und seiner Arbeitschaft*, Verlagsdruckerei Schmidt GmbH, 1992, p 30.
10 H Pauly, *Verfahren zür Herstellung künstlicher Seide aus in Kupferoxyd-ammoniak gelöster Zellulose*, Deutsche Patent (Kaiserliches Patent) 98 642, December 1897; British Patent 28 631, 1897.
11 W E Wicht, *Glanzstoff zür Geschichte der Chemifaser, eines Unternehmens und seiner Arbeitschaft*, Verlagsdruckerei Schmidt GmbH, 1992, p 28.
12 K Kamide, *History of Textile Industry*, Society of Textile Machinery of Japan, 1993, p 236, Fig. 7.5.
13 K Kamide, *History of Textile Industry*, Society of Textile Machinery of Japan, 1993, p 237, Table 7.5.
14 Glanzstoff, *25 Jahre Glanzstoff*, Ecksteins Biographischer Verlag, Berlin, 1924, p 46.
15 W E Wicht, *Glanzstoff zür Geschichte der Chemifaser, eines Unternehmens und seiner Arbeitschaft*, Verlagsdruckerei Schmidt GmbH, 1992, p 30.
16 J Foltzer, *Artificial Silk and its Manufacture*, translated by T. Woodhouse, 4th edition, Sir Isaac Pitman & Sons, 1928, p 31.
17 A Hard, *The Story of Rayon*, United Trade Press, 1939, p 44.
18 *25 Jahre Glanzstoff*, Ecksteins Biographischer Verlag, Berlin, 1924, p 51.
19 K Kamide, *History of Textile Industry*, Society of Textile Machinery of Japan, 1993, p 242.

20 *25 Jahre Glanzstoff*, Ecksteins Biographischer Verlag, Berlin, 1924, p 75.
21 See, for example, A Hard, *The Story of Rayon*, United Trade Press, 1939, pp 40–42.
22 D C Coleman, *Courtaulds II*, Clarendon Press, 1969, p 58.
23 K Kamide, *History of Textile Industry*, Society of Textile Machinery of Japan, 1993, p 243, Table 7.6.
24 Y Kami, *Recent Artificial Silk*, Meibundo, 1927, p 141.
25 K Kamide, *History of Textile Industry*, Society of Textile Machinery of Japan, 1993, p 245, Table 7.8.
26 W E Wicht, *Glanzstoff zür Geschichte der Chemifaser, eines Unternehmens und seiner Arbeitschaft*, Verlagsdruckerei Schmidt GmbH, 1992, p 35.
27 *25 Jahre Glanzstoff*, Ecksteins Biographischer Verlag, Berlin, 1924, p 41.
28 W E Wicht, *Glanzstoff zür Geschichte der Chemifaser, eines Unternehmens und seiner Arbeitschaft*, Verlagsdruckerei Schmidt GmbH, 1992, pp 39, 41.
29 E Thiele, Deutsche Patent 154 507, January 1901.
30 E Thiele, British Patent 8 083 (1902), April 1902.
31 E Thiele, Deutsche Patent 157 157, January 1901.
32 E Thiele, Deutsche Patent 173 628, June 1902.
33 E Thiele and R Linkmeyer, Deutsche Patent 179 772, August 1905; British Patent 16 088, 1906.
34 J P Bemberg Akt.-Ges., Deutsche Patent 220 051, May 1907.
35 J P Bemberg Akt.-Ges., Deutsche Patent 413 791, Dec. 1923.
36 J P Bemberg Akt.-Ges., Deutsche Patent 413 790, Nov. 1923.
37 J P Bemberg Akt.-Ges., Deutsche Patent 415 798, Nov. 1923.
38 J P Bemberg Akt.-Ges., British Patent 229 638 (1925), Jan. 1925, Shweizerische Patente 112 784.
39 J P Bemberg Akt.-Ges, British Patent 235 853 (1925), May 1925.
40 J P Bemberg Akt.-Ges., Deutsche Patent 162 866, September 1900.
41 J P Bemberg Akt.-Ges., Deutsche Patent 178 508, February 1905.
42 J P Bemberg Akt.-Ges., Deutsche Patent 408 889, October 1923.
43 K Kamide, *History of Textile Industry*, Society of Textile Machinery of Japan, 1993, p 315, Table 8.9.
44 Y Kami, *Recent Artificial Silk*, Meibundo, 1927, pp 51–55.
45 A Hard, *The Story of Rayon*, United Trade Press, 1939, p 49.
46 See, for example, Y Kami, *Recent Artificial Silk*, Meibundo, 1927, pp 136–159.
47 K Kamide, *History of Textile Industry*, Society of Textile Machinery of Japan, 1993, p 313, Table 8.5 (A Hard, *The Story of Rayon*, United Trade Press, 1939, pp 24, 42; D C Coleman, *Courtaulds II*, Clarendon Press, 1969, pp 13, 24, 27, 61, Y Kami, *Recent Artificial Silk*, Meibundo, 1927, p 173).
48 K Kamide, *History of Textile Industry*, Society of Textile Machinery of Japan, 1993, Chapter 8, pp 305–317.
49 K Kamide, *History of Textile Industry*, Society of Textile Machinery of Japan, 1993, p 308, Table 8.2.
50 K Kamide, *History of Textile Industry*, Society of Textile Machinery of Japan, 1993, p 166, Table 5.1.
51 K Kamide, *History of Textile Industry*, Society of Textile Machinery of Japan, 1993, p 310, Fig. 8.3.
52 K Kamide, *History of Textile Industry*, Society of Textile Machinery of Japan, 1993, p 311, Fig. 8.4.

53 See, for example, A Munekata, *Cellulose Handbook*, ed. K Sobue and N Migita, Asakura Publishing, 1959, Chapter 6, pp 213–214; and A Munekata, *Chemical Fibers*, eds R Oka, A Munekata and M Wadano, Maruzen, 1956, Chapter 2, pp 252–256.
54 See, for example, A Munekata, *Cellulose Handbook*, ed. K Sobue and N Migita, Asakura Publishing, 1959, p 214 and A Munekata, *Chemical Fibers*, Maruzen, 1956, Chapter 2, pp 256–257.
55 W Bonsdorf, *Ber Dtsch Chem Ges*, 1903 **36** 2322.
56 A Munekata, *Chemical Fibers*, eds R Oka, A Munekata and M Wadano, Maruzen, 1956, p 223.
57 A Munekata, *Chemical Fibers*, eds R Oka, A Munekata and M Wadano, Maruzen, 1956, p 222.
58 See, for example, C Nakayama, Memoirs of the Faculty of Education, Niigata University, 1989 **30**(2) 65.
59 A Munekata, *Chemical Fibers*, eds R Oka, A Munekata and M Wadano, Maruzen, 1956, pp 262–263; A Munekata, *Cellulose Handbook*, ed. K Sobue and N Migita, Asakura Publishing, 1959, p 215.
60 A Munekata, T Matsushima and H Tamura, Japanese Patent 181 338, 1949.
61 H A Neville and C T Ostwald, *J Phys Chem*, 1931 **35** 60.
62 E Posnjak and G Turnell, *J Phys Chem*, 1931 **35** 929.
63 O Binder, *Compt rend*, 1934 **198** 653.
64 O Binder, *Compt rend*, 1934 **198** 2167.
65 H B Weiser, W D Millinigon and E L Cook, *J Am Chem Soc*, 1940 **64** 503.
66 W Feitknecht and Z Kolloid, 1946 **92** 257.
67 Asahi Chemical Industry Co., *History of 30 Years, Asahi Bemberg Plant*, Asahi Chem Ind. Co., p 19.
68 A Munekata, *Chemical Fibers*, eds R Oka, A Munekata and M Wadano, Maruzen, 1956, p 225.
69 See, for example, A Munekata, *Cellulose Handbook*, ed. K Sobue and N Migita, Asakura Publishing, 1959, p 215.
70 Fr Billnheimer, *Ber Dtsch Chem Ges*, 1898 **31** 1453.
71 Fr Billnheimer, *Ber Dtsch Chem Ges*, 1899 **32** 2347.
72 W Traube, *Ber Dtsch Chem Ges*, 1911 **44** 3319.
73 W Traube, *Ber Dtsch Chem Ges*, 1921 **54** 3220; 1922 **55** 1899; 1923 **56** 268; 1930 **63** 2086; 1930 **63** 2094.
74 K Hess and E Messmer, *Ber Dtsch Chem Ges*, 1922 **55** 2441; 1921 **54** 834; Z Kolloid, 1925 **36** 260.
75 T Lieser and R Ebert, *Ann Chem*, 1925 **528** 281; 1925 **528** 532.
76 R E Reeves, *Science*, 1944 **99** 148.
77 S T K Bukhari, R D Guthrie, A Scott and A D Wrixon, *Chem Commun*, 1968 1580.
78 I Miyamoto, Y Matsuoka, T Matsui and K Okajima, *Polymer J*, 1995 **27**(11) 1113.
79 I Miyamoto, Y Matsuoka, T Matsui and K Okajima, *Polymer J*, 1995 **27**(11) 1123.
80 See, for example, A Munekata, *Chemical Fibers*, eds R Oka, A Munekata and M Wadano, Maruzen, 1956, pp 242–243.
81 H Suwa, 'Spinning and Fiber Structure of Cuprammonium Rayon,' in *Formation of Fiber and Emergence of its Structure (II)*, Kagaku-Dojin, 1970, p 87,

Chapter 5, Fig. 15. (See, also, A Munekata, *Chemical Fibers*, eds R Oka, A Munekata and M Wadano, Maruzen, 1956, p 249, Fig. II.9).
82 A Munekata, *Chemical Fibers*, eds R Oka, A Munekata and M Wadano, Maruzen, 1956, pp 247–249.
83 H Suwa, 'Spinning and Fiber Structure of Cuprammonium Rayon,' in *Formation of Fiber and Emergence of its Structure (II)*, Kagaku-Dojin, 1970, p 85.
84 H Suwa, 'Spinning and Fiber Structure of Cuprammonium Rayon,' in *Formation of Fiber and Emergence of its Structure (II)*, Kagaku-Dojin, 1970, p 87, Fig. 14.
85 See, for example, K Kamide, *Thermodynamics of Polymer Solutions – Phase Equilibria and Critical Phenomena*, Elsevier, 1990.
86 K Kamide, *Thermodynamics of Polymer Solutions – Phase Equilibria and Critical Phenomena*, Elsevier, 1990, Chapter 6, Fig. 6.5.
87 K Kamide, H Iijima and S Matsuda, *Polymer J*, 1993 **25**(11) 1113.
88 K Kamide, H Iijima and H Shirataki, *Polymer J*, 1994 **26**(1) 21.
89 H Iijima, S Matsuda and K Kamide, *Polymer J*, 1994 **26**(4) 439.
90 K Kamide, H Iijima and A Kataoka, *Polymer J*, 1994 **26**(5) 623.
91 H Iijima, A Kurata and K Kamide, *Polymer J*, 1995 **27**(10) 1033.
92 H Iijima, K Sogawa and K Kamide, *Polymer J*, 1996 **28**(9) 808.
93 H Iijima, M Iwata, M Inamoto and K Kamide, *Polymer J*, 1997 **29**(2) 147.
94 K Kamide, *Thermodynamics of Polymer Solutions – Phase Equilibria and Critical Phenomena*, pp 581–591.
95 K Kamide, *Thermodynamics of Polymer Solutions – Phase Equilibria and Critical Phenomena*, pp 563–581.
96 K Kamide, *Thermodynamics of Polymer Solutions – Phase Equilibria and Critical Phenomena*, p 452, Fig. 6.7.
97 K Kamide, *Thermodynamics of Polymer Solutions – Phase Equilibria and Critical Phenomena*, p 505, Fig. 6.35.
98 M Iwata, S Manabe and M Inoue, Japanese Patent 1 439 049, 1988; Japanese Patent 1 434 154, 1988; Japanese Patent 1 473 266, 1988; United States Patent 4 581 140, 1986; United States Patent 4 604 326, 1986; United States Patent 4 822 540, 1989.
99 K Kamide and M Saito, 'Cellulose and cellulose derivatives: recent advances in physical chemistry', *Adv Polymer Sci*, 1987 **83** 1–56.
100 K Kamide and M Saito, 'Recent advances in molecular and supermolecular characterization of cellulose and cellulose derivatives', *Makromol Symp*, 1994 **83** 233–271.
101 K Kamide, '17, Characterization of Chemically Modified Cellulose', *Wood and Cellulosic Chemistry*, eds D N-S Hon and N Shiraishi, Marcel Dekker, 1991, pp 801–860.
102 K Kamide and M Saito, *Polym J*, 1986 **18** 569.
103 S W Barker and R Alleston, 'The development of continuous rayon spinning', *J Text Inst*, 1948 **39** 1.
104 *25 Jahre Glanzstoff*, Ecksteins Biographischer Verlag, Berlin, 1924, p 69.
105 K Kamide, *History of Textile Industry*, Society of Textile Mach., Japan, 1993, Chapter 7, p 244.
106 W E Wicht, *Glanzstoff zür Geschichte der Chemifaser, eines Unternehmens und seiner Arbeitschaft*, Verlagsdruckerei Schmidt GmbH, 1992, p 36.
107 Sociėtė Françoise de la Viscose, Deutsche Patent 192 406, July 1906.

108 J Foltzer, Deutsche Patent 209 923, Oct. 1908.
109 W G Hagen and C E Ingham, *German Rayon Industry*, UK government report, 1945, p 40.
110 J P Bemberg Akt.-Ges., Deutsche Patent 763 735, 1940.
111 J Wichert, *Chemische Textilfasern Filme und Folien*, ed. R Pummerer, Ferdinand Enke Verlag, 1953, p 261.
112 See, for example, *Man-Made Fibers Vol. 2*, ed. H F Mark, A M Atlas and E Cernia), John Wiley, Chichester, 1968, p 1.
113 R W Moncrieff, *Man-Made Fibers*, 5th edition, Chapter 10, Heywood Books, 1970, pp 217–220.
114 VGF, Deutsche Patent 235 134, Oct. 1906; British Patent 16 495, 1907.
115 J Wichert, *Chemische Textilfasern Filme und Folien*, ed. R Pummerer, Ferdinand Enke Verlag, 1953, p 263.
116 T Miyazaki, E Omura, K Katakabe, M Makita, K Iwase, H Tsutsumiuti, T Yamamoto, F Ikeda, Y Takashima, H Sueyoshi and T Matsusaka, Asahi Chemical Industries, United States Patent 3 689 620, June 1972; Japanese Patent Publication No 29 926, Dec. 1969.
117 VGF, Deutsche Patent 236 584, Feb. 1910.
118 VGF, Deutsche Patent 239 822, May 1910.
119 W H Furness, United States Patent 1770 750, July 1930.
120 W H Furness, Japanese Patent 101 379, June 1933.
121 J Wichert, *Chemische Textilfasern Filme und Folien*, ed. R Pummerer, Ferdinand Enbe Verlag, 1953, p 264.
122 A Munekata, *Chemical Fibers*, eds R Oka, A Munekata and M Wadano, Maruzen, 1956, p 220.
123 H Hofmann, Beaunit Mills Inc., United States Patent 2 587 619, Mar. 1952; Japanese Patent 191 834, 1951.
124 S Ohhara, *Chemical Fiber Industry*, Tokyo Univ Press, 1961, pp 82–83.
125 Asahi Chemical Industry Co., *30 years History of Asahi Bemberg Plant*, Asahi Chemical Industries Co., 1962, pp 68–74.
126 A Munekata and R Maeda, Asahi Chemical Industries Co., Japanese Patent Publication No 2 132, 1950.
127 H Suwa, M Mihara and S Nagao, Asahi Chemical Industries Co., Japanese Patent Publication No 8 210, 1954.
128 S Nagao, Asahi Chemical Industries Co., Japanese Patent Publication No 115, 1956.
129 H Aizawa and F Ishida, Asahi Chemical Industries Co., United States Patent No 3 049 755, 1962.
130 Y Naniwa, Asahi Chemical Industries Co., Japanese Patent Publication No 22 204, 1969.
131 M Makita, E Komura and Y Takashima, Asahi Chemical Industries Co., Japanese Patent Publication 29 927, 1972.
132 I Moroe and K Iwase, Asahi Chemical Industries Co., Japanese Patent Publication 72 721, 1991. (Patent No 1 719 602).
133 Asahi Chemical Industries Co., *30 Years History of Asahi Bemberg Plant*, Asahi Chemical Industries, 1962; *50 Years History of of Bemberg Business*, Asahi Chemical Industries, 1981; Asahi Chemical Industries Co., *60 Years History of Bemberg Business*, Asahi Chemical Industries Co., 1991.
134 VGF, British Patent 1 283 (1905), July 1905.

135 W G Hagen and C E Ingham, *German Rayon Industry*, UK government report, 1945, p 43.
136 A Munekata and T Ootsuki, Japanese Patent 123 122, 1937.
137 A Haltmeier and I G Farben A.-G., Japanese, Patent 131 603, 1938.
138 A Haltmeier and I G Farben A.-G., Japanese, Patent 134 978, 1940.
139 A Haltmeier and I G Farben A.-G., Japanese, Patent 148 189, 1942.
140 H Yamazaki, *History of Japanese Chemical Fibers Industry*, Society of Chemical Fibres of Japan, 1974, p 47, Table 4.
141 H Yamazaki, *History of Japanese Chemical Fibers Industry*, Society of Chemical Fibres of Japan, 1974, p 46, Table 3.
142 *History Of Asahi Rayon Plant*, Asahi Chemical Industries, 1951, calculated using data in tables on pp 82, 96 and 101–102.
143 J Harrop, *Textile History*, 1969 **1**(2) 170.
144 K Kamide, *History of Textile Industry*, Society of Textile Machinery of Japan, 1993, pp 27–30; 58–60; 75–78; 111–118; 136–138.
145 H Yamazaki, *History of Japanese Chemical Fibers Industry*, Society of Chemical Fibers Japan, Part 1, Chapter 3, 4.2, p 151, Table 6 1974.
146 H Yamazaki, *History of Japanese Chemical Fibers Industry*, Society of Chemical Fibers Japan, 1974, p 150, Table 5.
147 T Ishii, H Iwasaki and S Komatsu, *History of Japanese Chemical Fibers Industry*, Society of Chemical Fibers Japan, p 1212.
148 Thr Lieser, *Chemische Textilfasern Filme und Folien*, ed. R Pummerer, Ferdinand Enke Verlag, 1953.
149 T Ishii, *et al.*, *History of Japanese Chemical Fibers Industry*, Society of Chemical Fibers Japan, 1974, p 1217.
150 *Year Book Textile Statistics*, compiled by Research and Statistics Department, Minister's Secretariat, Ministry of International Trade and Industry (Japan) 1999, pp 43, 44.
151 H Yamazaki, *History of Japanese Chemical Fiber Industry*, Society of Chemical Fibers Japan, p 217, Table 7.
152 *Chemical Fibers Handbook*, ed. Society of Chemical Fibers Japan, 1999, p 174.
153 A Munekata, *Chemical Fibers*, eds R Oka, A Munekata and M Wadano, Maruzen, 1956, p 281.
154 E Eguchi, Dissertation to Kumamoto University, 1999.
155 See, for example, K Kamide, *Thermodynamics of Polymer Solutions – Phase Equilibria and Critical Phenomena*, Elsevier, 1990, p 444, Table 6.1; K Kamide and H Iijima, *Cellulosic Polymers*, ed R D Gilbert, pp 189–190.
156 K Kamide and H Iijima, *Cellulosic Polymers*, ed R D Gilbert, pp 200–201, Fig. 10.8.
157 K Kamide and H Iijima, *Cellulosic Polymers*, ed R D Gilbert, 1994, p 191.
158 K Kamide, *Thermodynamics of Polymer Solutions – Phase Equilibria and Critical Phenomena*, Elsevier, 1990, p 610.
159 K Kamide and H Iijima, *Cellulosic Polymers*, ed R D Gilbert, p 202, Table 10.3.
160 K Kamide and S Manabe, *Polymer J*, 1981 **13** 459.

6
Fibres related to cellulose

JOHN W S HEARLE AND
CALVIN WOODINGS

This chapter is based on the accounts given by Moncrieff,[1] Coleman[2] and Raheel.[3] Other general references are by Cook[4] and Treiber.[5]

6.1 Cellulose acetate

6.1.1 Historical introduction

Paul Schützenberger, who heated cotton in a closed tube with acetic anhydride until it dissolved, first produced cellulose acetate in 1865.[6] When he diluted his solution with water, a white amorphous powder of highly degraded cellulose triacetate settled out. Over the next 25 years, important contributions to the science of cellulose acetylation were also made by Nadin,[7] Liebermann and Hoermann[8] and Franchimont,[9] but it was Cross and Bevan, of cellulose xanthate fame, who first described the industrial potential of the material in a series of patents and articles between 1889 and 1895.[10] The list of proposed uses, coming soon after their viscose patent, looks familiar: film, paints, varnishes, moulded articles, and when spun, filaments for electric lamps. Conversion of the cellulose ester into fibre was, however, a less straightforward proposition than the xanthate and it took a further 25 years before its fibre-making potential was realised on an industrial scale. Key developments of the fibre-making process were, in that period, made by George Miles, Arthur D. Little and the Dreyfus brothers, Camille and Henri.

In 1904 Miles discovered[11] that partially hydrolysing the cellulose acetate allowed it to be dissolved in relatively cheap and non-toxic solvents such as acetone, whereas the primary product of the acetylation reaction would only dissolve in 'difficult' solvents such as chloroform, nitrobenzene and epichlorhydrin. This conversion of the primary triacetate into the secondary diacetate was fundamentally important in allowing further development of practical fibre-making processes.

A D Little, working with W H Walker and H S Mork in Boston patented wet spinning of acetate fibres in 1905. Mork later became a Vice President of the Lustron Company of Boston, which, along with the Bayer Co in Germany pioneered small-scale commercial acetate fibre production. In 1907 the Bayer Co marketed cellulose acetate as 'Cellit' for the production of cinematograph film. The Swiss company Cellonite Co, run by the Dreyfus brothers, introduced a similar material a few years later. Count Donnersmark (see Chapter 1) had no more success with acetate than he had with viscose, and while the Chardonnet factory at Besançon (see Chapter 1) also experimented with dry spinning, none of the early developers succeeded in developing a commercial process. Overall, the acetate route looked more costly and more difficult than either the viscose process, or the nitrocellulose process that viscose was rapidly replacing. On top of this, acetate fibres had proved undyeable with those dyes commonly in use for viscose and cotton. Military interest in the use of the non-flam acetate dope to replace nitrocellulose dope nevertheless ensured developments continued.

The Dreyfus brothers also produced the fibre in a small way in Switzerland between 1910 and 1913, but during World War I their Cellonite company was the only company to respond to a British War Office tender for the UK manufacture of the acetate dopes used to coat the fabric skin of aeroplanes. They came to the UK, and with financial backing from Vickers Ltd and Pinchin Johnson & Co (Paints) Ltd, set up production in Spondon, near Derby as the British Cellulose Co Ltd.* As an interesting aside, they failed to deliver the product on time, and the War Office briefly encouraged Courtaulds to supply, before effectively granting exclusivity to the British Cellulose Co. Courtaulds nevertheless took what they learned about acetate seriously and convinced themselves that the problems of the fibre process would be solved. Acetate would then begin to threaten their dominant position in viscose.

The British Cellulose Co Ltd's unlisted shares rose from sixpence to £10 10s between 1916 and 1918 partly because they had the monopoly of aircraft dope and partly because they appeared able to enter the extremely lucrative artificial silk business. As war ended Camille Dreyfus set up the American Cellulose and Chemical Company to supply the US Army and Air Force with dope and both he and brother Henri, now running the British operation, successfully focussed on the artificial silk market. In 1921, when Spondon was capable of making about 1000 lb a day of acetate filament, Henri launched the product under the 'Celanese' brand. Camille commenced fibre production a year later in the USA. In 1923 the British

*They later changed the name to British Cellulose and Chemical Manufacturing Co Ltd.

company was renamed British Celanese Ltd, and the American operation later became The Celanese Corporation of America. Eastman Kodak produced acetate for films at about the same time, and their Tennessee Eastman subsidiary commenced fibre production as another outlet for the parent company's acetate in 1929. DuPont also started production of acetate fibre in 1929.

Meanwhile, in the UK, Courtaulds were becoming increasingly concerned by growing consumer interest in the new acetate yarns and decided to challenge the Celanese patents by setting up the Little Heath Works in Coventry to produce 'Seraceta' yarns. First deliveries were made in 1926, plans for US acetate production were laid, and a period of acrimonious relations between what were now the two leading rayon producers ensued. Briefly (for a full description see Coleman[12]), Celanese sued Courtaulds for infringements of their acetate patents in 1931, and in 1933 judgment favoured Courtaulds. Celanese appealed, lost again and took the case to the House of Lords where in 1935 they lost again and had their patents revoked. Amazingly in 1936 Dreyfus sued again (this time it was personal!) claiming that Glover at Courtaulds had given false evidence in a peripheral patent case, but in 1937 again lost and again appealed. The appeal was later withdrawn in what appeared to be the first move in talks leading to a merger of British Celanese and Courtaulds.*

The relative success of the upstart acetate against the established viscose can be judged by the fact that in 1930 the American Viscose Company (Courtaulds) commenced diacetate production and by 1940, 33% of all regenerated cellulosic filament output in the US was diacetate. In Japan, production commenced in 1919 in a joint venture between Snia Viscosa (see Chapter 1) and Teikoku, but it was the formation of Asahi with technology from VGF in 1925 that led to rapid expansion (see Fig. 6.1).

World production of acetate fibres, all made in the more easily dissolved diacetate form, reached 90 000 tonnes in 1939. Fibre for textile uses peaked in 1971 at about 500 000 tonnes, this being the combined figure for diacetate and triacetate which had been reintroduced in the 1950s. As for viscose, the decline in textile uses of acetate was caused by the rapid expansion and price cutting of synthetic fibres, but unlike viscose the losses were more than compensated by the development of a massive new industrial market for

* A deal which would have left Courtaulds in full control of the joint company was actually agreed with British Celanese in 1939. However war intervened, tax changes making the deal disadvantageous to both parties. Merger discussions were not renewed until Henry Dreyfus died in 1944. In 1947 they were again abandoned because of the risk of a combined company being nationalised by the Labour government. In 1957 Courtaulds finally acquired British Celanese in an agreed takeover.

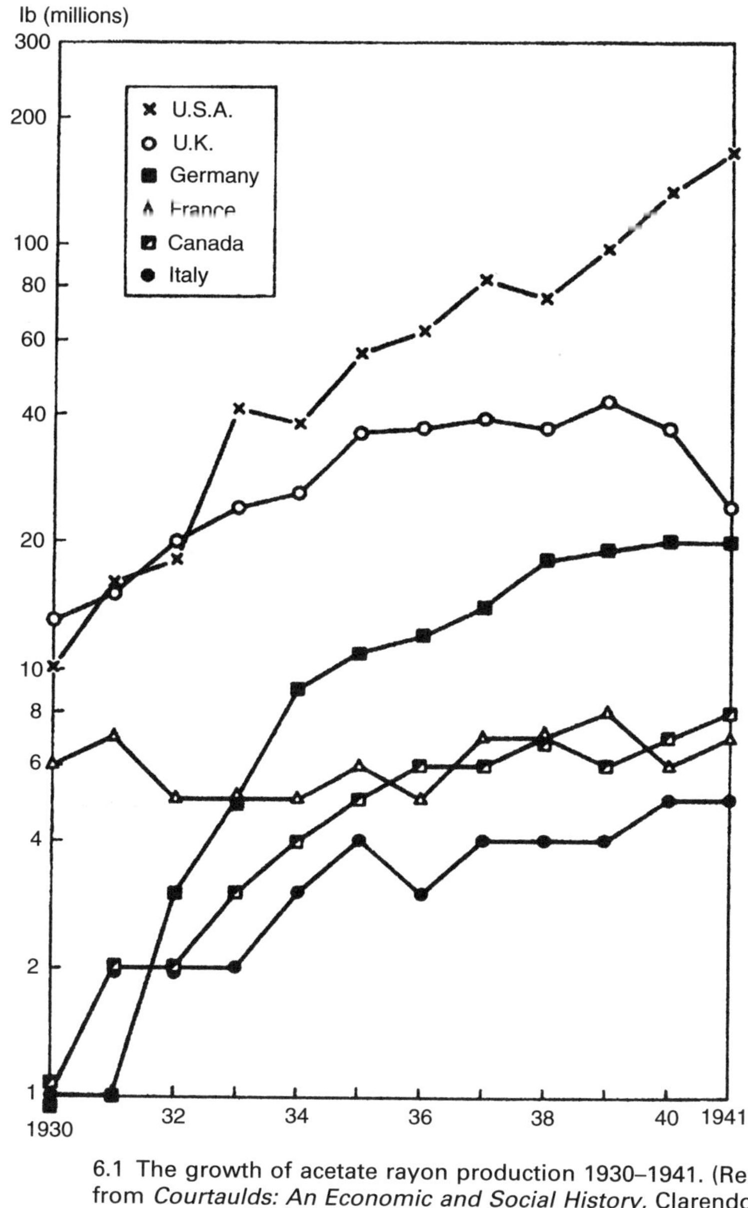

6.1 The growth of acetate rayon production 1930–1941. (Reproduced from *Courtaulds: An Economic and Social History*, Clarendon Press, 1969.)

the fibre: cigarette filter tips based on large crimped tows of diacetate. At the time of writing, despite many attempts by both synthetic and competitive cellulosic producers to displace it, diacetate tow reigns technically supreme in cigarette filters and worldwide production is now estimated to exceed 700 000 tonnes (see below and Chapter 10). Surprisingly in view of Western attitudes to smoking, the market continues to grow as the majority of the worlds smokers (in Asia) are persuaded to change to tipped cigarettes.

During the 1939–1945 war, very strong cellulose fibres, known as *Fortisan*, were made by highly stretching secondary acetate yarns and then regenerating the cellulose. The converse of this has been the production of acetate fibres by acetylation of cotton or viscose rayon.

The Lustron Company, from 1914 to 1924, had spun cellulose triacetate fibres in small quantities from chloroform, but the toxic solvent hazard made this an unsatisfactory operation. In the 1950s, an alternative solvent, methylene dichloride, became available cheaply and triacetate fibre production started again. Celanese introduced 'Arnel' in the USA in 1952 and 'Tricel' from the Spondon plant in the UK. Courtaulds UK followed in 1955 with 'Courpleta' from the Little Heath plant but abandoned this in favour of Tricel after the takeover of British Celanese in 1957.

Low moisture absorption and heat settability gave triacetate fibres some of the easy-care properties of the synthetic fibres, though not the durability, and at that time, the fibre was cheaper than nylon or polyester. The reduction in the price of the synthetics caused the triacetate market to shrink, and problems with the health and safety aspects of the solvent once again led to the closures of the US and UK operations in the late 1980s.

Other fibres mentioned briefly in this chapter are alginate fibres, derived from alginic acid, which is closely related to cellulose, and sodium carboxymethyl cellulose fibres.

6.1.2 Secondary acetate fibres

6.1.2.1 Manufacture

The starting material for acetate fibres was formerly cotton linters, the short fibres attached to the cotton seed, but wood pulp is now used. Thorough purification is necessary as a first step. The purified fibres are steeped in glacial acetic acid, so that they become more reactive, and are then thoroughly mixed with an excess of glacial acetic acid and acetic anhydride. In these conditions, acetylation would be slow, but it is speeded up by the addition of sulphuric acid, which forms sulpho-acetic acid as a fast acetylating agent. The reaction is strongly exothermic, and in order to avoid degradation of the polymer molecules, it is necessary to cool the vessels. After an hour at 20°C and about seven hours at 25–30°C the reaction is complete, and the fibre/liquid dispersion has turned into a viscous gelatinous mass. The cellulose has been converted to cellulose triaceate, which has the chemical formula shown in Fig. 6.2:

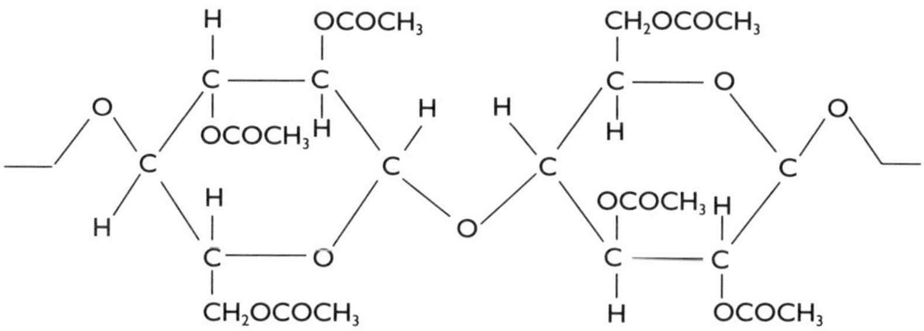

6.2 Fully acetylated cellulose. This primary product of the acetylation reaction, cellulose triacetate, is partially hydrolysed to the diacetate before fibre is produced.

It is necessary to sample the mixture in order to determine when this point has been reached, as indicated by the disappearance of swollen but undissolved fibres from microscopic observation and by solubility in chloroform. Degradation is monitored by viscosity measurement.

In order to make the secondary acetate, sufficient water is added to the mixture, which contains excess acetic acid and acetic anhydride, to give a 95% solution of acetic acid. Acid hydrolysis takes place over a period up to 20 hours at a higher temperature. The reaction is stopped when 1/6th of the acetate ($CH_3.COO–$) groups have been randomly changed to hydroxyl (–OH) groups. It is essential that any sulphate radicals are removed during hydrolysis. Excess water is added and the secondary acetate is precipitated. The liquor is removed to a recovery plant to extract the residual acetic acid. The polymer flakes are thoroughly washed, centrifuged and dried. Each batch is analysed and mixed with other batches in order to ensure a uniform product and avoid later troubles caused by differential dyeing or lustre.

It takes about 24 hours, with powerful stirring, to dissolve the secondary acetate in about three times its weight of acetone mixed with some water or alcohol. Titanium dioxide is added when dull yarns are being made. Black or other coloured pigments may also be added. After more blending, the spinning dope, which will be clear and colourless (unless pigments have been added), is filtered, de-aerated and passed to a feed tank (see Fig. 6.3).

Figure 6.4 illustrates the essential arrangement of dry-spinning equipment used to produce acetate fibres. The viscous dope, which contains 25–30% of secondary acetate, is pumped from the feed tank at a metered rate and through another filter to the spinneret, where the solution is extruded into the spinning cabinet. The spinneret is a metal plate, in which holes have been drilled typically between 50 and 100 μm in diameter. The

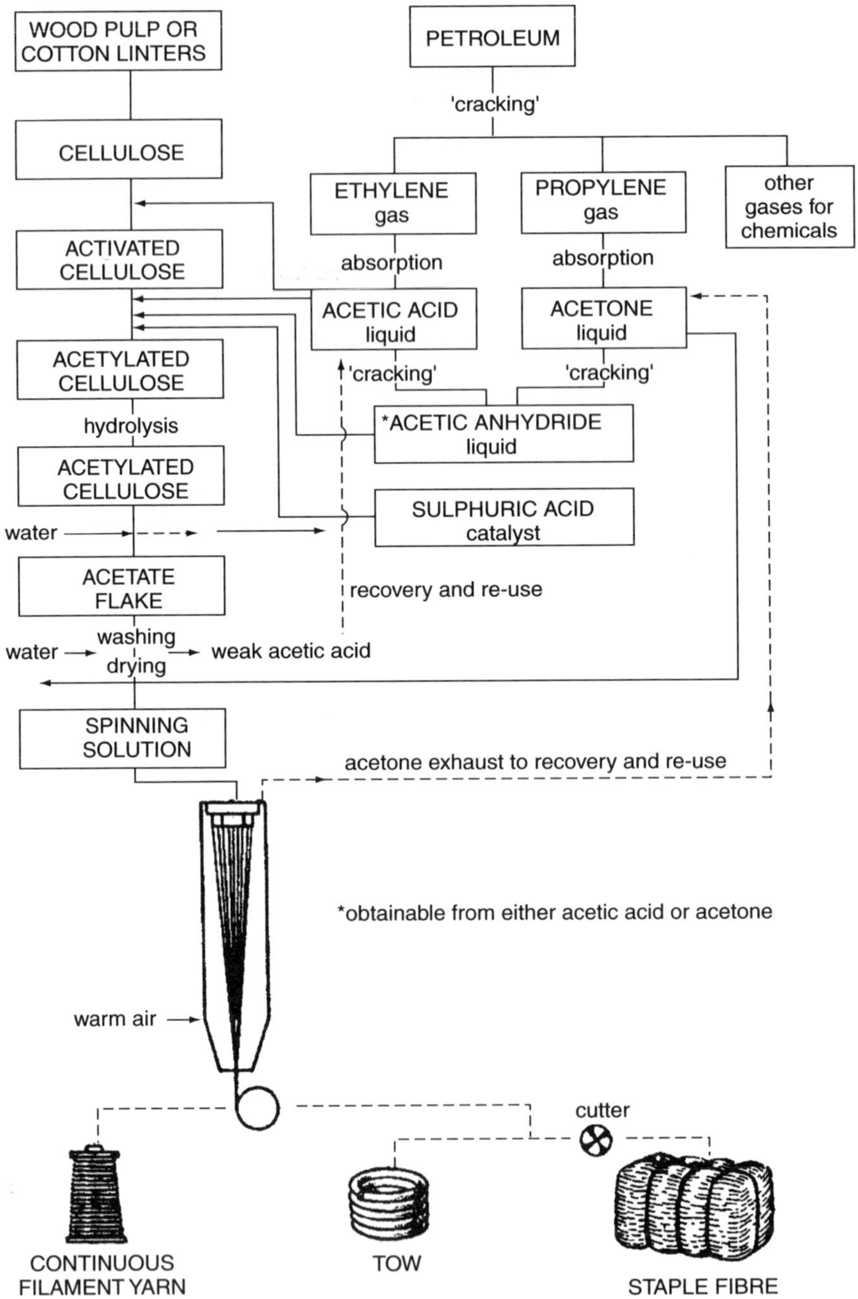

6.3 Diacetate fibre production flow chart. (Reproduced from Cook[4]).

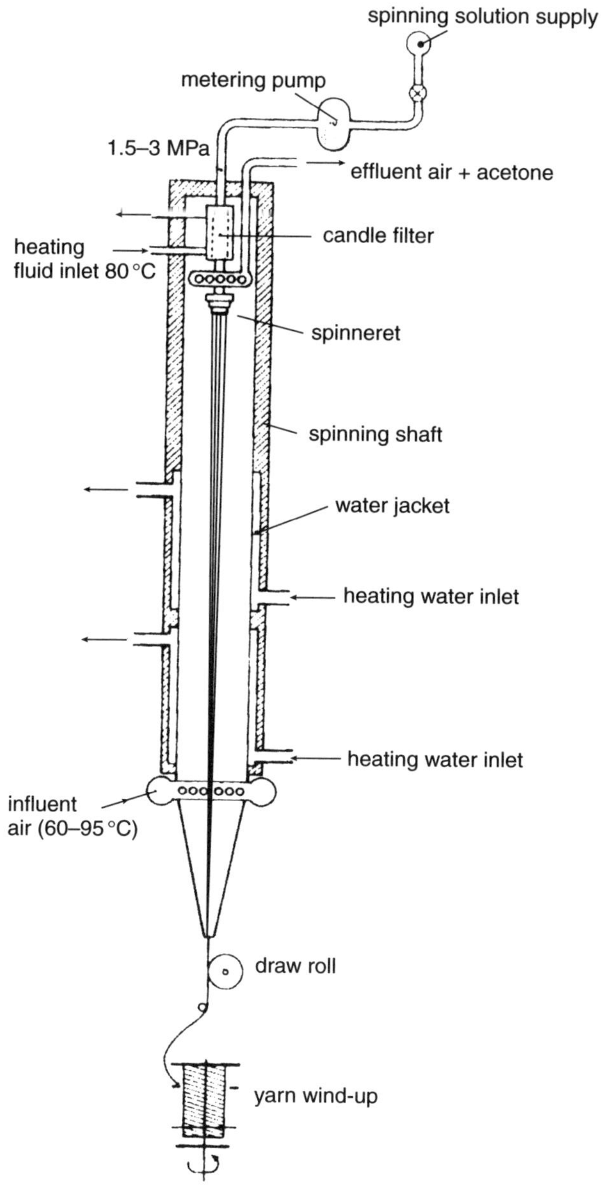

6.4 Diagrammatic representation of a spinning plant.

emerging dope descends vertically for 2–5 m against a countercurrent of hot air which enters the bottom of the tube. The temperature, moisture content and velocity of the air are important parameters. A guide roll at the bottom of the cabinet directs the fibres out to a collection device. For yarns, a take-up bobbin is arranged to insert a small amount of twist. Take-up speeds are typically between 200 and 400 $m\,min^{-1}$, but may be as high as 1000 $m\,min^{-1}$. A small amount of finishing oil may be applied to the yarn before wind-up. A stretch of around 100% is imposed on the yarn as it is drawn down from the spinneret. This stretch causes molecular orientation to develop in the plastic filaments in the upper part of the tube and provides the comparatively low strength needed for the textile applications. An additional drawing step is not required.

The linear density (tex or denier) of the filaments produced depends on the rate of mass flow through the spinneret holes, which is controlled by the metering pump, and the speed of take-up. The extent of drawdown is influenced by the hole diameter, since it depends on the linear flow rate through the spinneret and the take-up speed. The yarn linear density is given by the filament linear density, determined as stated, multiplied by the number of holes in the spinneret. Celanese Acetate, the world's largest producer of acetate yarns, produces about 16 different combinations of filament and yarn sizes in variants adapted to different applications. The yarns range from 55 to 600 denier (60.5–660 dtex). The commonest filament sizes are around 3.7 denier (4.1 dtex), but go down to 1.67 denier (1.84 dtex) in the 100/60 yarn and up to 6.25 denier (6.9 dtex) in the 200/32 yarn.

Tows, for cigarette filters or for cutting into staple fibres, are obtained by combining the output from a number of spinning positions.

Generally, dry spinning is a cleaner and faster process than wet spinning and requires less labour. Each kilogram of acetate fibre produced needs about 0.65 kg of cellulose, 1.5 kg of acetic anhydride and 4 kg of acetic acid, of which about 0.4 kg of the cellulose and 0.6 kg of the acetic compounds make up the fibres. In addition, 0.05 kg of sulphuric acid, 3 kg of acetone and 45 kg of water are used. Economy dictates efficient recovery of the acetone evaporated from the solution and of excess acetic acid from the earlier stages.

6.1.2.2 Properties and processing

Secondary acetate fibres, which do not show any evidence of crystallinity in X-ray diffraction, have a density of 1.32 $g\,cm^{-3}$ and a lobed cross-section. Different-shaped fibres can be made by using shaped spinneret holes. The fibres are comparatively weak, extensible, and have low stiffness, with poor elastic recovery above about 5% extension. Melting, which is accompanied by decomposition, is at about 230°C, but sticking occurs at lower temperatures, so that care must be taken in ironing. Because of the thermoplastic nature of the fibres, it is easy to emboss acetate fabrics. Moisture absorption is much lower than in viscose rayon, but the rate of change of moisture content is about the same as for cotton in the middle range of humidities. Cellulose acetate is an excellent insulator, which was one of the reasons for making acetylated cotton. Secondary acetate is soluble in a number of organic solvents and swollen by others, but is unaffected by ether; it is ultimately biodegradable, but much more resistant to biological attack than viscose rayon and moderately resistant to degradation by light. Acetate is attacked by some concentrated acids and is saponified by alkalis. Alkaline scours should be avoided.

Because of the lack of reactive groups, dyeing was initially a problem. However, new dyestuffs, notably disperse dyes, suitable for acetate were developed and a wide range of colours can be produced. Boiling water impairs the lustre of acetate fibres, so that dyeing temperatures must be kept below 85°C.

6.1.3 Cigarette filter tow

Filter tow was introduced in the mid-1950s, initially as a better looking, better tasting and more wet-collapse resistant alternative to the semi-crêpe filter paper used on the few brands sold with filter tips at that time. It has since become the preferred filter material for most of the world's cigarette producers and, thanks to changes in smoking habits, tipped cigarettes are now by far the most important sector of the world tobacco market.

The tow is produced in much the same way as diacetate textile yarns, with the output from many spinnerets being combined to form the tow rather than being wound individually on bobbins. There are, however, some key differences. Most cigarette tow filaments now have a Y-shaped cross-section achieved by spinning the dope through jets with triangular orifices. As the acetone solvent evaporates in dry spinning, the sides of the triangular dope stream collapse inwards to give the shape illustrated in Fig. 6.5.

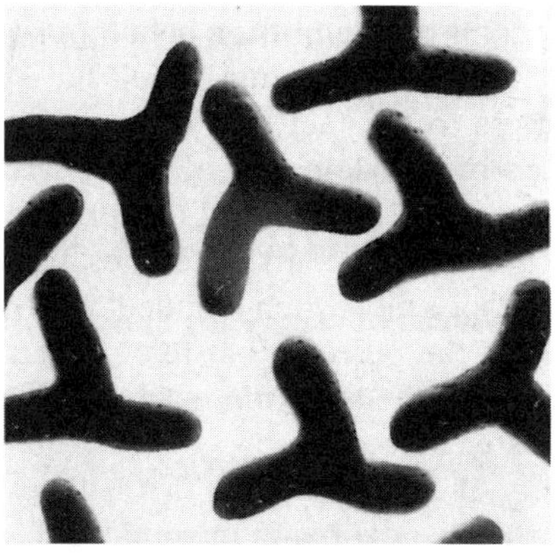

6.5 Typical Y cross-section tow.

This shape increases the bulk of the tow and increases the surface area available for intercepting smoke particles. Of even greater importance to the overall performance of the filter tow is the crimping and baling process. After the individual yarns have been lubricated with a mineral oil/surfactant mixture, they are 'dried' to remove most of the acetone solvent and combined into tow form for crimping in a stuffer box. The precise nature of the crimp has a significant effect on both the ability to make filter tips at the high speeds required by the tobacco industry and on the filtration and hardness characteristics of the filter tips. After crimping, the tow goes through a further drying stage to remove any residual acetone and to arrive at a precisely controlled moisture level in the region of 6%.

Filter tows are produced in sizes ranging from 18000 to 50000 total denier but most cigarette makers specify between 30000 and 40000 denier. The individual filaments can be as fine as 1.6 denier or as coarse as 7 denier, and while, as mentioned above, fibre cross-sections are mainly Y-shaped, I-shaped filaments and roundish filaments are also made. The tows are very carefully laid and packed into cardboard cartons which can contain up to a tonne of fibre. Precise presentation of long lengths of highly consistent tow is essential for the rod-making stage of the process of making filter tips. In this process the tows are 'bloomed' to deregister the crimp, sprayed with a partial solvent such as triacetin, wrapped in cigarette paper, and heated to fuse the filaments together at speeds that allow the production of some 6000 rods per minute. Each of these rods will have to be cut to form tips for 4 or 5 cigarettes at the next stage in the process, and each of these tips must meet exacting standards of diameter, porosity, hardness and pressure drop. They must also be capable of removing precise amounts of tar and nicotine from the smoke to allow the current wide range of advertised target tar and nicotine levels to be achieved with great consistency.

As mentioned earlier, the cigarette filtration market for diacetate tows has proved to be a 'fortress' that legions of attempts on behalf of polypropylene and other cheaper fibres including viscose and lyocell have failed to storm. It owes its continuing success to the fact that because it has been driven for the best part of 50 years by the demands of the tobacco industry, the process has evolved a quality and consistency of tow production that eludes those dealing mainly with the requirements of textiles. Its success is also a matter of taste. Diacetate, especially when bonded with triacetin, appears to have a beneficial effect on the perceived taste of filtered smoke: other fibres can perform just as well on the smoking machines, but human expert smokers always seem to prefer diacetate.

One possible future vulnerability of diacetate filter tips is their inability to filter down to the ultra-low tar levels now being introduced in response to consumer demands for 'safer' cigarettes. Tar levels below about 4 mg

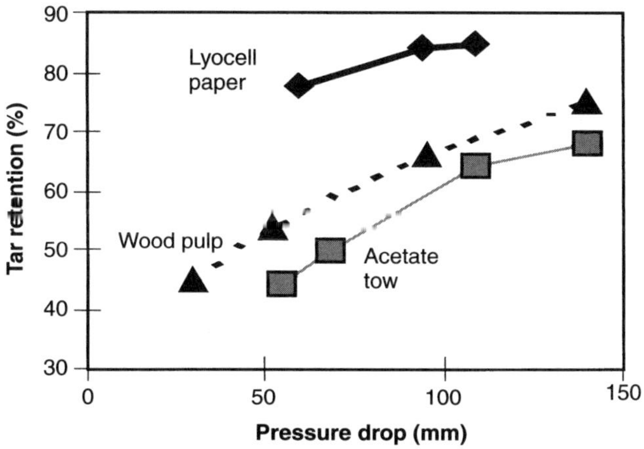

6.6 Filtrona Crest™ filter tips. Tar retention levels achieved by acetate tow, wood pulp paper and lyocell paper.

require filtration efficiencies that cannot be achieved with diacetate tow while maintaining acceptable pressure drops (draw resistance). Current commercial products therefore use laser perforation of the wrapping paper at the tobacco end of the tip which allow fresh air to dilute the smokestream entering the test equipment on the smoking machines used to test tar and nicotine yield. Ultra-low tar perforated diacetate cigarette tips can dilute the smoke with up to 70% of fresh air, a fact which human expert smokers detect very easily, and, allegedly, counteract easily by subconsciously or otherwise covering the perforations with their fingers. New 'Crest' tips have been developed by Filtrona using fibrillated lyocell fibres, and these, with numerous fine cellulosic hairs in the spaces between the filaments, allow ultra-low tar yields to be achieved without recourse to smoke dilution, and with acceptable pressure drops (see Fig. 6.6).

Hoechst Celanese addressed the need for ultrafine diacetate fibres that can intercept more tar and nicotine in cigarette filters with their 'Fibrets' product. These acetate fibrids* were manufactured by precipitating an acetate dope in water under conditions of high shear such that the precipitate was fibrous in nature. These fibrids could be made with very high specific surface area and could be made into papers, or attached to acetate tows for incorporation in filter tips.

* Fibrid: a term coined apparently by DuPont, which in the late 1960s developed the Fibrid–Textril approach to producing synthetic fibres (including viscose rayon) with paper-making properties. Fibrids were the pulp-like fibrillar microfibres that could be mixed with the short-cut regular fibres (Textrils), the combination being self-bonding enough to be converted on paper machines.

6.1.4 Cellulose triacetate

The advent of the synthetic fibres showed that there was a market for hydrophobic fibres, which could be heat set. This stimulated the production of triacetate fibres from solutions in methylene dichloride, which was easy and safe to handle and had become cheaply available by 1930. Acetylation of purified cotton linters or wood pulp may be carried out as described above for secondary acetate, but precipitated into water without hydrolyis of acetate groups, though it is necessary to remove sulphuric ester groups. In an alternative process, acetic anhydride acts directly on the cellulose in the presence of a non-solvent such as benzene, which has a slight swelling action, and an acid such as sulphuric acid, which speeds up the reaction. The solid cellulose triacetate is dissolved in methylene dichloride, mixed with some alcohol, and then dry spun into fibres in a similar way to secondary acetate.

Because of the regularity of the polymer molecules, triacetate fibres are partially crystalline with a melting point of 290–300°C. Many properties are similar to those of secondary acetate, but fabrics have a crisper handle. The moisture regain is lower and is further reduced by heat setting. There are two methods for heat setting: either dry heat at 170–200°C or steaming under pressure at 110–130°C.

6.2 Alginate fibres

Unlike other plants, the fibrous parts of seaweed are not based on cellulose but on the closely related polymers, alginic acid or its salts. The acid exists in two forms and has the chemical formula shown in Fig. 6.7.

Alginates, chain-forming heteropolysaccharides made up of blocks of mannuronic and guluronic acids, occur in the cell walls of the brown algae (*Phaeophycota*). They are present in sufficient concentrations for commercial exploitation (30–45% weight of dry seaweed) in the larger kelps and wracks (*Laminariales* and *Fucales*).

Alginic acid was first isolated by E C C Stanford in 1860 as jelly precipitated when a sodium carbonate solution of the abundant brown (not red or green) seaweed was acidified. He commercialized his process in the western highlands of Scotland and developed a range of alginates, mainly for thickening food, drinks and printing pastes. Scotland remains a leading producer, along with Norway, China and the USA. Some 25 000 tonnes of alginic acid are extracted from seaweed per annum worldwide, but the total is declining slowly.

Alginate's fibre-forming potential was explored by Speakman at Leeds University and first exploited during World War II when green monofilament yarns of chromium alginate were spun, on viscose machinery, to make

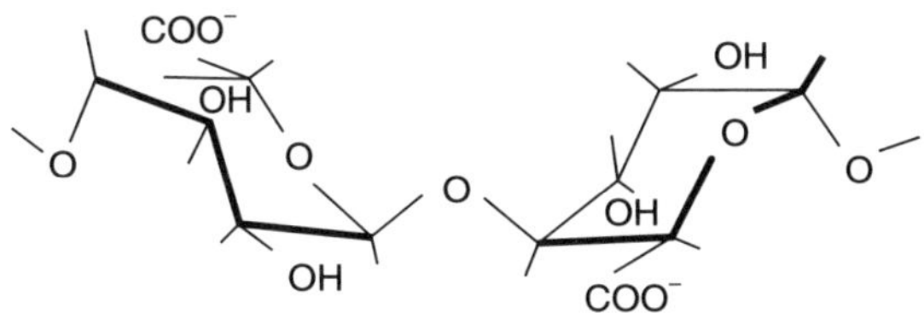

1–4 α L–Guluronic

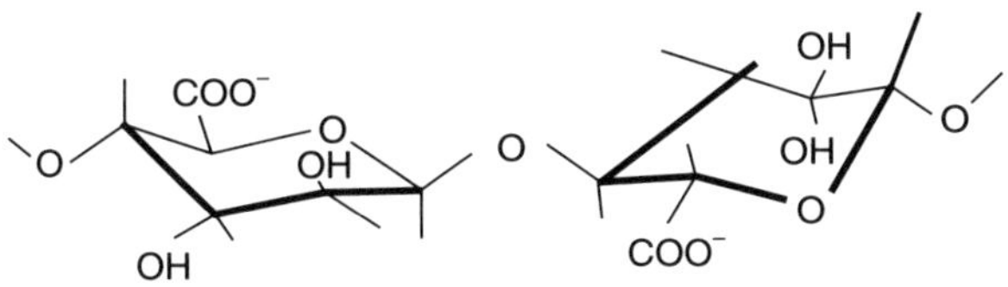

1–4 β D–Mannuronic

6.7 Two forms of alginic acid: guluronic and mannuronic acids.

camouflage netting. In order to make fibres, powdered and dried seaweed, usually the *Laminariae*, are treated with sodium carbonate and caustic soda to obtain a brown solution of sodium alginate. A series of purification steps, including conversion to alginic acid and back to sodium alginate, yield a dry white powder that can be dissolved in water to form a spinning solution. The fibres are formed by viscose-style wet spinning into a coagulating bath of 1 N calcium chloride solution and 0.02 N hydrochloric acid with a surface active agent. The filaments of calcium alginate are drawn, washed, lubricated, dried and wound up.

The non-flammability of calcium alginate fibres led to some applications. However, the main use found for alginate fibres derived from a property which rendered them useless for most purposes. They are soluble in hot soapy water and were used as 'disappearing threads' when fibres were required to be present in an intermediate stage of production but removed later. One example was to provide a scaffolding thread for the production of very lightweight fabrics. The commonest use was, however, as a linking thread during continuous knitting of socks, a market that the leading producer, Courtaulds, appears to have got into in order to help major customers to use viscose yarns in hosiery manufacture.

Polyvinyl alcohol, synthesized from acetic acid and acetylene via saponification of the polyvinyl acetate intermediate, became cheaper and more available than alginate during the 1970s and has now replaced alginate fibre in these markets. However, there was one small volume but very high value alginate product for which polyvinyl alcohol was not suitable: a knitted alginate gauze used as a haemostat – a dressing used to encourage blood to clot, for example a nasal cavity packing for treating otherwise unstoppable nose bleeds.* Collaboration between the fibre maker and the dressing manufacturer led to the development of alginate nonwoven dressings that were not only able to replace the old knitted product, but also capable of being used in a wider range of medical applications. In numerous clinical trials since 1990 it has been confirmed that alginate allows wounds to heal more quickly and less painfully than traditional cotton dressings.

The multimillion pound business now enjoyed by alginate fibres in advanced wound-care nonwovens is based on the special properties of mixtures of sodium and calcium alginate in fibre form. Whereas sodium alginate will dissolve in water, the calcium form is insoluble. A mixture of the two shows intermediate behaviour and swells in water to form a gel. Calcium alginate will also form a gel in fluids containing sodium ions – wound exudates for example – and as we learnt above will dissolve completely in alkaline soapy water. These properties have allowed Courtaulds, Innovative Technologies Ltd and others to engineer dressings that can provide the ideal healing environment for hard to heal wounds. When these dressings need to be changed they can be removed without trauma simply by dissolving the dressing in water containing sodium ions.

The differing response to sodium and calcium ions arises because alginate polymer is in fact a mixture of two monomers, mannuronic (M) and guluronic (G) acids, and in the polymer these tend to be arranged in blocks. Depending on the types of seaweed from which the alginate is extracted, fibre can be made rich in one monomer or the other and those with more of the mannuronic form (high in M units) exchange calcium for sodium more readily. High-M fibres form gels more rapidly than high-G fibres. In fact when any mixed fibre swells, it does so because the calcium crosslinking in the M units is exchanged first and the gel is prevented from dissolving by the links between the G units (see Fig. 6.8).

Alginate fibres have dry properties similar to viscose and can be converted into a variety of dry-laid nonwovens with ease. They are typically bonded by needlepunching, but can also be produced in the spun-laid form and as such can be self-bonded or bonded after drying using water as the solvent. The nonwovens will absorb up to 20 times their own weight of test fluid when tested according to the Pharmacopoeia procedure.[13] High-M products, for example Sorbsan®, can be distinguished from High-G products such as Kaltostat® by this test (see Fig. 6.9).

*The wound-healing properties of seaweed have apparently been appreciated by sailors for centuries.

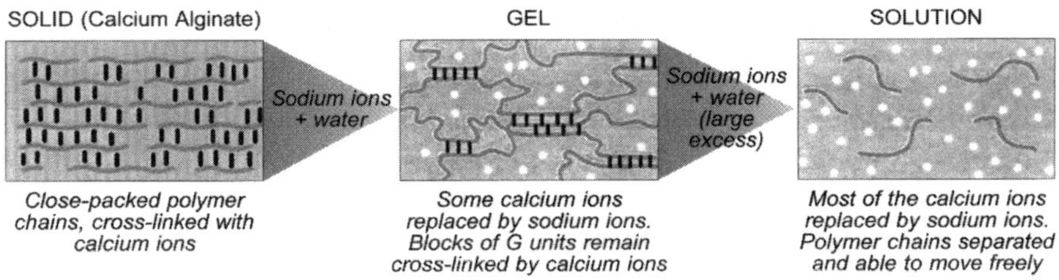

6.8 Principles of gel dressings based on alginates. (Reproduced by courtesy of Acordis Speciality Fibres.)

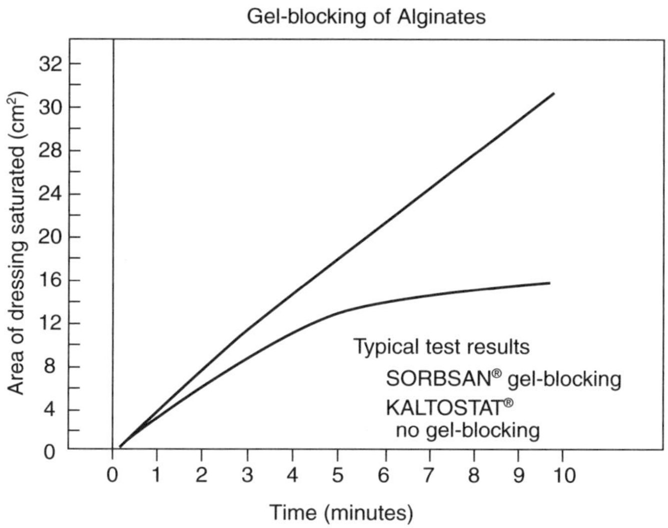

6.9 High-M dressings such as Sorbsan® (lower curve) resist the spreading of wound exudates better than high-G dressings such as Kaltostat® (upper curve). (Reproduced by courtesy of Acordis Speciality Fibres.)

In a wound-care context, gel blocking, as the restriction in lateral transfer of fluid is known, prevents healthy tissue surrounding the wound from being damaged by maceration.

6.3 Sodium carboxymethyl cellulose (CMC) fibres

The carboxymethylation of cotton and wood cellulose has been practised for many years to produce thickening agents for food, drinks and drilling lubricants. Alkaline cellulose is reacted with monochloracetic acid, Equation [6.1]:

$$(C_6H_9O_4 \cdot O^-Na^+)_n + nCH_2ClCOOH \rightarrow (C_6H_9O_4 \cdot OCH_2COOH)_n + nHCl \quad [6.1]$$

While these products were usually more powdery than fibrous, short-fibre versions could be produced from cotton linters and these could be blended with rayon fibres to enhance the absorbency of hygienic disposables, especially tampons. (Hercules in the USA produced 'Aqualon' for such end-uses.) Normally water soluble, these products could be crosslinked to limit the extent of swelling in water and as such they were early examples of superabsorbent fibres. Buckeye Cellulose produced 'CLD2' (Cross-Linked Derivative) in the USA for Procter and Gamble and the 'Rely' tampon. In Japan, Asahi developed and commercialized a superabsorbent version of their cuprammonium rayon spun-bond nonwoven fabric as 'Super AB' for use in medical dressings.

Viscose rayon can be carboxymethylated to give products ranging from highly absorbent to water soluble depending on the degree of substitution and crosslinking, but the fibres are generally too harsh and brittle to be used as textile or nonwoven fibres. Less costly products with useful absorbency benefits were however commercialized by Avtex and Enka in the USA, based on dissolving CMC in sodium hydroxide and injecting it into the viscose dope just prior to spinning. These so-called 'alloy fibres' were not true CMC fibres but a practical way of bringing the functionality of CMC into regenerated cellulose fibres.

Lyocell fibres (see Chapter 4) are much stronger than viscose, a feature which led Courtaulds (now Acordis) to investigate continuous carboxymethylation processes on lyocell tows and batch processes on both tow and staple fibres. Like the alginate products mentioned above, nonwoven dressings made from these true CMC fibres provided the ideal healing environment for difficult wounds, and the product has now been commercialized by the Speciality Fibres division of Acordis Fibres.

For a fibre that gels instantly in water, imbibing over six times its own weight while remaining coherent enough in the wet state to be handled, the physical properties of this CMC fibre are outstanding. Air dry (i.e. at 12% moisture content) 2 decitex fibres have a tenacity of $29\,cN\,tex^{-1}$ and an extensibility of 13%. Needlepunched nonwovens are easily manufactured and freely absorb $58\,g\,g^{-1}$ of water, or $35\,g\,g^{-1}$ of 0.9% saline. Furthermore they absorb $22\,g\,g^{-1}$ of 0.9% saline under loads of 3 kPa (0.45 psi).

References

1 R W Moncrieff, *Man-made Fibres*, 6th edition, London, Newnes-Butterworth, 1975 Chapters 11, 12, 16.
2 D C Coleman, *Courtaulds: an Economic and Social History*, Oxford, Clarendon Press, 1980.

3 M Raheel, History of cellulose acetate fibers, *Man-Made Fibers: Their origins and development*, eds R G Seymour and R S Porter, Oxford, Elsevier, 1991, pp 142–168.
4 J G Cook, *Handbook of Textile Fibres. II Man-made Fibres*, 5th edition, Shildon, Co. Durham, Merrow, 1984, Chapters 2, 4.
5 E E Treiber, Formation of fibers from cellulose solutions, *Cellulose Chemistry and its Applications*, eds T P Nevell and S H Zeronian, Chichester, Ellis Horwood, 1985, Chapter 18.
6 P Schützenberger, *Comp. Rend* 1865 **61** 485–486.
7 L Nadin and P Schützenberger, *Comp Rend* 1869 **68** 814.
8 C Liebermann, and O Hoermann, *Berichte* 1878 **11** 1618.
9 A Franchimont, *Rec Trav Chim* 1889 **18**(2) 465, 472.
10 C Cross, E Bevan and C Beadle, *Proc Chem Soc* 1895 **31**.
11 G Miles, German Patent M-28289, Sept 30th 1905.
12 D C Coleman, *Courtaulds: an Economic and Social History*, Oxford, Clarendon Press, 1969, Vol. II.
13 Thomas and Loveless, *Pharm J* 1992 **248** 850–851.

7
Other processes

ALBIN TURBAK

7.1 Historical review

Ever since the discovery of the xanthate process by Cross and Bevan in 1892 and its commercialization in the early 1900s there have been many efforts to dissolve cellulose directly in non-aqueous solvents which would be both easy to use and readily recoverable. The viscose process produces good quality films and fibers. However, by the mid-1970s pollution problems and price competition from synthetic films and fibers had seriously eroded the cellophane and rayon markets. In order to remain commercially competitive it was necessary for rayon and cellophane producers to seek simpler, less polluting processes.

In the mid-1970s ITT Rayonier decided that to preserve their pulp markets it would be in their best interest to take the lead in trying to find new ways to make rayon. To their credit, Rayonier's management authorized a multimillion dollar research effort which ran for a period of 7 years at the ITT Rayonier Eastern Research Division Laboratories. The overall goals were to develop new direct cellulose solvent systems which would require lower investment costs and be readily recovered and recycled in closed loop non-polluting systems.

In order to identify potential solvent candidates that had been examined previously, an extensive literature search was made covering 80 years of cellulose solvent reports. A close examination and evaluation of these reports from a physical organic viewpoint led to a review paper which, for the first time, delineated all cellulose solvent systems into four simple categories.[1] Subsequently, several other solvent system reviews have been reported.[2,3] Rayonier reported some of their preliminary results at an American Chemical Society symposium in New Orleans in 1977.[4] As a result of Rayonier's stimulus other companies and research universities undertook cellulose solvent research projects.

Concurrently, Franks, McCorsley and Varga at the American Enka Corp were deeply involved in evaluating *N*-methyl morpholine-*N*-oxide

(NMMO).[5] The Enka NMMO technology was subsequently pursued by Courtaulds in England. Courtaulds' Tencel® NMMO solvent spun cellulose fiber technology is now a major commercial success and solvent spun cellulose films for food packaging are currently under intense investigation. Tencel® lyocell is actually not a regenerated cellulose fiber but, rather, is a *reconstituted* cellulose fiber since it was never derivatized.

7.2 Thermodynamic requirements for dissolution

No chemical reaction, including dissolution of any polymer, can proceed unless the free energy requirement for the overall solution process is negative, that is, $\Delta G_{solution}$ must have a negative value. There are several factors involved in the dissolution process and while any one of them may have a positive value, such a value must ultimately be overcome by a larger negative value of the other factors if solution is to occur, Equation [7.1]:

$$-\Delta G_{solution} = -\Delta G_{fusion} + \Delta G_{reaction/complexing} + \Delta G_{mixing} \qquad [7.1]$$

In the case of liquids the situation is simpler than it is for solids and particularly simpler than it is for polymers. Since liquids do not have sufficient molecular interaction to form solid crystals the free energy of crystallization (fusion) can be ignored and then the overall solution process can be fairly well defined by considering only the free energy of interaction (or complexing) and the free energy of mixing.

With many solids, the free energy of interaction (complexing) or wetting and swelling of the solid by the incoming liquid is sufficiently negative to weaken the crystal forces greatly and lower the free energy of fusion. The solid then dissolves if the free energy of mixing is negative. With polymers, however, the free energy of fusion is so high owing to the many points of molecular interaction along chains that the process of dissolution becomes much more difficult. This is especially true for cellulose and protein polymers which have very large positive values for ΔG_{fusion}, and do not melt.

A good example of the importance of overcoming the ΔG_{fusion} is found in the dissolution of polypropylene. There is no known solvent which will dissolve polypropylene directly at room temperature. However, once the polypropylene is heated near its melting point then many liquids that have a solubility parameter similar to polypropylene will readily dissolve the polymer. The same is true for many other thermoplastic polymers.

Therefore, in order to dissolve cellulose, the proposed solvent system must either directly or indirectly lower the large positive value of ΔG_{fusion} or raise the negative values of $\Delta G_{complexing}$ and ΔG_{mixing} to allow these other two factors to yield an overall negative result. It is obvious that $\Delta G_{complexing}$ will always have a negative value and, once complexed, will always present a new surface for solvent interaction to improve the negative value of ΔG_{mixing}.

In trying to explain why certain solvents dissolve polymers, some authors have emphasized the requirement that the solubility parameter (δ) of the solvent must be in the same range as the solubility parameter of the polymer in question. While such statements are directionally correct, it is obvious from the above discussion that they are fundamentally incomplete. For example dimethylformamide (DMF), dimethylsulfoxide (DMSO) and several other liquids have solubility parameters that are in the range calculated for cellulose yet they do not dissolve cellulose directly. All free energy values (ΔG) are dependent upon their corresponding enthalpy (ΔH) and their entropy (ΔS). In the case of mixing this becomes, Equation [7.2]:

$$\Delta G_{mixing} = \Delta H_{mixing} - T\Delta S_{mixing} \quad [7.2]$$

or the free energy of mixing = the heat of mixing – $T \times$ (the entropy of mixing).

Since the entropy of mixing (ΔS) will normally have a positive value it is necessary that the ΔH_{mixing} has the smallest possible value if the ΔG_{mixing} is to be negative. Hildebrand relates the solubility parameters (δ) of the solute and solvent to the heat of mixing by Equation [7.3]:

$$\Delta H_{mixing} = k(\delta_1 - \delta_2)^2 \quad [7.3]$$

Therefore, ΔH_{mixing} will be at its lowest positive value when the solubility parameters for the solvent and the polymer are as identical as possible. This defines the extent to which the solubility parameters (δ) will contribute to the overall dissolution process.

7.3 Cellulose solvent systems

In each and every instance of dissolution of cellulose, the above described thermodynamic factors can be delineated and shown to have a significant and controlling effect on the success of the dissolution process. In each case the solvent system swells the cellulose and then modifies the molecule either by derivatizing, proton weakening or removal, or by complexing, to give a new intermediate species that can more readily react with the liquid solvent system. Thus, all cellulose solvent systems can be categorized under four mechanisms:

1 cellulose as an acid
2 cellulose as a base
3 cellulose unstable derivatives
4 cellulose stable derivatives.

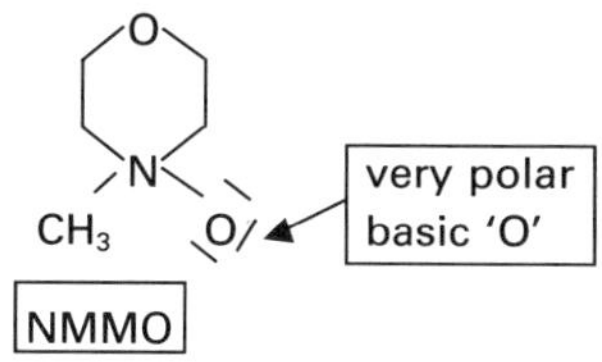

7.1 Structure of N-methyl morpholine-N-oxide.

This chapter will review selected solvent systems involving areas (1), (2), and (3). Stable derivatives (4) will not be discussed. All compounds can be considered to be acids or bases relative to some other stronger species. For example the following acid/base relationships:

(A) is an Acid	relative to	*Base (B)*
HCl, H_2SO_4		Carboxylic acids
Phosphoric		Acetic, cellulose
Phenol		NaOH, sodium carbonate
Alcohol		NaOH, sodium metal
Cellulose		NMMO, N_2N_4
$Li^{(+)}$/DMAc (dimethylacetamide)		Cellulose
Cellulose-OH		Unsolvated anhydrous Cl^-
Hydrocarbons, (isobutane)		$NaNH_2$

It may seem odd to realize that cellulose can act either as an acid or a base, yet this is exactly the case, especially in anhydrous or limited aqueous systems.

Water is an excellent swelling liquid for cellulose and opens millions of internal pores in the cellulose microfibrillar structure. However, H_2O is itself both an acid or a base depending on the other molecular species. Therefore, water normally reacts more rapidly than the cellulose with incoming molecules and this totally disrupts the effectiveness of the incoming potential dissolving moiety.

As examples of this, consider that NMMO has a very polar basic oxygen atom which will seek to abstract hydrogens (Fig. 7.1) wherever it can find them. Thus, water can be used to swell the cellulose. However, since it is a stronger acid than the cellulose (OH), the excess water must first be removed before this polar oxygen can exert its influence in abstracting or loosening the 'H' atoms from the (OH) positions on cellulose to begin the dissolution process. Similarly, the Li^+ of complexed LiCl/DMAc requires essentially anhydrous conditions before it can interact effectively as an acid toward the basic electrons on the 'O' of the (OH) on cellulose. In each case it is important to recognize that the incoming reactant has so modified the original cellulose that a negative $\Delta G_{complexing}$ has resulted and this will give the solvent a new surface species for interaction.

7.4 Unstable cellulose derivatives

Cellulose forms many unstable derivatives and complexes. Except for cuprammonium and xanthate very few unstable derivatives have been spun into fibers. While the literature reports include cellulose carbamates and formates as possible candidates, no fiber spinning studies have been reported from these systems. Only two extensive studies exist where good fibers were spun from solutions of cellulose nitrite in DMF and also from methylol cellulose dissolved in DMSO.

7.4.1 Cellulose nitrite

7.4.1.1 Reaction chemistry

The reaction of cellulose with nitrogen dioxide (N_2O_4) was initially studied by Kenyon, Yackel and Unruh[6,7] who found that it produced 6-carboxy cellulose in good yields. Fowler and McGee[8] then made a detailed study using 40 different solvents and found that the degree of cellulose oxidation was related to the polarity of the solvent employed – the higher the polarity, the lower the degree of oxidation. Although N_2O_4 has an $O_2N\text{-}NO_2$ structure it reacts homolytically or heterolytically depending on the polarity of the solvent.[9]

$$N_2O_4 \xrightarrow{\text{non-polar solvent}} 2NO_2 \text{ paramagnetic (oxidizer)} \quad [7.4]$$

$$\text{Cellulose} - \text{6-OH} + NO_2 \longrightarrow \text{Cellulose} - \text{6-Carboxy} \quad [7.5]$$

$$N_2O_4 \xrightarrow{\text{polar solvent}} NO^+ + NO_3^- \quad [7.6]$$

$$\text{Cellulose} - OH + NO^+ \longrightarrow \text{Cellulose} - ONO + HNO_3 \quad [7.7]$$

Equations [7.4]–[7.7] show the formation of cellulose nitrite using N_2O_4.

While good solutions of cellulose nitrite can be prepared, the temperature of the reaction is important in minimizing the amount of aldehyde that is formed. Temperatures lower than 5°C must be employed during dissolution or significant amounts of aldehyde are formed.

$$\text{Cellulose} - CH_2OH + N_2O_4 \begin{cases} \xrightarrow{\text{non-polar solvent}} \text{6-carboxy} \\ \xrightarrow[\text{room temp.}]{\text{polar solvent}} \text{aldehydes} \\ \xrightarrow[\text{0–5°C.}]{\text{polar solvent}} \text{cellulose nitrite solutions} \end{cases} \quad [7.8]$$

Equation [7.8] shows the effect of N_2O_4 reaction conditions on cellulose reactivity.

Higher dissolving temperatures ultimately lead to fibers that have poor resistance to alkaline washing cycles. Solubility in 6.5% NaOH at 20°C ($S_{6.5}$) is an acceptable test for predicting fiber losses in alkaline wash cycles. Early attempts to find the aldehydes were not successful because the standard oxidation analytical method employed was not sufficiently sensitive. However, Hergert *et al.*[10] reported that when the aldehyde content was determined by the alkaline copper number method, the alkaline conditions enhanced the aldehyde sensitivity owing to the well-known alkaline peeling reaction. There is a direct correlation of $S_{6.5}$ and dissolving temperature. At 0–5°C reaction temperature only 10–15% of the fiber would dissolve in 6.5% caustic soda, while at 20°C reaction up to 80% of the fiber would subsequently dissolve in 6.5% caustic soda. Such high levels would be totally unacceptable commercially.

7.4.1.2 *Fiber spinning and properties*

The cellulose nitrite/DMF solutions can be readily regenerated by a wide variety of protonic liquids. In most cases the regeneration is very rapid and care must be taken to obtain some molecular alignment and stretching before the polymer is completely regenerated. Water is particularly fast, and simply exposing the nitrite solution to air will result in formation of a surface skin. To obtain good quality fibers it is necessary to have high jet stretch and high initial stretch on the first godets. Addition of bases to the alcohol regeneration bath slows the regeneration process and allows for greater stretch.

In addition to high mechanical speeds of extrusion, the nature of the protonic regenerating liquid is important. Extensive studies by Hammer and Turbak[11] demonstrated that as the molecular weight of the regenerating alcohol bath is increased, alcohol diffusion inward is decreased and the cross-sectional shape of the fiber is significantly affected. While methanol gives a round fiber, amyl alcohol and benzyl alcohol give a crenulated fiber and octanol actually gives a hollow segmented fiber. The hollow fiber results from the fact that octanol can readily form an outer skin but then cannot readily penetrate this skin to regenerate the fiber internally. Faced with this situation the lower molecular weight internal DMF diffuses towards the outer skin and the cellulose nitrite solution densifies from the inside out to give a hollow structure (Fig. 7.2).

It should be noted that this is one of the few instances, if not the only one, where a hollow fiber could be spun by controlling inside and outside diffusion without relying on the use of some type of a gaseous blowing agent.

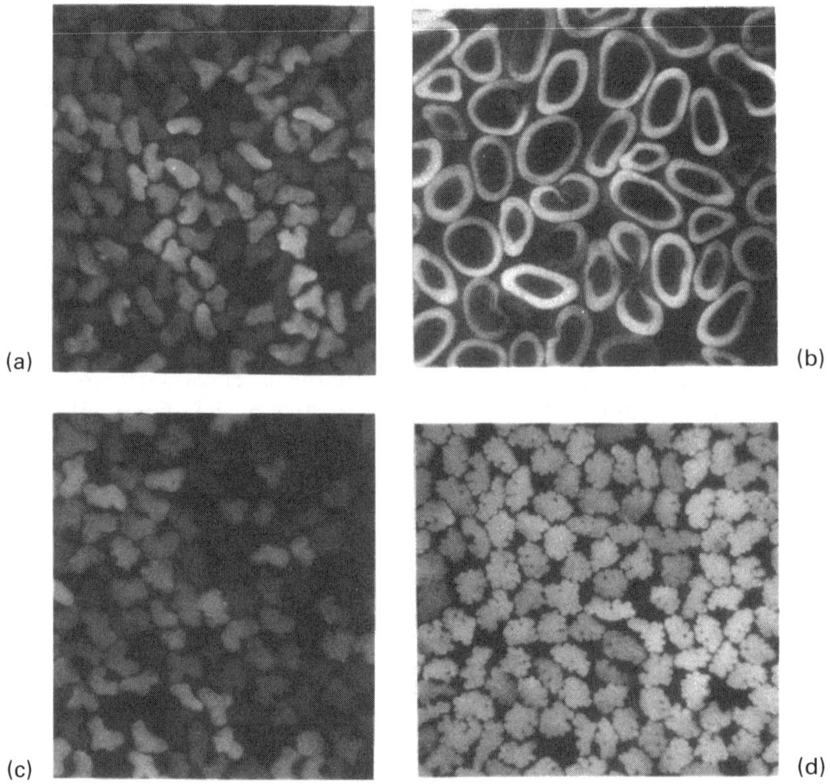

7.2 Cross-sections of N_2O_4/DMF spun fibers. (a) Precipitated in isoamyl alcohol; (b) precipitated in octanol; (c) precipitated in benzyl alcohol; (d) regular rayon.

With cellulose nitrite, as essentially with all solvent spun fibers, outer skin formation is very rapid giving rise to fibers that have intermediate wet strength properties without the need for any regeneration bath additives. The comparative fiber physical properties are given in Fig. 7.3.

7.4.1.3 Recovery and recycle

One of the most important factors in deciding the fate of all cellulose solvent systems is the ability to recover and recycle all of the reactants in good yields in closed loop systems. In the case of cellulose nitrite this is a difficult and expensive sequence. McDonald[12] has proposed spinning cellulose nitrite solutions into DMF/aqueous ammonium nitrate baths then subsequently concentrating the ammonium nitrate and converting it back to

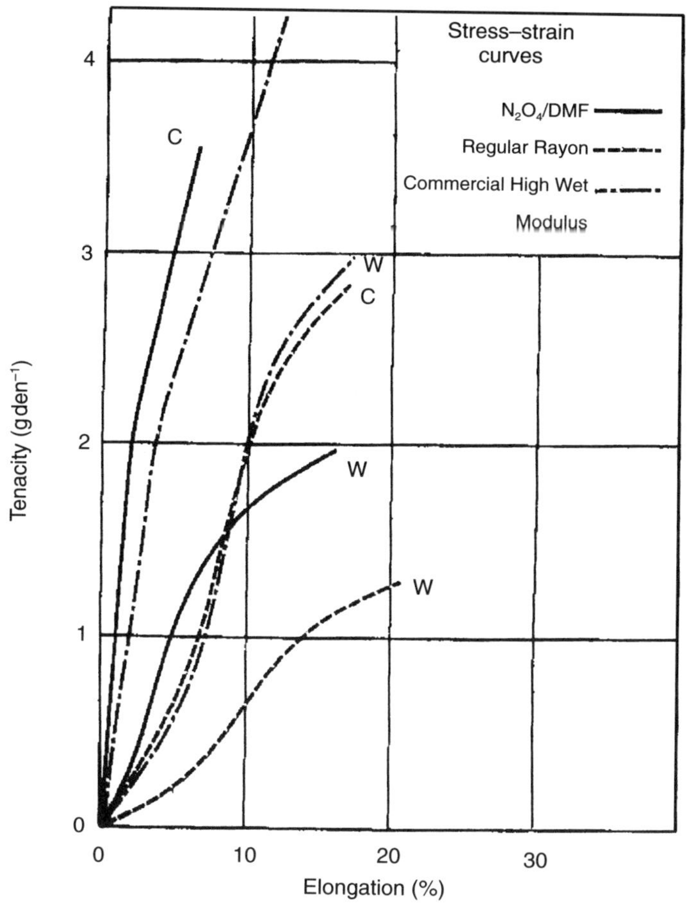

7.3 Fiber properties of N_2O_4/DMF spun fibers: Conditioned (C); Wet (W).

NO_2. However, no details are given about how such a conversion could be achieved.

Portnoy and Anderson[13] undertook a major detailed study on the recovery and recycle of NO_2 from the spinning of cellulose nitrite solution into an isopropanol regeneration bath. Their detailed scheme is given in Fig. 7.4 and demonstrates that recovery and recycle of this system would be quite complex.

7.4.2 Methylol cellulose

7.4.2.1 Reaction chemistry

The reaction of cellulose with formaldehyde to form a reasonably stable methylol derivative in the presence of DMSO was discovered and first

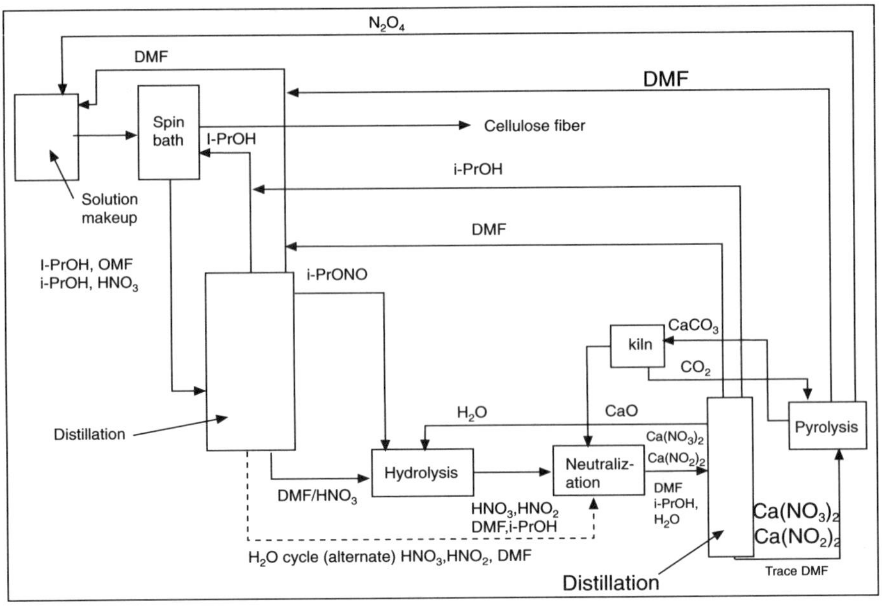

7.4 Total recovery and recycle scheme for the N_2O_4/DMF fiber system.

reported by Johnson, Nicholson and Haigh[14] at the Institute of Paper Science in 1976. This was a most interesting discovery since formaldehyde had previously been used almost exclusively to crosslink cellulose and the supposed methylol cellulose was always reported as a hypothetical non-isolatable intermediate.

Several studies have confirmed that excess formaldehyde is needed to effect the initial dissolution and that paraformaldehyde – being a polymeric form $(CH_2O)_x$ – is the best material to supply the required excess rapidly when it decomposes at higher temperatures. The paraformaldehyde (PF) is preferably used as a solid suspension. Methylation of the cellulose in the DMSO/PF solution demonstrated that the methylol formation initially occurred mostly in the C6 position of cellulose.[15–16] Figure 7.5 shows the reaction sequence for formation of methylol cellulose.

Baker, Schroeder and Johnson then reported that solvents other then DMSO could also be used in this reaction,[17,18] at low dissolving temperatures (80°C), where the amount of molar substitution (MS) could be as high as 24:1, depending on the polarity, complex-forming ability and basicity of the specific solvent. Lower polarity solvents required more formaldehde to effect solution of the cellulose.

The high MS levels dropped significantly at higher temperatures as excess formaldehde was removed. The high initial MS was probably due to the fact

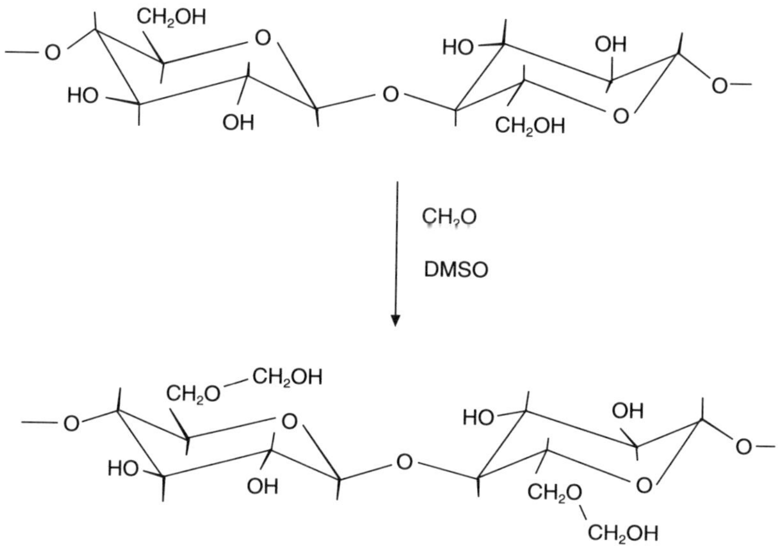

7.5 Formation of methylol cellulose in DMSO/PF.[16]

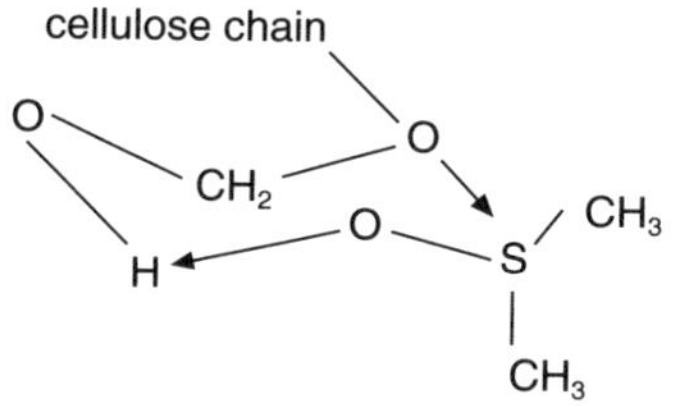

7.6 Complex formation of DMSO with methylol cellulose.[18]

that lower polarity solvents need the higher MS to act as a wedge to help keep the chains apart. Baker *et al.*[18] reported that even at high temperatures, over extended periods of time under vacuum, the DMSO to formaldehyde molar ratios in these solutions could not be lowered below a 1:1 level. This is strong evidence for the formation of a molar complex between the methylol content and the DMSO and they proposed the complex shown in Fig. 7.6.

Seymour and Johnson[19] studied this reaction and developed a chart showing the optimum conditions under which cellulose will dissolve in the DMSO/PF system and their chart is shown in Fig. 7.7.

All of these studies clearly indicate that cellulose dissolves readily in the presence of a polar solvent, that DMSO is the preferred polar solvent because it forms a strong 1:1 complex with the dissolved methylol

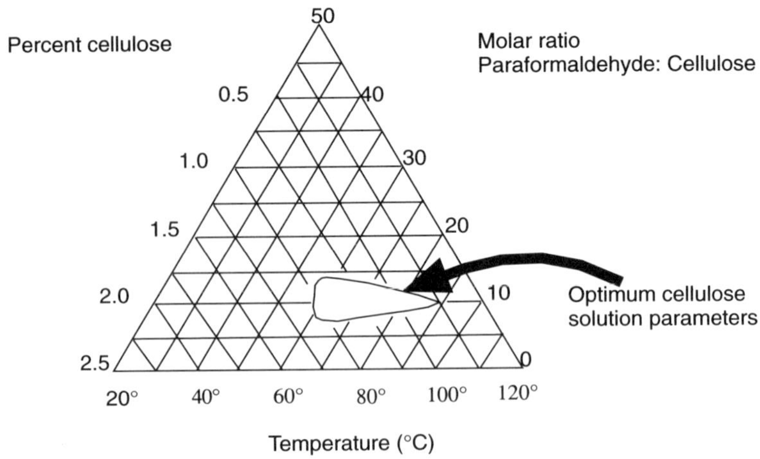

7.7 Optimum dissolving conditions for cellulose in DMSO and formaldehyde.[19]

cellulose and that excess formaldehyde is needed initially to help keep the chains apart. The formaldehyde favors formation of the 6-methylol and the large amount of formaldehyde that can be released from paraformaldehyde as it decomposes at temperatures above 120°C greatly accelerates the dissolution reaction.

7.4.2.2 Fiber spinning and properties

Hammer, O'Shaughnessy, Strauch, Portnoy and Turbak[20,21] made extensive spinning studies with clear particle-free polymer dopes containing from 4–15% cellulose prepared from pulps having a degree of polymerization (DP) from 1000 down to 300. Swenson[22] has reported the viscosity versus pulp DP relationships for DMSO/PF solutions. For most spinning studies 6–8% cellulose/CH_2O/DMSO solutions were prepared at 120°C using paraformaldehde. These solutions were spun and coagulated into protonic solvents such as water or methanol. Surprisingly, these fibers were coagulated but were not regenerated and could be readily redissolved in fresh DMSO at 50–70°C.

This demonstrated that the DMSO/methylol complex was much stronger than first expected and was not being broken down by simple dilution and coagulation with protonic liquids. In fact, this type of coagulation followed by redissolution in fresh DMSO could be repeated several times until the formaldehyde level of the cellulose solution fell below 1% which corresponds to the level required to maintain the stability of the methylol cellulose in the initial DMSO/PF solution.

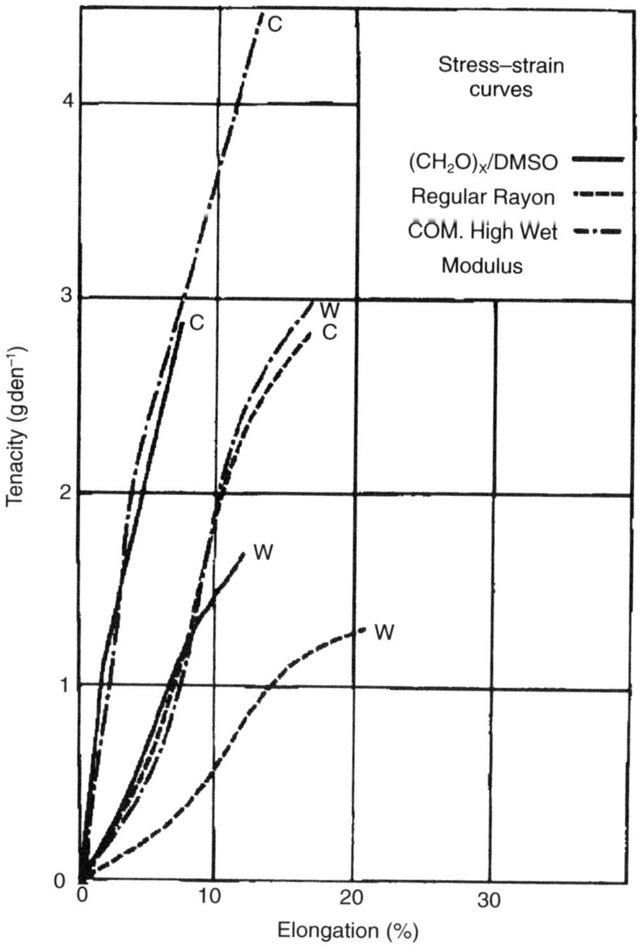

7.8 Fiber property comparisons for DMSO/PF spun fibers: Conditioned (C); Wet (W).

Figure 7.8 shows how the DMSO/PF fibers compare with known rayon fibers. One of the big advantages of the DMSO/PF fibers is their low $S_{6.5}$ levels of 3–15% which puts them in the range of high wet modulus (HWM) premium rayon fibers.

Various nitrogen compounds (urea, melamine, etc.) that are known to react with formaldehyde were added to the cellulose/DMSO/PF solutions to try to convert the excess formaldehyde into known crosslinking agents. While the expected compounds were formed, they all came out in the spinning baths and were not retained for possible further internal fiber reaction. A large range of synthetic fiber polymers which were soluble in DMSO were also added and spun into fibers but they acted mostly as fillers in the final fibers with no apparent tendency to give improved fiber properties.

In order to achieve more rapid regeneration of the cellulose/DMSO/PF dopes, various additives which are known to react with formaldehyde were added to the regeneration bath. These included ammonia, various amines, sodium sulfide, sodium sulfite and sodium thiosulfate. These coagulation bath additives did, indeed, increase the regeneration rate and in some cases gave improved fibers. However, the problem with such additives was that they would later have to be recovered and recycled. Rodier at Rhone Poulenc[23,24] has issued several US and British patents dealing with the area of fiber spinning into various coagulation baths. Lenoni at Snia Viscosa[25] has taken advantage of the coagulated fibers and then stretched them extensively before they were regenerated, thereby improving the fiber properties.

7.4.2.3 Recovery and recycle

With the DMSO/PF solvent system, as with all solvent systems, the overall potential commercial success is fundamentally determined by how simply and completely the dissolving materials can be removed, recovered and recycled.

The complex recovery system noted for the N_2O_4/DMF system essentially eliminated it from commercial consideration.

For the DMSO/PF system the recovery of the formaldehyde in a readily reusable form is absolutely mandatory. Unfortunately, the strong complex formation of the DMSO to the formaldehyde greatly limits the ease of CH_2O recovery. Further, the formaldehyde that is recovered is not in the paraformaldehyde form, but rather in some other solid form that does not lend itself to ready use for redissolving purposes. If some way could be found to overcome this recovery and recycle problem, this system could hold significant promise in the future.

7.4.3 Cellulose/urea systems

While the solutions of cellulose heated with urea apparently form a derivative with above average stability, they are included under this heading of unstable derivatives since they ultimately can be broken and spun under viscose-like conditions.

7.4.3.1 Reaction chemistry

The reaction of excess urea with cellulose preswollen in liquid ammonia was first promoted in the early and mid-1980s in Finland by Turunen *et al.*[26]

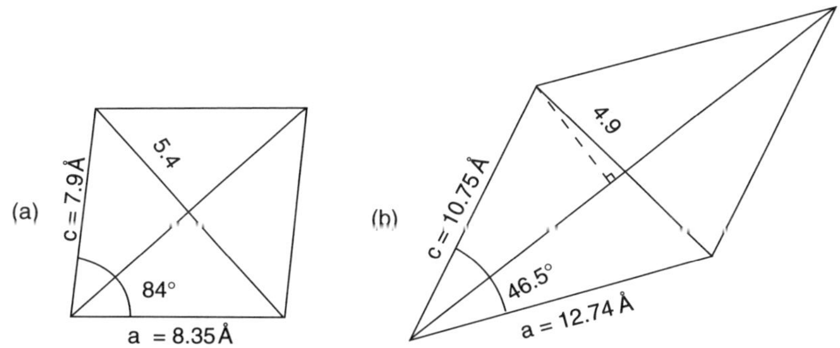

7.9 (a) Unit cell of cellulose I; (b) unit cell of ammonia–cellulose I.

and by Huttunen *et al.*[27] who subsequently filed a series of patents for their work. Evidently, one of the major factors in achieving a satisfactory product lies in having a high degree of swelling of the cellulose in order for the urea to obtain good and even penetration. The ability of liquid ammonia to swell the cellulose and to alter the crystal structure is vividly demonstrated in the comparison of cellulose crystal structures shown in Fig. 7.9.

The swollen cellulose is impregnated with a high concentration of urea also dissolved in liquid ammonia. Following evaporation of the liquid ammonia, the cellulose/urea material is heated to 120°C, a temperature sufficient to cause decomposition of the urea to isocyanic acid (HN=C=O) which then reacts with the cellulose (OH) supposedly to form a cellulose carbamate. The carbamate when washed free of excess urea (and by-products) and dissolved in 6% caustic soda gives a spinnable solution. Since most carbamates that have terminal NH_2 groups are quite unstable, this author questions whether the supposed carbamate is, indeed, the derivative formed. Most terminal carbamates readily undergo further reaction with more isocyanic acid to form much more stable 'allophanates'.

$$\text{Cell–OH} + \text{HN=C=O} \longrightarrow \underset{\text{Carbamate}}{\text{Cell–O–}\overset{\overset{\displaystyle O}{\|}}{C}\text{–NH}_2} \qquad [7.9]$$

$$\underset{\text{Allophanate}}{\text{Cell–O–}\overset{\overset{\displaystyle O}{\|}}{C}\text{–NH–}\overset{\overset{\displaystyle O}{\|}}{C}\text{–NH}_2} \xleftarrow{\text{HN=C=O}} \text{Carbamate}$$

Equation [7.9] shows reactions of cellulose with isocyanic acid.

Since there obviously is an excess of HN=C=O available for further reaction one must wonder why such reaction would not occur in this case. Also, HN=C=O should react with liberated NH_3 and with urea.

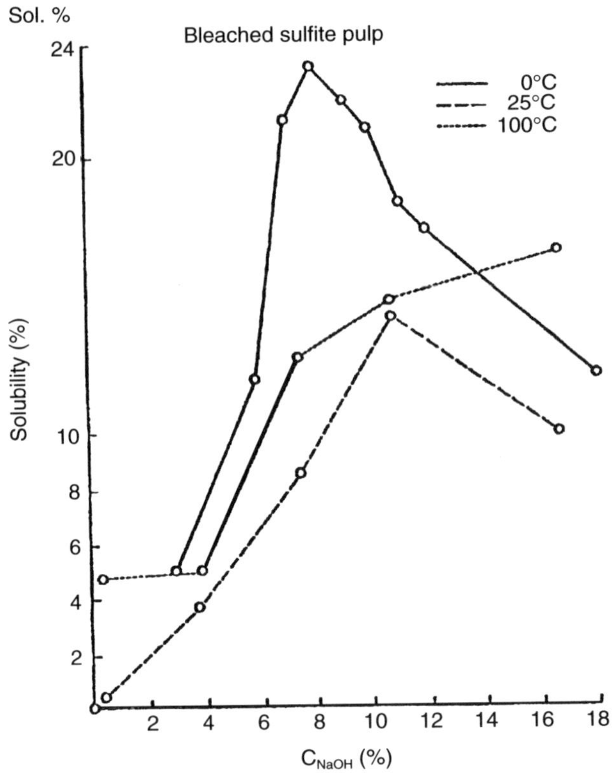

7.10 Swelling solubility of pulp versus NaOH and temperature.

The possibility of running this reaction without the need for liquid ammonia was reported by Struszczyk in Poland.[28] Mercerizing strength caustic (18%) was used to swell the cellulose in the presence of excess urea. The excess caustic was removed and the dry urea saturated cellulose was heated to generate the desired HN=C=O. The derivative was then dissolved in more dilute caustic soda and spun into an acid coagulation/regeneration bath similar to viscose.

Bridgeford and Rahman[29,30] realized that 18% caustic soda was not necessary to obtain good swelling of cellulose and filed several patents showing how much lower levels of caustic soda could be used if the temperature of swelling was kept cold. The use of cold temperatures to achieve maximum swelling is well known in the pulping industry and has been used extensively to remove low ends of cellulose along with hemicelluloses in the S_{10}–S_{18} tests and to prevent mercerization during pulping operations. Figures 7.10 and 7.11 demonstrate the theoretical soundness of the Bridgeford/Rahman approach.

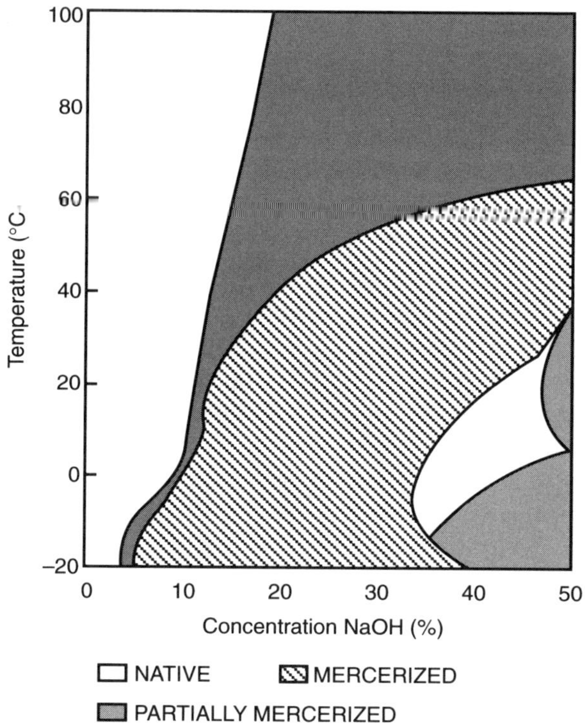

7.11 Caustic temperature and concentration to achieve mercerization.

7.4.3.2 Fiber properties and recovery and recycle

For the cellulose/urea system very little has been published regarding the quality of the fibers that might be expected and even less regarding the possible economics of recovery and recycle for the reagents. Struszczyk[28] has reported spinning 1.6 dtex fibers having strengths of 15–22 cN tex^{-1}, with 12% elongation. No overall recovery scheme has been made available but removal and handling and recovery of liquid ammonia would involve significant investment costs. The recovery and separation of purified urea and by-products along with the standard recovery of acid and salts for the alkali swelling method would also be of some consequence. Obviously much more needs to be done in the recovery/recycle area if this approach is to become competitive.

7.5 Cellulose as an acid or a base

7.5.1 Lithium chloride/dimethyl acetamide

7.5.1.1 Reaction chemistry

In this system the complexed positive Li^+ ion exerts a strong interaction with the (O) electrons of the cellulose (OH) while simultaneously releasing a strongly negative Cl^- ion to pull the (H) atom away from the (OH) on cellulose. While Cl^- ions are not normally considered to be strong bases, in anhydrous media they have sufficient electron density to exhibit strong base behavior. This is supported by the fact that the much larger bromide ion in lithium bromide does not dissolve cellulose under these identical conditions, nor do other lithium salts.

Lithium, being a small atom, has a very high positive charge density. In aqueous electrolyis measurements this small lithium ion moves much more slowly than the much larger sodium ion. This has been shown to be due to the six associated water molecules that Li^+ carries with it while sodium carries none. The LiCl/DMAc system has received considerable research attention.

LiCl/DMAc was first used as a polymer solvent system by DuPont for dissolving and spinning synthetic fibers. Later, Austin at the University of Delaware[31] used this solvent system to dissolve chitin. Subsequently, McCormick[32] reported the use of low levels of cellulose (less than 3%) to run various derivative reactions on cellulose in Li/DMAc. However, it was not until the research team at Rayonier[33] discovered how to get higher amounts of cellulose into solution that this solvent system could possibly be considered for commercialization. This group also demonstrated that *N*-methylpyrrolidone could also be used in place of the DMAc.

The Rayonier group found that activating the cellulose was a critical step prior to dissolution. If the cellulose was not preactivated, then temperatures of 150°C were required, at which temperature the solutions become brownish in color indicating significant decomposition. The best and simplest way to improve the dissolution process was to preswell the cellulose with water (or steam) and then maintain the swollen state by exchanging the water down to less than 2% with DMAc prior to adding the LiCl/DMAc mixture.

Liquid ammonia preswelling could also be used. If the water level was too high the cellulose did not dissolve because the LiCl would be interacting with the excess water. In this way clear solutions containing up to 16% cellulose/12% LiCl/72% DMAc at DP 400–600 could be readily obtained within 4–6 hours at temperatures of 85°C. At DP levels of 800–1700 up to 4% cellulose could be dissolved with less than 30 DP unit losses. Ekmanis[34] at Waters Labs used the LiCl/DMAc solvent system to develop a totally

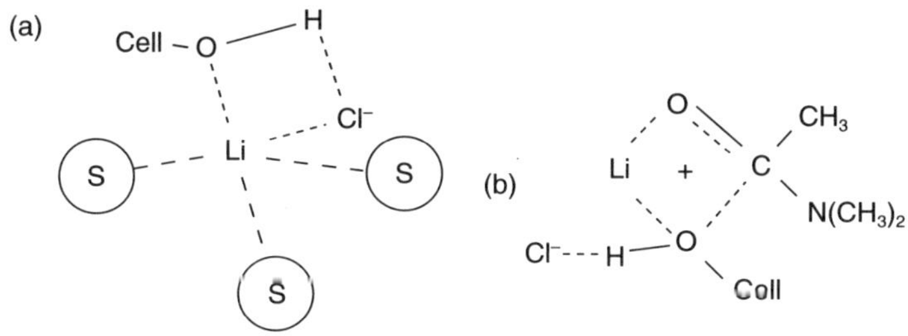

7.12 Proposed LiCl/DMAc/cellulose complex, (a) by Morgenstern and (b) by El-Kafrawy.

new simple and rapid GPC technique for determining molecular weight distributions of cellulose. This method has been adopted by many industrial laboratories.

The exact nature of the dissolving species has been investigated. El-Kafrawy[35] reported a ^{13}C-NMR study which demonstrated that the LiCl/DMAc did not form a derivative, meaning it was a complex with the cellulose. Morgenstern and Werner-Kammer[36] have also studied the nature of the complex formation and dissolution mechanism using ^{7}Li-NMR. Both studies agree that a complex is formed between the lithium ion and the DMAc. They differ, however, about what happens to this complex in the presence of cellulose. The two model proposals are given in Fig. 7.12.

Morgenstern reports that a large ^{7}Li-NMR shift occurs when the LiCl/DMAc complex first encounters cellulose suggesting that the cellulose (OH) is replacing the DMAc. Yet, one must wonder if this observed shift isn't exactly what might be expected when the Li^+/DMAc complex moiety starts to share its charge with the cellulose (OH) without break-up of the Li^+ complex structure. Further, if close continued DMAc complex interaction was not crucial at this contact stage, then many other solvents should be just as effective as DMAc once their Li^+/solvent complex met the cellulose (OH). This is simply not the case. Regardless of which concept of interaction is more nearly correct, the Li/DMAc system is capable of producing excellent quality fibers.

7.5.1.2 Fiber spinning/properties

Extensive spinning studies were made with the Li/DMAc system dopes by Hammer, Snyder, Kafrawy and associates.[37] They spun fibers by all three fiber spinning methods, that is, (a) wet spinning, (b) dry spinning and (c) dry jet-wet spinning (or air gap spinning). The effects of jet stretch, godet stretch, temperature, residence time, dope concentration and regeneration bath variations were evaluated for each spinning system. In general

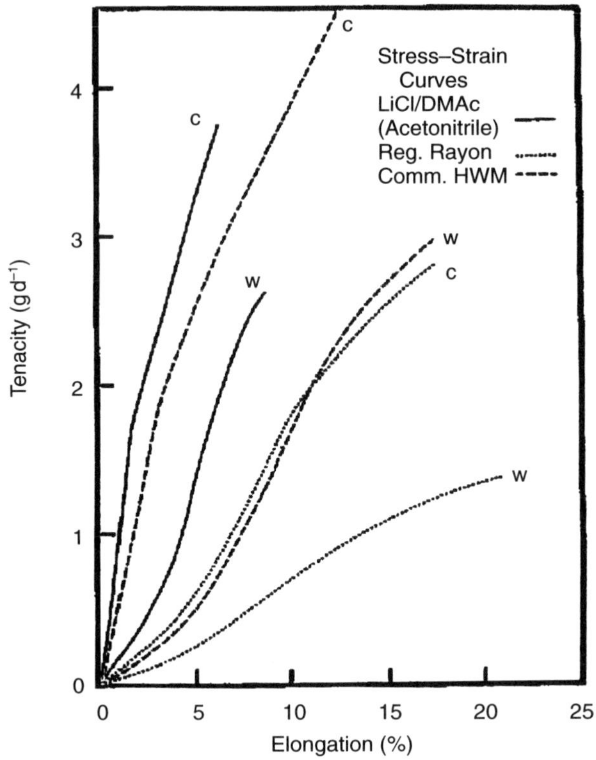

7.13 Stress-strain curves for LiCl/DMAc fibers and rayon fibers: Conditioned (C); Wet (W).

6–10% cellulose dopes gave good spinning performance. With polar regeneration baths good quality intermediate wet strength fibers could be spun at over 200 m min^{-1}. Physical properties of fibers spun into an acetonitrile bath are shown in Fig. 7.13.

Fibers spun into organic liquids like methanol, acetone or acetonitrile had round shapes. Fibers spun into saturated aqueous salt baths containing NaCl or LiCl had a 'peanut' shape or a 'C' shape. Fibers spun into organic polar liquids typically had slightly higher conditioned tenacities and about equivalent conditioned elongations.

7.5.1.3 Recovery and recycle

In this case, as is the case with all solvent spun rayon systems, the ease of complete recovery and recycle of reactants and solvents is the controlling factor for possible commercial success. The N_2O_4 recovery was very complex. The CH_2O/DMSO system presented significant problems in the recovery of re-usable paraformaldehyde owing to possible strong com-

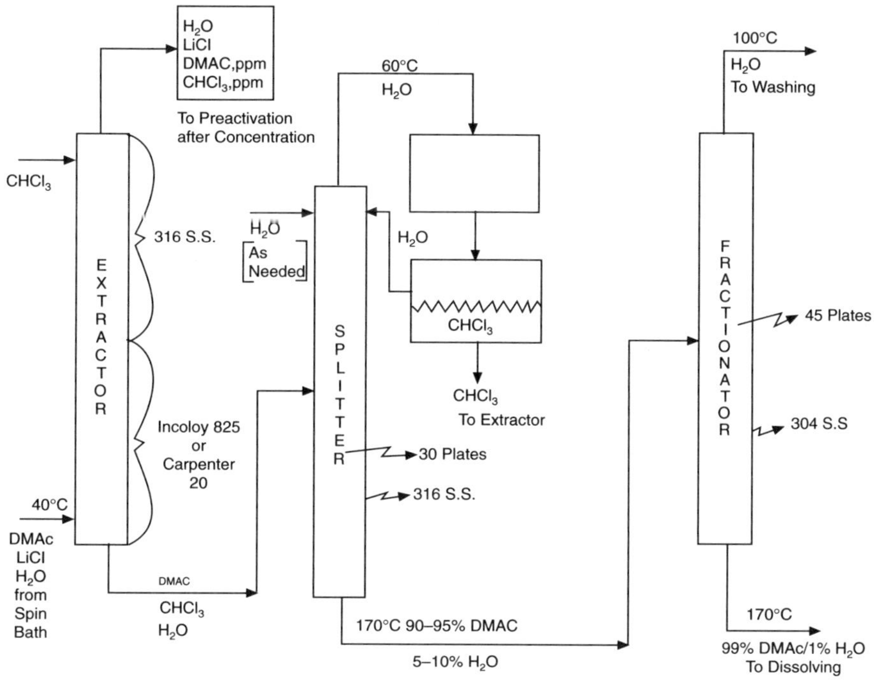

7.14 LiCl/DMAc recovery/recycle process for spinning cellulose into concentrated aqueous LiCl coagulation bath.

plexing with the DMSO. The LiCl/DMAc system is much more friendly. The recovery and recycle of DMAc has been quite thoroughly worked out by the DuPont company and they were more than willing to help anyone interested in using their DMAc as a potential solvent.

Figure 7.14 is a diagram of the overall recovery process for spinning 12% cellulose dopes into an aqueous saturated LiCl bath followed by extraction of the DMAc with $CHCl_3$, fractionating, recovery and recycle of all starting materials. This system produced very good quality fibers. A detailed analysis was made of the overall costs of such a system with the following achievable process parameters:

1 £100 million/year greenfield plant
2 99.50% DMAc recovery/recycle
3 99.85% LiCl recovery/recycle
4 99.90% $CHCl_3$ recovery/recycle.

For the above parameters, the overall manufacturing costs would be 11% lower than the viscose rayon process and would be non-polluting. This system appears to have possible commercial potential.

Films were also cast from these regeneration baths. Films from LiCl/DMAc had significantly better tenacity, elongation and tear strength than cellophane. For example, the LiCl/DMAc films cast into methanol had 34% higher conditioned tenacity, 62% higher conditioned elongation and 162% higher tear strength compared to cellophane. When properly cast and coagulated these films had a dense dye resistant skin indicating that such films may have potential as membranes for various uses.

7.5.2 Cellulose/hydrazine system

7.5.2.1 Reaction chemistry

Hydrazine is a rather unstable chemical and has been used as a rocket fuel. Therefore, extreme caution is necessary when running reactions with a carbon source like cellulose. Hydrazine has a boiling point of 113.5°C, a flash point of 52°C, an auto ignition temperature of 270°C in air and glass; 156°C in air and stainless steel and 23°C in air and rust. A special composition of stainless steel must be used to prevent explosion at higher temperatures. The vapors are very toxic at 1 ppm with continued exposure. Hydrazine has a strong irritating ammonia-like odor. Hydrazine is a strong enough base to dissolve cellulose at higher temperatures. Such solutions suffer significant DP loss.

Dissolution of cellulose in hot hydrazine has been reported by Litt and Kumas[38] and studies on the morphology of cellulose regenerated from hydrazine solutions have been reported by Kolpak, Blackwell and Litt.[39] Owing to the overall danger of try to spin such solution, no fiber data is available.

7.5.3 Cellulose/phosphoric acid

Cellulose readily dissolves in 85% phosphoric acid at room temperature with little loss of DP. At this high acid concentration the (OH) of cellulose is protonated and dissolution results. These clear dopes are readily prepared and they could form the basis of a potential commercial spinning system if some way could be found to recover the phosphoric acid easily without neutralizing the H_3PO_4.. In this respect, water is a sufficiently strong base to make distillation of the water to reconcentrate the H_3PO_4 too expensive. A previously reported approach tried by this writer was to coagulate a 6% cellulose solution in H_3PO_4 into a weaker base such as strong acetic acid. This worked fairly well and acetic acid recovery was reasonably effective, but the coagulation rate was quite slow. Finding the proper volatile weak acid might well develop into a novel system. Simple systems such as this are worth more investigation.

[Editors note: Patents[40,41] published in February 1996 describe anisotropic cellulose solutions at concentrations up to 40% in a 'solvent containing from 72 to 79 wt% of phosphorous pentoxide'. Claiming a highly economical route for the production of tyre yarns, the process described therein appears to be simple and elegant.

A twin-screw extruder is fed with dry fluff-pulp and the phosphoric acid solvent. The first cooled zone mixes the cellulose into the solvent and reduces the particle size, the second hotter zone starts the dissolution and the third zone provides further blending and dissolution time, its temperature being controlled to achieve the desired reduction in cellulose DP. A final low-pressure zone allows deaeration and dewatering (or the injection of additives) as necessary, and filters remove particles before spinning.

While 'all available types of cellulose' can be used, dissolving pulps are preferred for best spinning and fibre properties. The solvent preparation involves mixing orthophosphoric acid with polyphosphoric acid, in 80/20 ratio at around 50°C and then cooling and storing for several hours to arrive at the desired concentration of anhydride.

After less than 15 min in the extruder, the cellulose solution, typically containing 20% cellulose, is spun like lyocell, through a similar-sized air gap, but into cold acetone, followed by water washing, sodium carbonate or soda neutralization, and a final water washing. Tyre yarn with a breaking tenacity of 76 cN tex^{-1}, an elongation at break of 6.5%, capable of sustaining a load of 20 N tex^{-1} without extending by more than 2%, was obtained at 120 m min^{-1} spinning speed.

A similar process[41] from Michelin Research appeared in March 1996.

Further work at Akzo-Nobel[42] showed that if extra water is injected with the solvent, or added to the anisotropic solutions described above, isotropic solutions of cellulose could be obtained at concentrations of 7–20%. A 7.6% cellulose solution could be spun to give fibres with tenacities of 18–38 cN tex^{-1} (extension at break declining from 17.4% to 8.4%) as draw ratios were increased from 60–150%.

One of the key requirements for making high quality high modulus fibres using the phosphoric acid approach is to keep the level of phosphorous bound to cellulose at a minimum. Allowing it to rise from the <0.5% level preferred in the above process to more than 3% by increasing the phosphorous pentoxide level in the solvent, or by increasing the storage temperature of the solvent prior to introduction of the cellulose, resulted in weaker, hard-to-handle gel fibres with very high absorbency. Akzo-Nobel therefore postulated the production of superabsorbent fibres using a variation of the phosphoric acid route.[43] While the resulting patent covers conventional fibre production, perhaps in recognition of the difficulties of cleaning up such fibres, it also covers the production of fibrids by extrusion into a high shear mixer to obtain a fibrous slurry. Water absorbencies of up to 100 g g^{-1} are *claimed if the fibres that contain more than 6% bound phosphorous are crosslinked with ethylene glycol diglycidyl ether in a 70/30 ethanol/water mixture.*

Centrifugal spinning of phosphoric acid solutions of cellulose was also explored,[44] *and found capable of producing mixtures of filaments with linear densities ranging from 1 to 23, and tenacities of 4–90 cN tex*$^{-1}$ *and extensions at break up from 1–15%.*

In an attempt to remove the problems associated with the acetone coagulation bath, fibres were air-gap spun into aqueous salt solutions, for example up to 25% of diammonium hydrogen phosphate or potassium phosphate or 5% zinc sulphate.[45] *This patent also makes it clear that the corrosive nature of the solution is an issue in spinneret design.*

None of the above patents address the crucial solvent recovery issues, and at the time of writing, work at Akzo-Nobel (now Acordis) appears to have ceased. We can only conclude that once again the process economics, especially those related to solvent recovery, have compared unfavourably with the lyocell approach.

Boerstoel's work on fibres from liquid crystalline solutions of cellulose in phosphoric acid is also referred to below in Chapter 8.]

References

1 A F Turbak, R B Hammer, R E Davies and H L Hergert, 'Cellulose solvents', *Chemtech*, Jan 1980 **10** 51–57.

2 S M Hudson and J A Cuculo, 'The solubility of unmodified cellulose: a critique of the literature', *J Macromol Sci*, 1980 **18**(1) 1–82.

3 D C Johnson, *Solvents for Cellulose, Cellulose Chemistry and its Applications*, Ellis Horwood, Chichester, 1985.

4 A F Turbak, *Solvent Spun Rayon, Modified Fibers and Derivatives*, ACS Symposium Number 58, American Chemical Society, Washington, D.C., 1977.

5 N E Franks, C C McCorsley and J K Varga, 'Process for making a precursor for a solution of cellulose', US Patent 4142913 March, 1979; 'Process for making a shaped cellulose and shaped cellulose products', US Patent 4196282 April, 1980.

6 C C Unruh and W Kenyon, 'Investigation of the properties of cellulose oxidised by nitrogen dioxide', *J Am Chem Soc*, 1942 **64** 127.

7 W Kenyon and E Yackel, 'Reactions of cellulose with nitrogen dioxide', US Patent 2448892, 1945.

8 W Fowler, P McGee, C C Unruh and W Kenyon, 'Solubility of cellulose in mixtures of nitrogen tetroxide with organic compounds', *J Am Chem Soc*, 1947 **69** 1636.

9 P Gray and A D Yoffe, *The Reactivity and Structure of Nitrogen Dioxide*, Chemical Reviews, American Chemical Society, 1955 1069–1154.

10 H L Hergert, R B Hammer and A F Turbak, 'New methods for preparing rayon', *Tappi J*, 1978 **61**(2) 63–67.

11 R B Hammer and A F Turbak, *Production of Rayon from Solutions of* N_2O_4 *and DMF*, ACS Symposium Series Number 58, 1977, 40–51 and 71–80.

12 D M MacDonald, *Unconventional Cellulose Solutions*, ACS Symposium Series Number 58, 1977 25–39.
13 N A Portnoy and D P Anderson, *Chemistry of the Cellulose N_2O_4–DMF Solution, Recovery and Recycle of Raw Materials*, ACS Symposium Number 58, 1977 52–70.
14 D C Johnson, M D Nicholson and F C Haigh, *Dimethyl Sulfoxide/Paraformaldehyde: A Non Degrading Solvent for Cellulose*, Applied Polymer Symposium No 28, J Wiley and Sons, NY, 1976 911–943.
15 D C Johnson and M D Nicholson, *Solvent System for Cellulose*, US Patent 4 097 666 June 1978.
16 M D Nicholson and D C Johnson, 'Carboxymethylation and methylation of cellulose in DMSO/PF solvent system', *Cell Chem Tech*, 1977 **11** 349–359.
17 T J Baker, L R Schroeder and D C Johnson, 'Dissolution of cellulose in polar aprotic solvent via formation of methylol cellulose', *Carbohydrate Res*, 1978 **67** C4–C7.
18 T J Baker, L R Schroeder and D C Johnson, 'Formation of methylol cellulose and its dissolution in polar aprotic solvents', *Cellulose Chem Technol*, 1981 **15** 311–320.
19 R B Seymour and E L Johnson, 'The effect of solution variables on the solution of cellulose in dimethyl sulfoxide', *J Appl Polym Sci*, 1976 **20** 3425–3429.
20 R B Hammer, M E O'Shaughnessy, E R Strauch and A F Turbak, 'Process and fiber spinning studies for the cellulose/paraformaldehyde/dimethyl sulfoxide system', *J Appl Polym Sci*, 1979 **23** 3485–3494.
21 R B Hammer, A F Turbak and N A Portnoy, 'Solvent spun rayon, modified cellulose fibers and derivatives', *Production of Rayon Fibers from Solutions of Cellulose in $(CH_2O)_x$ – DMSO*, ACS Symposium Series Number 58, American Chemical Society, 1977, pp 71–80.
22 H A Swenson, *Solution Properties of Methylol Cellulose in Dimethyl Sulfoxide*, Applied Polymer Symposium Number 28, John Wiley & Sons, 1976, 945–952.
23 H Rodier, *Cellulose Fiber Process*, US Patent Nos. 4 145 391, 1979; 4 129 640, 1978; 4 173 613, 1979.
24 J Menault and H Rodier, *Spinnable Cellulose Solution and Process for Making Same*, US Patent 4 129 451, 1978.
25 R Lenoni, G Angelinin and A Baldini, *Regenerated Cellulose Filaments and Films*, British Patent 2 026 934, Feb 1980.
26 O Turunen, L Mandel, V Eklund, K Ekman and J Huttunen, 'Process for producing soluble cellulose derivatives', US Patent No 4 486 585, 1984; 'Procedure for treating cellulose derivative fibers', US Patent 4 583 984, 1986; O Turunen, J Huttunen, J F Selin, J Fors and V Eklund, 'Cyclic process for producing an alkali solution of cellulose carbamate and precipitating the carbamate', US Patent 4 639 515, 1987.
27 J Huttunen, O Turunen, L Mandel, V Eklund and K Ekman, 'Method for producing alkali-soluble cellulose derivative', US Patent 4 404 369, 1983.
28 H Struszczyk, *Cellulose Carbamate and its Utilization*, ACS Meeting, New Orleans, 1990.
29 D J Bridgeford and M Rahman, 'Cellulose aminomethanate sausage casings', US Patent 5 295 514, 1994.
30 M Rahman and D J Bridgeford, 'Preparation of cellulose aminomethanate', US Patent 5 229 506, 1993.

31 P R Austin and C J Brine, 'Chitin films and fibers', US Patent 4029727, June 14, 1977; 'Solvent for purification of chitin', US Patent 4062921, Dec 13, 1977.
32 C L McCormick, 'Novel cellulose solutions', US Patent 4278790, July 14, 1981.
33 A F Turbak, A Kafrawy, F Snyder and A B Auerbach, 'Solvent system for cellulose', US Patent 4302252, Nov 22, 1981; 'Process for forming shaped cellulose product', US Patent 4352770, Nov 24, 1981.
34 J Ekmanis, *GPC Analysis of Cellulose*, Pittsburgh Analytical Conference, Atlantic City, 1986, paper number 251.
35 A El-Kafrawy, 'NMR studies of the cellulose/LiCl/DMAc system', *J Appl Polym Sci*, 1982 **27** 2435.
36 B Morgenstern and H Werner-Kammer, 'Solvation in cellulose-LiCl-DMAc solutions', *Trends Polym Sci*, 1996 **4**(3) 87–92.
37 A F Turbak, *Newer Cellulose Solvent Systems, Wood and Agricultural Residues*, Academic Press, 1983, pp 87–99.
38 M H Litt and N G Kumar, *Cellulose Solutions and Products Prepared Therefrom*, US Patent 40280132, June 1977.
39 F J Kolpak, J Blackwell and M H Litt, 'Morphology of cellulose regenerated from hydrazine solution', *Polym Let Edition*, 1977 **15** 655–658.
40 H Boerstoel, B M Koenders and J B Westerlink, WO9606208 to Akzo-Nobel, 29 Feb, 1996.
41 R Huston *et al.*, WO9609356 to Michelin, 28 March, 1996.
42 J B Westerlink, H Maatman and H Boerstoel, WO9728298, to Akzo-Nobel, 7 Aug, 1997.
43 J B Westerlink, H Maatman and H Boerstoel, WO9730090, to Akzo-Nobel, 21 Aug, 1997.
44 S J Rusticus and E Van Andel, WO9730196, to Akzo-Nobel, 21 Aug, 1997.
45 M Ypma, H Boerstoel and H Maatman, WO9830741 to Akzo-Nobel, 16 July, 1998.

The author would like to thank Dr D C Johnson and Dr D J Bridgeford for providing selected reference materials.

8
Physical structure and fibre properties

JOHN W S HEARLE

8.1 Introduction

Among all the textile fibres, cellulose fibres have the most diverse range of structures and properties. Even apart from the variety of natural cellulose fibres, with their highly crystalline fibrillar structures in various helical forms of lay-down, the less highly ordered regenerated cellulose fibres have many different structures, which lead to different properties and applications. The dominant method for manufacturing textile fibres, typified by nylon, polyester and polypropylene, is now melt spinning, in which the structural formation is controlled only by molecular weight, extrusion, cooling and stretching. In contrast to this, the manufacture of cellulose fibres is also controlled by two or three other factors: always by solvent and by the means of separation from solution, and, in some methods, by chemical reactions. The choice of manufacturing conditions among the large number of parameters involved leads to the diversity of structure and properties.

Table 8.1 summarises the various modes of formation of cellulose fibres that influence the resulting structure. Within each type, technical and economic factors lead to detailed differences both between different manufacturers and over time, so that the information in this chapter must be regarded as general and not particular. Some processes which are no longer commercial are mentioned for their historical relevance and will be referred to briefly below when there are features of particular structural or property interest. Cuprammonium rayon has had a recent revival and the lyocell fibres, such as Tencel®, spun from organic solvents are growing in importance. However, the various forms of viscose rayon remain the largest commercial types and will receive the greatest attention. Following the production of aramid fibres, such as Kevlar and Twaron, from liquid crystal solutions, there has been research on the formation of high-modulus and high-tenacity cellulosic fibres by similar methods. None of these has yet been commercialised, but there is potential for the future. In view of the detailed information in the thesis of Boerstoel,[1] the fibre spun from a liquid

Table 8.1 Cellulosic fibres

Type	Process and structure-determining factors
Nitrocellulose[a]	Regeneration of cellulose nitrate
Cuprammonium	Solution in cuprammonium hydroxide; formation of a copper complex of cellulose; coagulation in water
First viscose spinning in laboratory	Sodium cellulose xanthate, dissolved in caustic soda, coagulation and regeneration in acid bath
Viscose rayon: regular[d]	Zinc ions added to acid bath, giving skin-core structure
Lilienfeld[a]	Viscose into 65% sulphuric acid with concurrent stretch
Fortisan[a]	Regenerated from highly stretched cellulose acetate fibres
High-tenacity yarn[d]	High zinc in bath and modifiers in viscose solution, 'all-skin'
Modal staple[e]	Modifications to viscose and weak acid bath
Polynosic[e]	High viscosity viscose, modifed bath, allowing gel formation
Viscose: crimped staple	Modification of viscose process to cause skin bursting
Cordenka EHM HT yarn	Addition of formaldehyde to viscose process
Lyocell (e.g. Tencel)	Solution in an amine oxide; coagulation in weak aqueous solution of the amine oxide
Liquid crystal solutions of cellulose derivatives[b]	Cellulose esters and ethers in organic solvents or inorganic acids, with liquid crystal orientation, coagulation and with or without regeneration
Liquid crystal solution of cellulose: Fiber B[c]	Solution in phosphoric acid, spun with liquid crystal orientation; coagulated in acetone bath
Acetate	Secondary cellulose acetate dry spun from solution in acetone
Triacetate	Cellulose triacetate dry spun from solution in methylene chloride

[a] Not now commercial. [b] Various patented processes, not commercialised. [c] Laboratory process described by Boerstoel.[1] [d] (Coagulation + regeneration) → stretch. [e] (Coagulation + stretch) → regeneration.

crystalline solution of cellulose in phosphoric acid will be described as an example of this group of cellulosic fibres, and following the thesis, will be referred to as Fiber B.

In addition to the effects of the major process differences, the products of any of the methods can, within limits, be varied by the molecular weight distributions, the concentrations of cellulose in the spinning solution and the stretch in the stages following extrusion. Additives, such as delustrants, pigments to give colour, carbon black to give electrical conductivity, or barium salts for opacity to X-rays, may be incorporated in the fibres.

8.2 Fibre forms

All regenerated cellulose fibres may be produced as continuous filament yarns, containing up to a few thousand filaments, or as heavier tows with millions of filaments, which are cut into staple fibres of suitable length for short or long staple spinning. Typical fibre fineness would be in the range 0.1–0.5 tex (10–20 μm in diameter), though various finer and coarser fibres, up to around 5 tex (60 μm), have been available. At one time, fancy yarns were made by periodic variations in thickness along the fibres.

The diversity of 'regular' viscose rayon yarns is shown by the following extract from the *Fibre Data Summaries*, edited by Ford:[2]

Types available
Continuous-filament yarns in wide range of deniers; bright, matt, and spun-dyed; available in cake form or continuously spun on bobbin. Also special types, e.g., tow, textile Tenasco (higher tenacity); coarse, crimping yarn; slub yarn; Ribbonfil; Visca straw yarn; monofilaments (up to 60 000 denier). Staple fibre (Fibro) from 1.5 to 50 denier; bright, matt, spun-dyed (Fibro-Duracol), and tow. Also crimping types Sarille 2.2 to 4.5 denier, and Evlan 8 and 15 denier.

In the basic viscose process, regeneration and coagulation start at the surface of the fibre and form a skin. Subsequent loss of solvent from the interior of the fibre causes a collapse into a strongly serrated form, as shown in Fig. 8.1(a). The difference in fine structure, which arises from the different diffusion rates of hydrogen and zinc ions, between skin and core is discussed in Section 8.3.2. The modifications to the viscose process, which led to the high-tenacity tyre-cord yarns, progressively increased the proportion of skin, Fig. 8.1(b) and (c), until the final forms were 'all-skin' fibres. These changes led to a reduction in the serration, with the all-skin fibres being bean shaped. In certain process conditions, the core can be caused to burst out of the first-formed skin. This leads to an asymmetric fibre, Fig. 8.1(d), which will crimp on drying. The skin is thicker on the side that has the initial skin and thinner on the portion that has burst out.

Cuprammonium, modal, lyocell, and liquid crystal fibres, such as Fiber B, are circular in cross-section. Acetate and triacetate fibres have a serrated cross-section, caused by solvent being lost from the interior after the surface has solidified by evaporation.

The shapes referred to above result from effects during the solidification processes after extrusion through circular holes. Other shapes can be made by special procedures at extrusion. Hollow fibres, such as Viloft, are

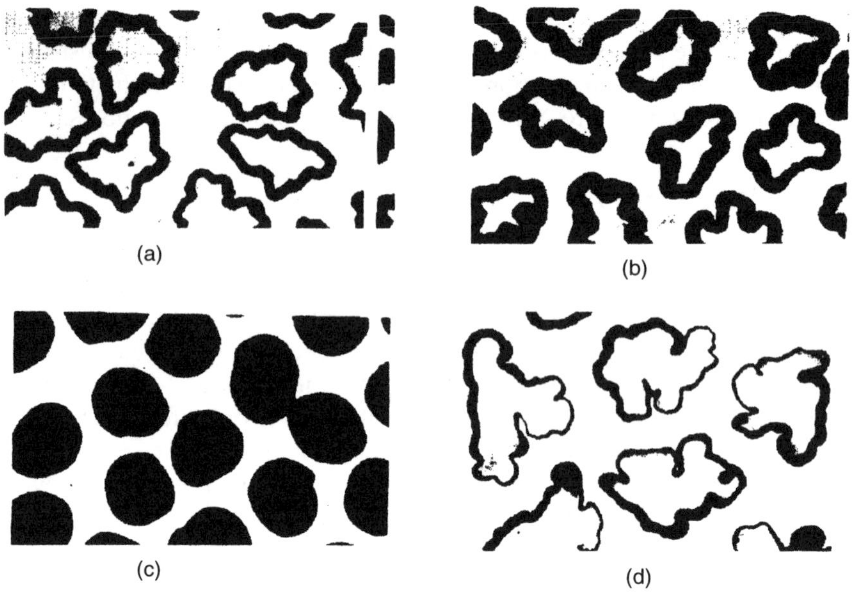

8.1 Cross-sectional shapes of cellulose fibres, with differential dyeing of skin. (a) Regular viscose rayon; (b) Tenasco, early high-tenacity rayon with thicker skin; (c) Tenasco Super 105, an 'all-skin' fibre; (d) crimped staple viscose rayon with asymmetric skin. (From Woodings.[3])

produced by inflation techniques. These can collapse into flat or multilimbed fibres. Subsequently when spinneret engineering had improved to enable the punching of small enough slots, it was found to be more effective to extrude through shaped spinneret holes. Rectangular slots give flat fibres. Y- and X-shapes give multilimbed forms, which match the inflated fibres in bulk and handle.

8.3 Fine structure

8.3.1 The cellulose molecule: order and disorder in assemblies

In natural cellulose, the molecules are extremely long with degrees of polymerisation (DP = number of anhydro-glucose rings in the chain) over 10000, but extraction reduces the length. Even purification will reduce the DP to about 2500, and industrial processing is more severe. The molecular weight distributions shown in Table 8.2, which were measured by Harland[4]

Table 8.2 Degree of polymerisation (from Harland)[4]

DP	Relative frequencies		
	Viscose rayon	Tenasco rayon	cuprammonium rayon
200	0.195	0.102	0.095
400	0.107	0.102	0.104
600	0.072	0.102	0.095
800	0.038	0.072	0.078
1000	0.020	0.042	0.050
1200	0.012	0.023	0.026
1400			0.015
1600			0.013

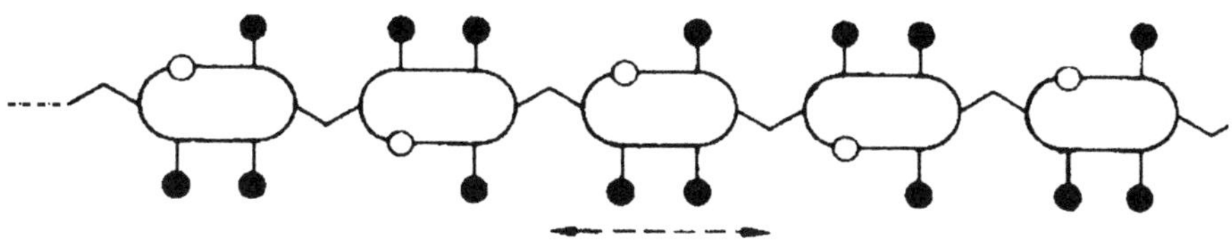

8.2 Essential physical features of the cellulose molecule. The rings consist of five carbon atoms and one oxygen; the bridges are single oxygen atoms; the paired projections are —OH groups; the single projections are $—CH_2OH$ groups; the remaining valencies are taken by hydrogen atoms. (From Morton and Hearle.[5])

in the 1950s, illustrate that short molecules are reduced in the better rayons and the distribution is moved to higher DPs.

In relation to fibre structures and properties, the important features of the molecule are shown in Fig. 8.2. The cellulose chain is directional, which allows for parallel and antiparallel packing. The anhydro-glucose rings are relatively rigid, but there is more freedom at the oxygen bridges. The ribbon shape of the molecule allows for twisting and for bending in the direction out of the plane, so that the molecule is moderately flexible. There is relatively strong interaction between neighbouring cellulose molecules in dry fibres owing to the hydroxyl (–OH) groups, which stick out from the chain and form intermolecular hydrogen bonds. Absorbed water competes for hydrogen bonding and reduces the lateral cohesion.

Cellulose can crystallise in a variety of forms: cellulose I to IV. The common crystal lattice in regenerated celluloses is cellulose II, which is presumed to have antiparallel chains, in contrast to the parallel chains of cellulose I in natural cellulose. There has been controversy[6] because of the difficulty of explaining the change from parallel to antiparallel when natural fibres are mercerised, but it is now generally accepted that this is due to a mixing from neighbouring microfibrils. The fine details of the crystalline lattice are not important in determining properties, and a range of values of unit cell dimensions are quoted in the literature. At right angles to the chain direction, the unit cell is a parallelogram with an angle of about 62° and sides of about 0.8 nm (*a*-axis) and 1 nm (*b*-axis). The *c*-axis is about 1 nm. The rings are roughly in line with the long diagonal of the parallelogram. Since each unit cell dimension corresponds to two glucosidic repeats, the distance between rings along the chains is about 0.5 nm, and the distances between chain centres are about 0.8 nm in the plane of the rings and 0.45 nm perpendicular to the chains. Figure 8.3 shows views of the conformation of the chain molecule in cellulose II and looking along the chain axis, as used in the calculation of chain modulus by Kroon-Batenburg *et al.*[7]

Most regenerated fibres are less than 50% crystalline. Wooding[3] quotes values from 32 to 53%, but the numbers depend on the interpretation of measurements as well as the differences in fibre formation. The size, shape and orientation of crystalline regions and the paths of cellulose molecules in the amorphous regions constitute the diversity of structure, which gives rise to the range of properties. Table 8.3 shows values reported by Boerstoel[1] for the birefringence Δn, which is a measure of orientation, and the crystal dimensions of several cellulose fibres. The widths and heights quoted should not be taken to imply that these are sharp boundaries; they merely represent the limits of coherence as shown by diffraction. As Howsmon and Sisson[8] pointed out with particular reference to rayon, there is a whole sequence of forms from high amorphous disorder to perfect crys-

Table 8.3 Axial birefringence and crystal width and height derived from $1\bar{1}0$ and 004 reflections (from Boerstoel)[1]

Fibre	$10^4\Delta n$	Width (nm)	Height (nm)
Enka viscose	260	2.9	9.0
Cordenka 660	330	3.6	10.3
Cordenka 700	390	3.7	9.6
Cordenka EHM	510	4.3	15
Fortisan	480	5.5	15.6
Fiber B	502	3.9	17.8

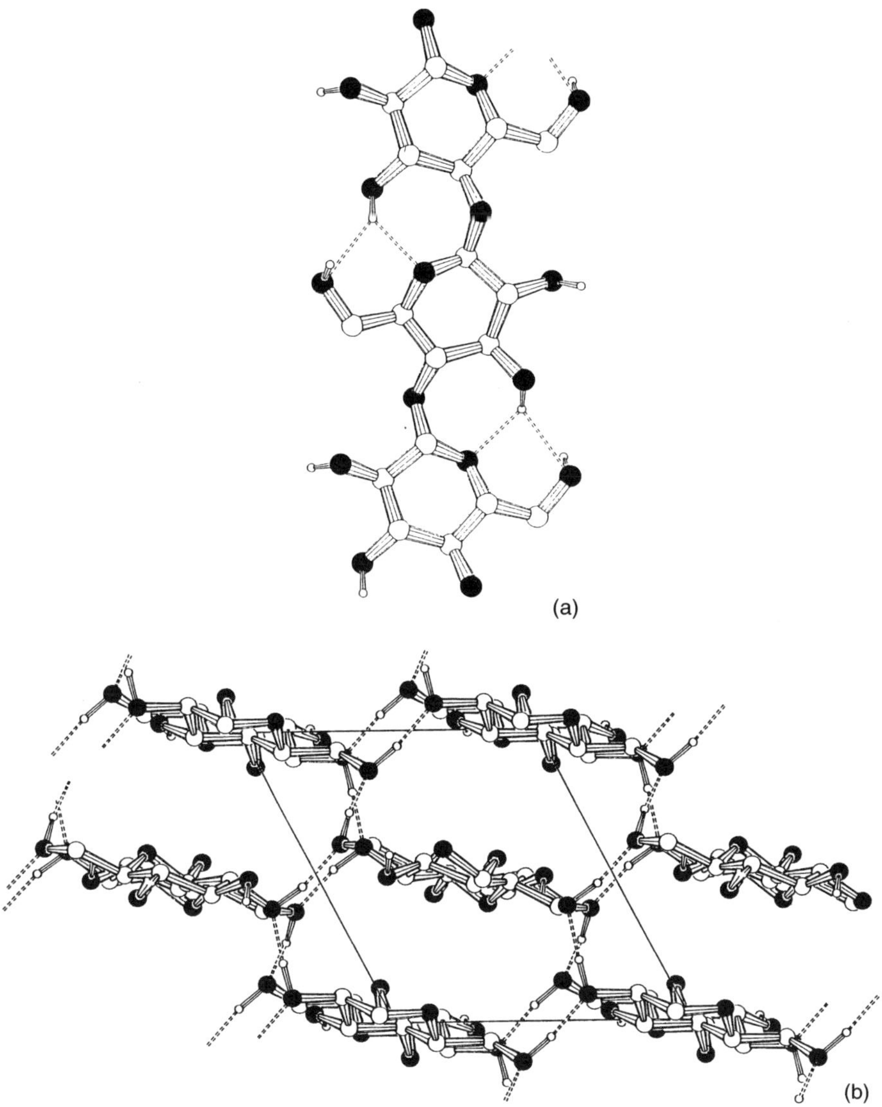

8.3 (a) Chain conformation in cellulose II. (Courtesy of M G Northolt, based on reference.[7]) (b) Crystal lattice of cellulose II viewed along the chains. (Courtesy of M G Northolt, based on reference.[7])

talline order. A variety of combinations of order and disorder, or even intermediate states of continuous partial order, can give rise to 'crystallinities' between 30 and 50%, as indicated by density of regenerated cellulose fibres. Elucidation of details of these complexities of structure in all manufactured fibres remains uncertain,[9] and it is beyond the scope of this chapter to review all the experimental evidence and interpretive argument. The informed view given below is derived in three ways: qualitative scenarios of how the change from dispersed molecules in solution to solid semi-crystalline fibres might occur; the results of optical and electron microscopy, X-ray diffraction, spectroscopy and other specialised techniques; and the interpretation of physical properties.

8.3.2 Viscose rayons

Direct coagulation of cellulose, as in regular viscose rayon, leads to a fringed micellar structure of brick-like crystallites linked by tie molecules in amorphous regions. If the reader imagines conversion into a three-dimensional array of chains with forms like those in Fig. 8.2, two early pictures give an idea of the likely structure, though that by Bunn,[10] Fig. 8.4(a), which was drawn for synthetic polymers, is too densely crystalline, and that by Hearle,[11] Fig. 8.4(b), which is intended to show the intercrystalline links, has too much amorphous material. Single crystals of cellulose derivatives, grown in the laboratory, do probably have crystallographic folds, but the stiffness of the cellulose molecule makes it likely that any folding back in fibres will be in amorphous regions, as shown in Fig. 8.4(a) and (b). The degree of orientation, both of crystallites and of amorphous segments, will depend on the stretch imposed in manufacture.

There are differences in structure between skin and core. Taking into account various sources of evidence, Wooding[3] concluded that 'it is not unreasonable to suggest that the skin region is generally composed of cellulose possessing more, smaller crystalline regions, whereas the core contains fewer, larger crystallites'. It is also possible that the core is generally more disorganised. There is a difference in lateral orientation of the planes of the cellulose molecules. As shown in Fig. 8.5, the collapse leads to a radial orientation in the skin, but the core has no preferred lateral orientation. The 'all-skin' high-tenacity yarns, used in tyre cords, will consist entirely of the finer, more uniform texture.

When coagulation and stretch occur together, before regeneration and crystallisation of the cellulose, as in modal and polynosic fibres, the structure has a fibrillar texture. The fringed fibril structure, suggested by Hearle[12] and illustrated in Fig. 8.4(c), may be a reasonable representation of the structure. Alternative though extreme possibilities would be a discontinuous assembly of separate long crystals, without fringing along their length, or a continuous oriented paracrystalline structure, with local ordered packing but no identifiable crystallites.

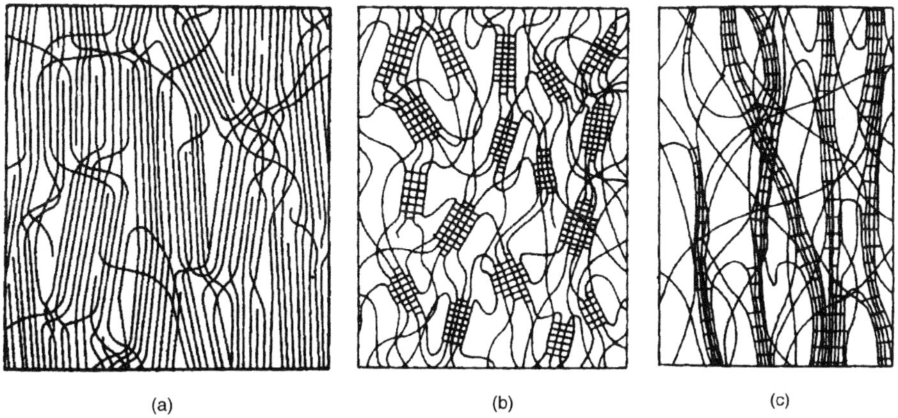

8.4 (a) Fringed micelle structure. (From Bunn.[10]) (b) Fringed micelle structure. (From Hearle.[11]) (c) Fringed fibril structure. (From Hearle.[12])

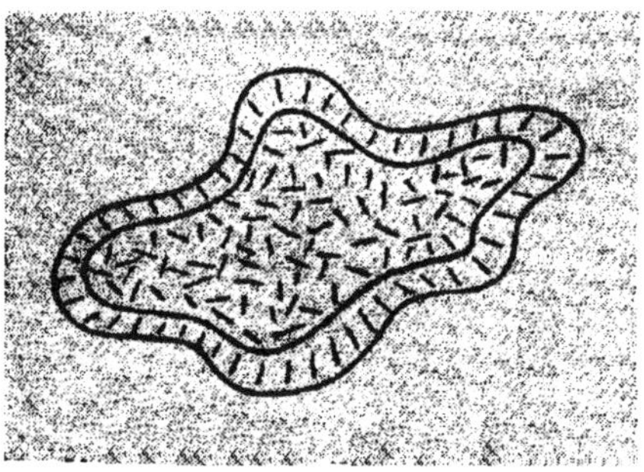

8.5 Lateral orientation in skin and core in regular viscose rayon. (From Hearle.[13])

8.3.3 Newer cellulose fibres

Information on the structure of lyocell fibres, such as Tencel, is given by Coulsey and Smith.[14] They state that the 'dry-jet/wet-spinning' method used to produce Tencel has similarities to the extrusion of lyotropic rigid-rod systems. This leads to a microfibrillar structure. Wide-angle X-ray diffraction data indicates that Tencel has a degree of crystallinity of about 60%, compared to 40% in a high-tenacity viscose rayon. Scherrer analysis gives a value of about 4 nm for the lateral dimension of crystals, which is similar to that in high-tenacity rayon, but the meridional peaks are very sharp, which shows that the crystals have a high aspect ratio. The fine structure may be paracrystalline. There is no evidence of differences in orientation across the fibres. An important benefit of the lyocell process is that much higher molecular weight solutions can be formed and spun.

Boerstoel[1] reports that Fiber B, spun from a liquid crystal solution in phosphoric acid, shows a much sharper wide-angle X-ray diffraction pattern than tyre-cord viscose rayon. This indicates a high crystallinity and a high orientation, which is similar to that in para-aramid fibres. As noted in Table 8.3, the crystalline aspect ratio is high. An optical micrograph between crossed polars shows positive lateral birefringence, which indicates a radial texture.

8.3.4 Acetate and triacetate fibres

Because of the irregular location of the residual unacetylated hydroxyl groups, secondary cellulose acetate fibres are poorly crystalline. Typical textile fibres are not highly oriented. However, the low crystallinity does enable acetate fibres to be drawn to a high degree of orientation. This technique was used to make the precursor fibres for the highly oriented cellulose fibre, Fortisan.

Because of the regularity of the chemical repeat, triacetate fibres are more crystalline and behave somewhat like semi-crystalline polyamide or polyester fibres.

8.4 Physical properties

8.4.1 Comparative values

The variety of cellulose fibre types, and the minor differences between fibres of nominally the same type from different manufacturers or from different dates, lead to a wide range of values of physical properties. Although major differences are shown up in reported values, differences in experimental techniques mask smaller differences between different fibres. A comparative view from a single source is provided by the fibre data summaries from 1966, edited by Ford.[2] Values are shown in Table 8.4, but must be regarded as indicative and not as applicable to all fibres of the same designation. The viscose fibres were from Courtaulds, the cupro from Bayer, and the acetates from British Celanese. The regular viscose rayon is from continuous filament yarn, with additional entries for staple Fibro and for crimped Evlan staple. The high-tenacity yarn is Tenasco Super 105, with staple entries for Fibro-Durafil. The modal is Vincel staple. The cupro is continuous filament Cupresa, with additional entries for Cuprama staple. The acetates are from Dicel and Tricel continuous filament yarns. Another set of comparative values, including more recent fibres, is given in Table 8.5.

Table 8.4 Properties of cellulosic fibres (from Ford)[2]

Property	Viscose			Others		
	regular (see notes in text on types)	high ten	modal	cupro	secondary acetate	triacetate
Density ($g\,cm^{-3}$) dry	1.51	1.51	1.53	1.53	1.33	1.32
Moisture regain (%), 65% rh	12.5	14.1	11.8	12.5	6.9	4.5
Water imbibition (%)	101	79	75	97	26	13
Tenacity ($mN\,tex^{-1}$): 65% rh, 20°C	177	416	265	159	111	115
wet, 20°C	88	265	186	97	62	71
wet, 95°C	80	239	177	88	27	40
65% rh, 20°C, after wet, 95°C	168	327	265	150	115	115
staple 65% rh, 20°C	212	310		132		
crimped 65% rh, 20°C	97					
Break extension (%): 65% rh, 20°C	19	12	7.0	10	35	30
wet, 20°C	38	24	8.5	17	48	38
wet, 95°C	40	30	9.0	20	95	68
65% rh, 20°C, after wet, 95°C	22	23	7.5	10	33	30
staple 65% rh, 20°C	17	30		26		
crimped 65% rh, 20°C	36					
Work of rupture ($mN\,tex^{-1}$): 65% rh, 20°C	25	28	11.5	11.5	28	18
wet, 20°C	19	22	7.1	9.7	18	19
wet, 95°C	19	27	7.1	11.5	13	18
65% rh, 20°C, after wet, 95°C	25	36	11.5	8.8	26	17

Table 8.4 (cont.)

Property	Viscose			Others		
	regular (see notes in text on types)	high ten	modal	cupro	secondary acetate	triacetate
staple 65% rh, 20°C	21	49		23		
crimped 65% rh, 20°C	19					
Initial modulus (mN tex^{-1}):	8850	8850	13300	8850	3100	3100
65% rh, 20°C						
wet, 20°C	442	177	1060	619	531	1770
wet, 95°C	442	133	885	531	35	531
65% rh, 20°C, after wet, 95°C	4420	4420	8850	7080	2480	2210
staple 65% rh, 20°C	4420	6190		2650		
crimped 65% rh, 20°C	221					
Extension (%), 8.85 mN tex^{-1}: 65% rh, 20°C	0.1	0.1	0.07	0.1	0.3	0.3
wet, 20°C	2.5	4	0.8	1.4	2.5	0.6
wet, 95°C	2.5	5	1.0	1.6	26	3.0
65% rh, 20°C, after wet, 95°C	0.2	0.2	0.1	0.12	0.3	0.4
staple 65% rh, 20°C	0.2	0.15		0.3		
crimped 65% rh, 20°C	4					
Refractive index: *n* parallel	1.542	1.544	1.551	1.553	1.476	1.469
n perpendicular	1.520	1.505	1.513	1.519	1.473	1.469
birefringence	0.022	0.039	0.038	0.034	0.003	0

rh = relative humidity.

Table 8.5 Properties of cellulosic fibres (from Courtaulds)[15]; see also Appendix

Property	Viscose				Others	
	regular	improved	modal	polynosic	cupro	lyocell
Water imbibition (%)	90–100	90–100	75–80	55–70	100	65–70
Tenacity ($mN\,tex^{-1}$): dry	200–240	240–300	340–360	350–400	150–200	400–440
wet	100–150	120–160	190–210	270–300	90–120	340–380
Break ext (%): dry	20–25	20–25	13–15	10–15	7–23	14–16
wet	25–30	25–35	13–15	10–15	16–43	16–18
Initial modulus ($mN\,tex^{-1}$): wet	400–500	400–600	1000–1200	2000–2500	300–500	2500–2700
Degree of polymerisation	250–350	250–350	300–600	500–600	450–550	550–600

Where property values given below are not specifically attributed, they are taken from Morton and Hearle.[5]

8.4.2 Density and moisture absorption

The density of the cellulose II crystal is $1.58\,g\,cm^{-3}$, but the amorphous material has a lower density. Careful measurements under the same conditions can be used to estimate comparative crystallinities. Morton and Hearle[5] quote densities of $1.52\,g\,cm^{-3}$ for dry viscose fibres and $1.49\,g\,cm^{-3}$ at 65% rh (relative humidity), but values up to $1.55\,g\,cm^{-3}$ can be found in the literature. Cellulose acetate fibres have a density of $1.32\,g\,cm^{-3}$.

The values in Table 8.4 show small differences between different regenerated cellulose fibres in the moisture regain (expressed as a percentage of dry weight) under standard conditions. Because there are fewer or no hydroxyl groups in acetate fibres, the moisture absorption is less. Water retention values are for fibres that have been wetted and then centrifuged at 9.25 N for 7 minutes. The values in Table 8.5 confirm the lower water imbibition of fibres with round cross-sections. Fibrillated lyocell fibres show a larger take-up of water.

The variation in moisture absorption of cellulose fibres with humidity, which is shown in Fig. 8.6, follows a typical sigmoidal curve, increasing

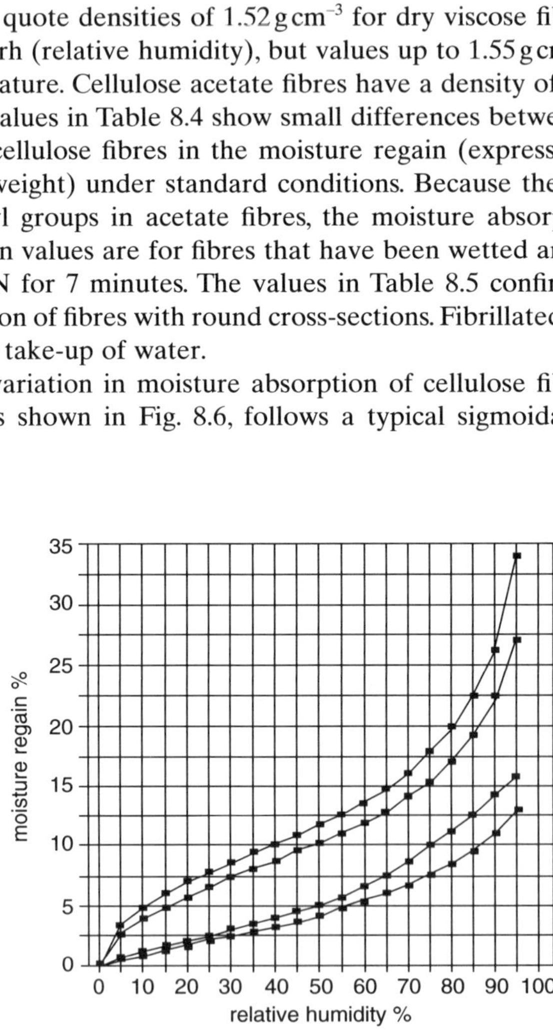

8.6 Moisture absorption in viscose rayon (upper pair of curves) and acetate (lower pair), showing hysteresis between absorption (lower curves in each pair) and desorption (upper curves). (From Urquhart and Eckersall.[16])

rapidly from 0% rh, turning down and becoming almost linear from about 5% rh and 6% regain to 70% rh and 14% regain, and then increasing more rapidly to 80% regain at at 100% rh. Hysteresis leads to a difference between the absorption and desorption curves. Urquhart and Eckersall,[17] in a paper on cotton, stated that the hysteresis persisted up to saturation (in other words, the moisture content on drying from the wet state was higher than from 100% rh), but Ashpole[18] disputed this view and showed that, in rayon as well as in cotton, there was a rapid increase at high humidities. At higher temperatures, there is a small reduction in moisture absorption at a given relative humidity.

Absorption of water is accompanied by an increase in volume, as shown in Fig. 8.7. At high water contents, the volume increase equals the volume of water, but at the start the combined (fibre + water) volume decreases as

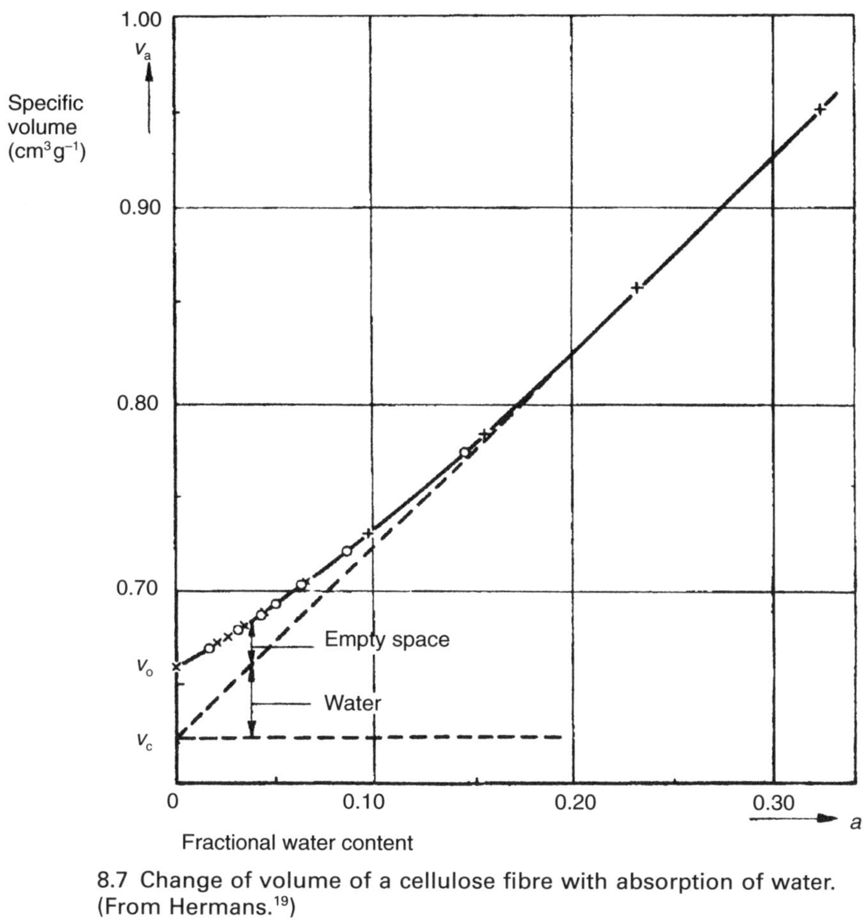

8.7 Change of volume of a cellulose fibre with absorption of water. (From Hermans.[19])

the molecules pack more closely together. The density initially increases and then falls to about 1.40 at 30% moisture content. The volume change is mainly due to transverse swelling with a small axial swelling. Preston and Nimkar[20] found values in the literature of 50–114% for the area increase on wetting and 3.7–4.8% for the length increase.

The absorption of water generates heat. The heat of wetting from zero moisture content to wet is 106 J g^{-1}. The differential heat of sorption falls from 1.17 kJ g^{-1} at zero relative humidity to 0.24 at 75% rh. A typical atmospheric change, such as indoors to outdoors, from 40–70% rh produces 168 kJ kg^{-1} of viscose rayon.

8.4.3 Tensile strength and extensibility

The first viscose fibres in 1904 had the low strength of about 100 mN tex^{-1}, but by the 1920s this had been doubled. The values in Table 8.4 and the plots in Fig. 8.8 illustrate the tensile properties of cellulosic fibres from the 1950s and 1960s. Some more recent comparative values, which show superior properties, are in Table 8.5 and Fig. 8.9. Many factors contribute to this wide range of mechanical properties.

Values of work of rupture, which is a measure of energy to break, are included in Table 8.4. If the stress–strain curve is linear, work of rupture equals ½ (tenacity × break strain), with higher factors for curves above a straight line. The highest value in Table 8.4 is given by the staple high-tenacity fibre which combines moderately high strength with high break extension. Although the strength is greater, the more highly oriented tyre-cord yarn has a lower work of rupture. The high extension and the shape of the curve compensate appreciably for the lower strength of regular rayon, but the low extension of modal rayon leads to low values of work of rupture.

Degree of polymerisation affects the strength of all fibres. Figure 8.10 indicates that a certain minimum DP is needed to achieve any strength. The strength increases for DPs above 100, but levels off beyond 500, which is typical of DPs in the stronger fibres in Table 8.5. The theoretical predictions of Cumberbirch and Mack[21,22] are based on the distributions of lengths of tie molecules between crystalline micelles, as given by polymer chain conformations. Statistical analysis takes account of the fact that short segments will break first and longer ones later. The low strengths at low DP are due to the fact that the molecules are not long enough to bridge many gaps and many free ends will emerge from crystallites.

Figure 8.10 also illustrates how strength increases with degree of orientation, as shown by birefringence. A more detailed picture of the effects of increasing orientation is shown in Fig. 8.11 for a range of cellulose acetate fibres subject to increasing stretch and then regenerated into cellulose. The

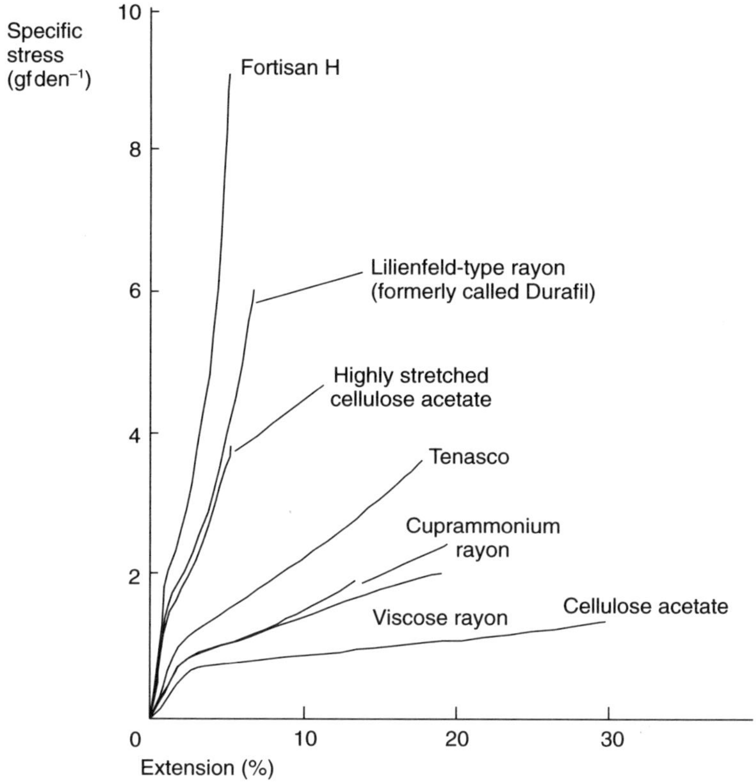

8.8 Stress-strain curves for cellulosic fibres 10 gf den^{-1} = 0.88 N tex^{-1} (gf = grams force). (From Wooding.[3])

highest curves would be typical of Fortisan, the high strength cellulose fibre, which was important before the advent of high-tenacity synthetic fibres. Although the shape of the stress–strain curves differs in different cellulose fibres, the same pattern of change with orientation is found. This is illustrated schematically in Fig. 8.12, which shows strength increasing and break extension decreasing on hyperbolic loci.

Figure 8.12 also demonstrates another trend. Strength and extension increase together when the structure is improved, as shown in Table 8.6 by the improvements in properties of tyre-cord yarns between the 1930s and 1950s. The strengths are higher than those shown for the corresponding yarns in Table 8.4, because the yarns in Table 8.6 had been post-treated by stretching wet and drying under strain. In addition to any changes in molecular packing, this reduces the linear density and so gives higher values

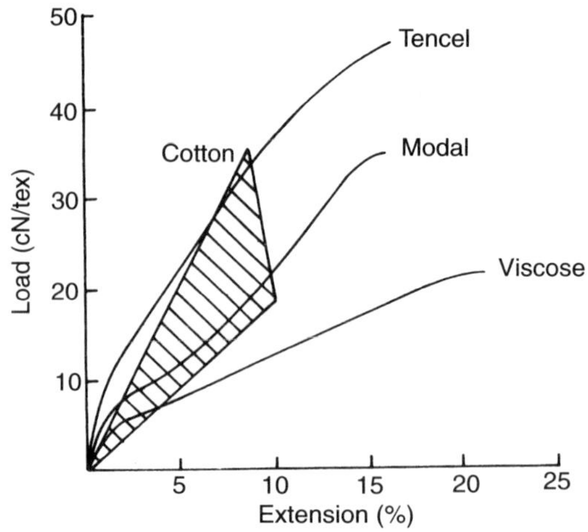

8.9 Stress-strain curves of newer cellulose fibres. (From Courtaulds.[15])

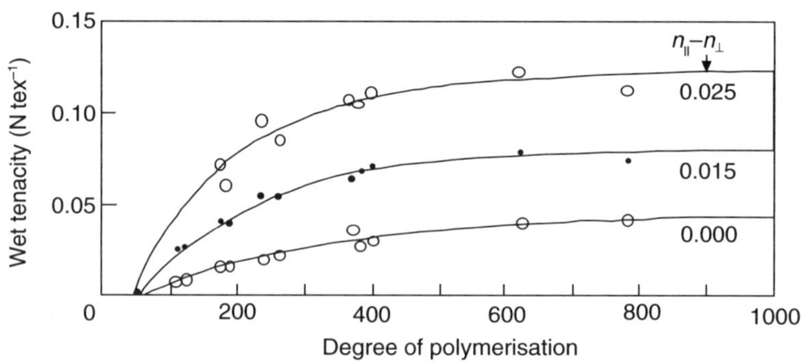

8.10 Variation of strength of wet rayon with degree of polymerisation reported by Cumberbirch and Mack.[21] Circles are experimental points and lines are theoretical predictions. (From Morton and Hearle.[5])

Table 8.6 Structure and strength (from Wooding)[3]

Tyre-yarn type	Tenacity (mN tex^{-1})			Break extension (%)		
	0% rh	65% rh	wet	0% rh	65% rh	wet
Tenasco	398	283	168	7.0	9.5	20.0
Tenasco 35	442	345	221	8.0	11.5	22.0
Tenasco Super 70	460	363	257	9.0	12.0	11.5
Tenasco Super 105	531	442	354	9.0	11.0	24.0
Development yarn	602	496	398	11.0	13.0	26.0

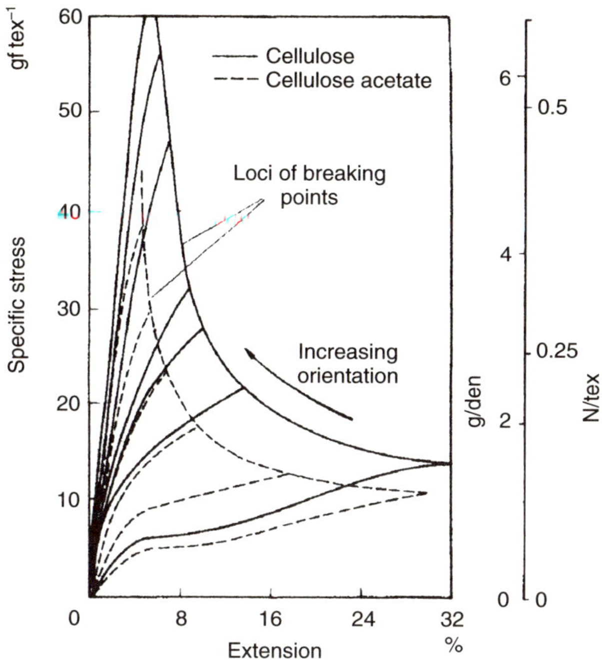

8.11 Effect of orientation on tensile properties reported by Work.[23] Dotted curves are for cellulose acetate at increasing degrees of stretch. Solid curves are after regeneration of cellulose. (From Morton and Hearle.[5])

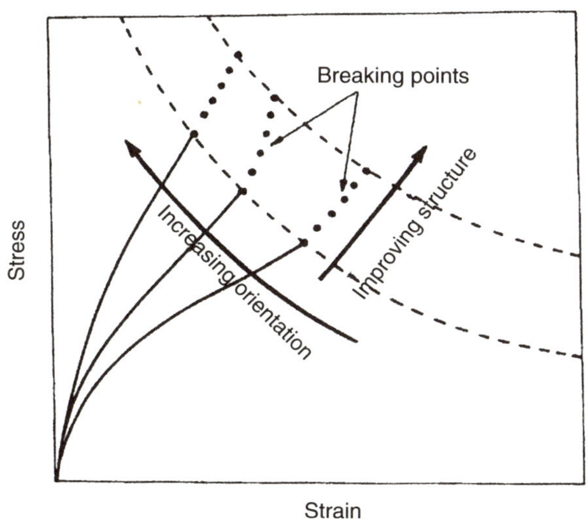

8.12 Changing tensile properties by increasing orientation and improving structure. (From Morton and Hearle.[5])

of stress. The first Tenasco yarns had a skin-core structure. The core structure is weak, because the crystallites are large and widely spaced. Bridging by tie-molecules is less effective and more of the chains emerging from micelles are free ends, which do not contribute to strength. The finer structure of the skin is stronger. Most of the improvement in properties is due to process changes which led to 'all-skin' fibres, although DP and orientation were also addressed. The weakness of the core in regular rayon was confirmed by the experiments of Chamberlain and Khera,[24] who showed a reduction in tenacity as outer layers of the fibre were removed.

Tables 8.4–8.6 all show the major influence of moisture. The weakest yarn in Table 8.6 is almost 2½ times stronger oven-dry than wet, though the gap narrows to 1½ times for the strongest yarn. Intermediate values are found in standard atmospheric conditions. Typical changes between wet and dry (65% rh) over the whole stress–strain curve are shown in Fig. 8.13, which also illustrates the difference in shape between regular and modal rayons.

The production of regular rayon, and of all the other types that have coagulation and regeneration followed by orientation, leads to fibres with

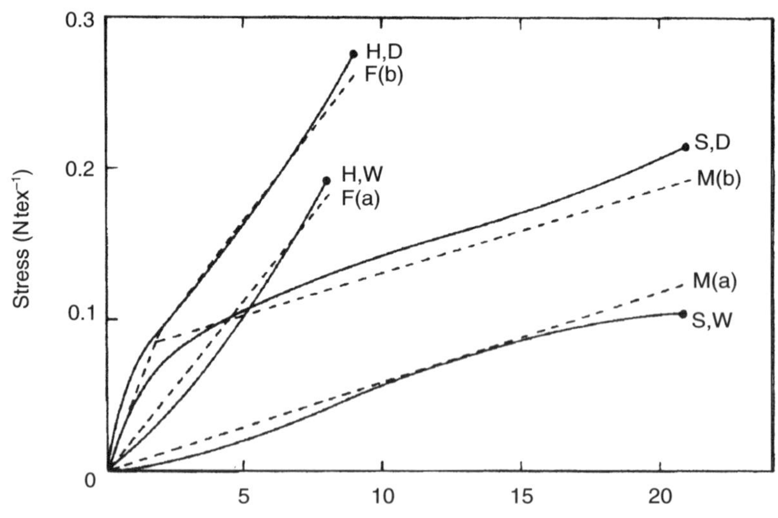

8.13 Stress–strain curves (solid lines) for dry (D) and wet (W) regular (S) and modal (H) rayon, with theoretical curves (dotted lines) from Fig. 8.14 for fibrillar (F) and micellar (M) structures. (From Hearle.[25])

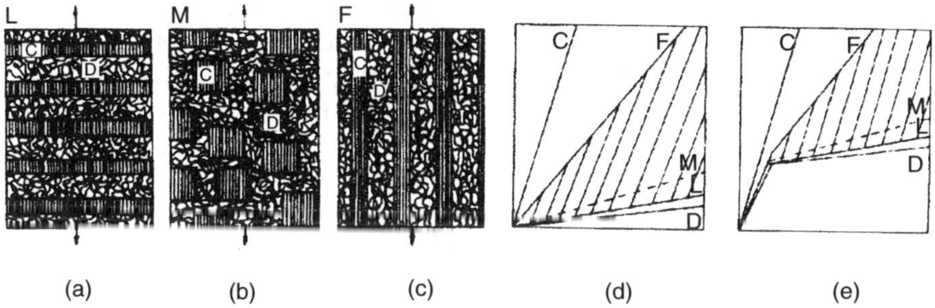

8.14 (a) Lamellar (L), (b) micellar (M), and (c) fibrillar (F) structures. Predicted tensile responses for the three structures in (d) wet and (e) dry states, with assumed properties for crystalline (C) and disordered (D) regions. (From Hearle.[25])

low wet modulus. In contrast to this, the modal process, and others in which coagulation and stretch occur together and are followed by regeneration, leads to high wet modulus. Hearle[25] explained this as being due to differences between a micellar texture in regular rayon and fibrillar in modal rayon.

Figure 8.14 shows three idealised assemblies of crystalline and amorphous material and predictions of stress–strain response. The lamellar form follows an averaging of strain values, since both components are under the same stress, whereas the fibrillar form has equal strains and stress must be averaged. Predictions for the micellar form, which cannot be simply analysed, are somewhat arbitrarily placed at strains midway between lamellar and fibrillar values at the same stress. Linearised stress–strain relations are assumed for dry and wet crystalline and disordered material, and the averaging is weighted for $\frac{1}{3}$ crystalline and $\frac{2}{3}$ disordered material. The predictions from Fig. 8.14(d) and (e) are shown as dotted lines in Fig. 8.13 and agree well with the experimental results.

In this model, the crystalline regions are assumed to be stiff and to be unaffected by absorbed water. In the dry state, amorphous segments are crosslinked by hydrogen bonds, which give a relatively stiff structure. The initial modulus is not as high as for crystals, because of the irregular packing. At about 2% extension, the stress reaches a level at which crosslinks break. In a way typical of glassy polymers, this leads to a yielding and a lowering of modulus. In the wet state, the direct hydrogen bonding between cellulose molecules is broken by mobile water molecules. The amorphous matrix behaves like a rubber with a low initial modulus, which remains constant at the same value as the dry yield modulus. Having formulated the model, the numerical values for the crystalline and disordered stress–strain relations were chosen to fit the experimental curves for regular rayon; the fit for the high wet modulus rayon then has no adjustable parameters. The differences in properties are most marked in the wet state, in which the micellar form can be compared to hard glass beads in a soft rubber, allowing for easy extension, whereas in the fibrillar form the stiff rods would have a stronger influence. As shown in Table 8.5, the beneficial effects are greater in polynosic than in modal fibres, and greater still in lyocell.

In the above model, the effects of orientation are taken into account empirically in fitting the values of the parameters for regular rayon, with the tacit assumption that the effect is the same in the high wet modulus rayon. As shown in Figs. 8.11 and 8.12, increasing orientation causes marked stiffening of fibres. In the micellar form, this is due partly to crystalline orientation, which is shown by the X-ray diffraction patterns. If the structure is formed under low tension, the micelles will be appreciably misoriented. Extension takes place as they swing round towards the fibre axis. However, the major influence is amorphous orientation. Under the high tensions of stretching processes, hydrogen bond formation locks the matrix into a more oriented, less extensible form.

In low orientation, high-wet modulus forms, the fibrils will be following paths over a range of directions to the fibre axis, though whether there is local parallelism or not is an open question. If there is variation in orientation along individual fibrils, extension will occur as they are pulled straight. Alternatively, if the fibrils are initially straight, but at angles to the fibre axis, extension would cause them to be pulled into alignment. In either case, major resistance will come from distortion of the matrix between the fibrils.

The production of high-performance aramid fibres by spinning from liquid crystalline solutions led to similar approaches to cellulosic fibres. There are a number of reports in the patent literature of spinning of fibres from liquid crystalline solutions of cellulose derivatives.[26] At low degrees of orientation (e.g. 35° orientation angle), fibres of ethylcellulose and cellulose triacetate have been made with strengths ranging from 350 mN tex^{-1}, comparable to high-tenacity viscose rayon, to almost 950 mN tex^{-1}, comparable to high-tenacity nylon or polyester. Very high break extensions, ranging from 100–280%, mean that the fibres are extremely tough. Other process modifications lead to higher tenacity and lower break extensions in cellulose derivative fibres. High-performance cellulose fibres can be regenerated from derivative fibres which have been spun from liquid crystalline solutions, but the most interesting results are from spinning of cellulose itself.

Very high-modulus and high-tenacity are achieved when cellulose is spun from liquid crystalline solutions, owing to the resulting high crystallinity and high orientation of the fibres. The description given here relates to Fiber B, which is spun from a solution of cellulose in phosphoric acid. The information is taken from chapter 7 of the thesis by Boerstoel.[1] Figure 8.15 shows the stress–strain curves of Fiber B compared with regular viscose and with high-tenacity yarns made by modifications of the viscose process. The initial modulus of Fiber B is 44 GPa, which is lower than the initial modulus of about 80 GPa for the first para-aramid (PPTA) fibres, such as Kevlar and Twaron, which later reached around 120 GPa for high modulus versions. The conversion factor of 0.5 from the crystal modulus is similar to that of the first para-aramids. The limitation is the crystal modulus of cellulose II, which is calculated to be 90 GPa,[7,27] compared with 140 GPa for cellulose I and 240 GPa for PPTA. Nevertheless at 30 N tex^{-1}, the modulus is higher than that in earlier high-modulus cellulose fibres, such as Fortisan (21 N tex^{-1}) and Cordenka EHM (25 N tex^{-1}). The sonic modulus of Fiber B is 58 GPa, but this increases to 67 GPa after drying the fibre, owing to an estimated increase in shear modulus from 3.8–6 GPa. The strongest Fiber B filaments had a tenacity of 1130 mN tex^{-1}, and a break extension of 6.5%, though the average values were 870 mN tex^{-1} and 5.1%. Improvements in control of spinning could be expected to improve tenacity and break extension.

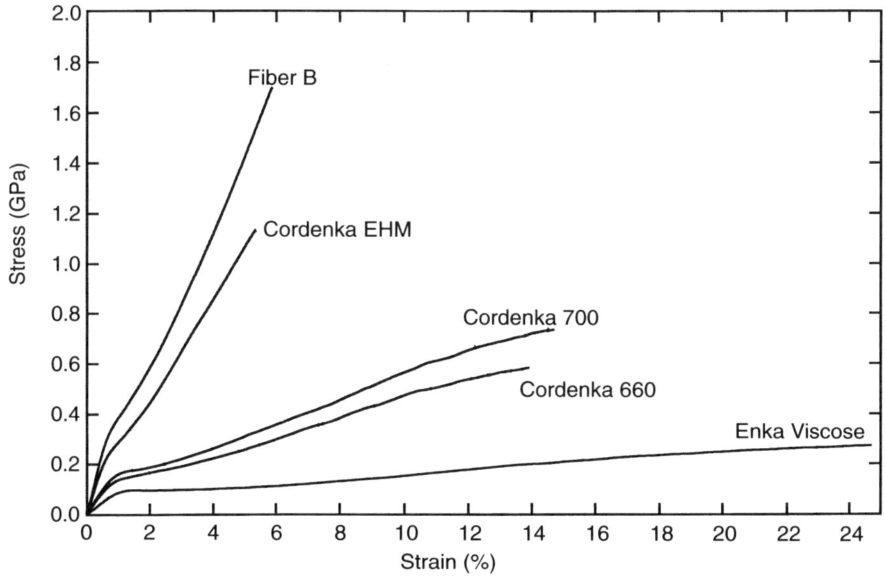

8.15 Stress–strain curves for Fiber B compared with cellulose II fibres.[1] (Courtesy of M G Northolt.)

As shown in Figs. 8.8 and 8.11 and in Table 8.4, cellulose acetate is weaker and more extensible than cellulose fibres. The lower initial modulus leads to limp fabrics, which are attractive in some fashions. In addition to the effects of the low crystallinity of secondary acetate, the bulky acetate groups contribute to volume but not to mechanical cohesion. The reduction in strength and increase in extension at higher temperatures, which is small in cellulose fibres, is substantial in acetate.

8.4.4 Elastic recovery and time dependence

Meredith[28] showed that both regular viscose rayon and secondary acetate fibres show poor elastic recovery after the stress reaches the yield stress at about 2% extension. The elastic recovery, as shown by percentage of imposed extension that is recovered, for viscose fell rapidly to about 40% and for acetate to about 15%. However, Guthrie and Norman[29] have shown that the magnitude of elastic recovery is dependent on the times involved. Figure 8.16 is an example of the influence of the time for which the fibre is held extended and the time allowed for recovery.

For Fortisan, Meredith[28] found much better elastic recovery, falling only to about 70%. This must reflect the influence of the fibrillar structure, and so would be expected to apply to modal, polynosic and lyocell fibres.

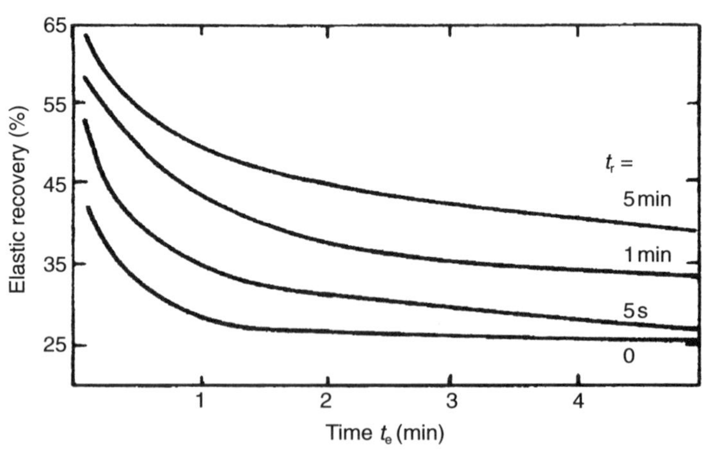

8.16 Elastic recovery of viscose rayon from 5% extension, plotted against time held extended, t_e for different recovery times, t_r, at zero load, after Guthrie and Norman.[29] (From Morton and Hearle.[5])

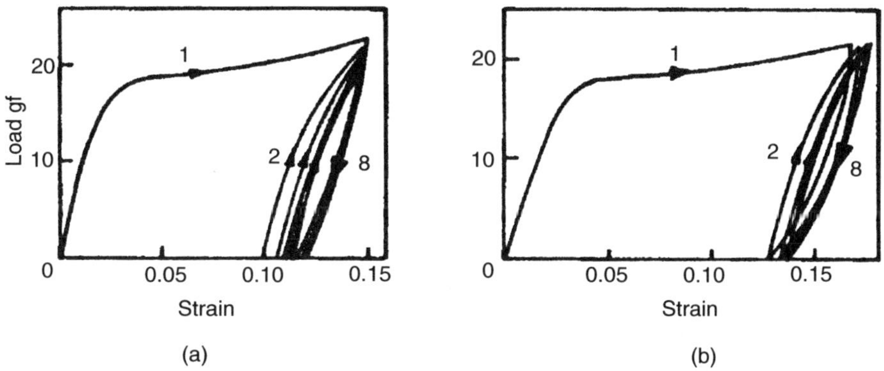

8.17 Cyclic loading of 2.5 tex acetate fibre, after Hearle and Plonsker[30] (a) in simple extension cycling; (b) in load cycling. (From Morton and Hearle.[5])

As illustrated for acetate in Fig. 8.17, cyclic loading beyond the yield point shifts the initial part of the stress–strain curve owing to the lack of recovery. The behaviour of viscose rayon is similar. These plots are for cycling to constant extended length and constant load. Another procedure is cumulative extension cycling, in which slack is removed after each cycle. Hearle and Plonsker[30] found that at 2% imposed extensions of acetate, the length levelled off at a constant value after a few cycles, but continued to increase indefinitely at higher imposed extensions, requiring only 20 cycles to reach the break extension at 5% imposed extension. Viscose rayon showed continuing extension even at 2% imposed extension.

Figures 8.16 and 8.17 show indirectly the influence of time on mechanical behaviour. The direct effect of load is shown by the curves in Fig. 8.18(a). The first point at 6s can be regarded as a combination of 'instantaneous' extension, caused by changes in covalent and intermolecular bond spacings, and rapid creep. The total change in length from 0.1–100 min can be divided into primary creep, which is recovered in similar times, and secondary creep, which is not recovered. At low stress, the primary creep shows a typical sigmoidal curve, which levels off at longer times. At high stress, only the upper part of the sigmoid is seen. Secondary creep becomes important beyond the yield point, which is between 1 and 2% extension. Another set of data[33] shows primary creep reaching 3.5% extension after 11.5 days at 71 mN tex^{-1} and secondary creep reaching 8%. The converse of creep is stress relaxation, namely decrease in tension at constant extended length, as shown in Fig. 8.18(b). Note that this plot is against time on a linear scale, which rapidly reduces in rate of fall, whereas a plot on a logarithmic scale shows an almost linear decrease of about 10 MPa per decade between 0.1 and 10^5 s (30 min) at 20% extension;[32] at lower extensions, the rate of fall is

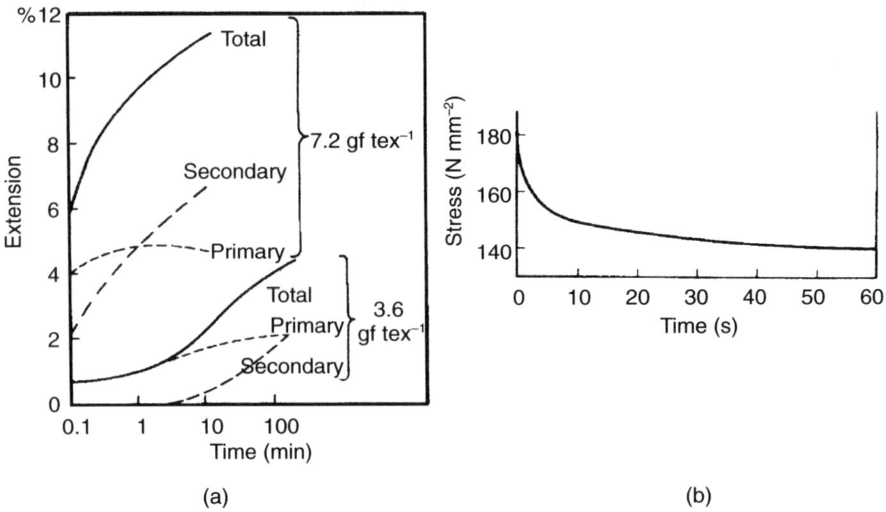

8.18 (a) Creep of viscose rayon at 60% rh, after O'Shaughnessy[31] (3.6 gftex^{-1} = 35.3 mNtex^{-1}; 7.2 gftex^{-1} = 70.6 mNtex^{-1}); (b) stress relaxation of viscose rayon.[33] (From Morton and Hearle.[5])

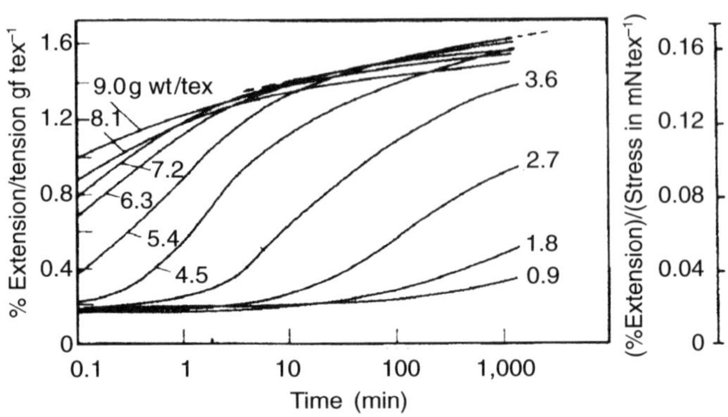

8.19 Generalised creep curves for viscose rayon at 60% rh[31] (9 gftex^{-1} = 88.3 mNtex^{-1}). (From Morton and Hearle.[5])

slightly lower and the sigmoidal tendency becomes apparent at short times. By varying test conditions and normalising extension in terms of stress, different parts of the sigmoidal curve can be shown as generalised creep or stress relaxation curves on one graph. An example is shown in Fig. 8.19. A shift leads to superposition of the curves.

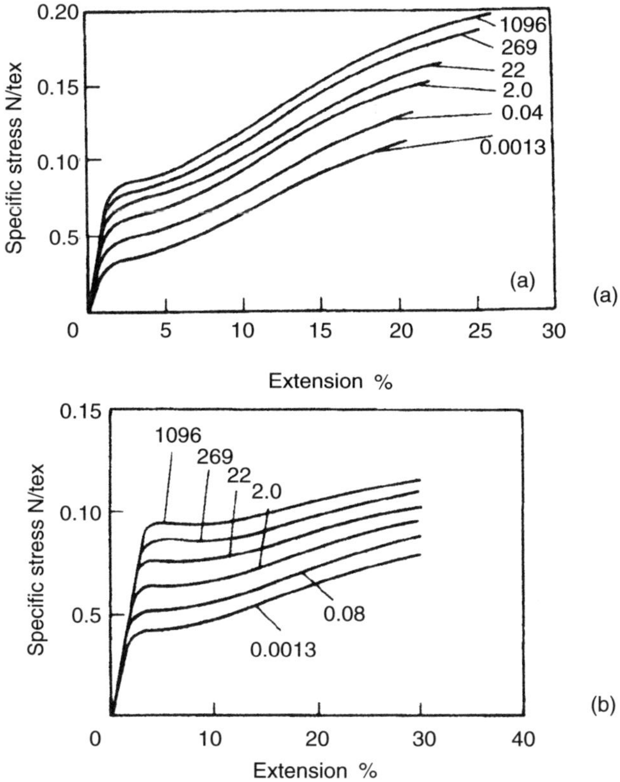

8.20 Stress–strain curves at various rates of extension in % per second.[24] (a) Viscose rayon; (b) acetate. (From Morton and Hearle.[5])

A third way of showing the influence of time is by measuring load–elongation curves at different rates. Figure 8.20 shows stress–strain curves for viscose rayon and acetate at rates from 0.0013–1096% per second. There are slight changes in moduli, as shown by the slopes of the curves, but the dominant effect is a rise in the yield stress. Meredith[34] found that the fall in strength with increase in time was given by equation 8.1 with strength–time coefficient k equal to 0.083 for viscose rayon and 0.060 for acetate:

$$F_1 - F_2 = kF_1 \log_{10}(t_2/t_1) \qquad [8.1]$$

where F_1 and F_2 are break loads in times t_1 and t_2.

For a highly oriented cellulose fibre, Fortisan, the changes in going from 0.017–2000% per second were from 560 to 800 mN tex^{-1} in tenacity and 14 to 22 N tex^{-1} in initial modulus, with a small drop in break extension from 5.4–5.2%.[35]

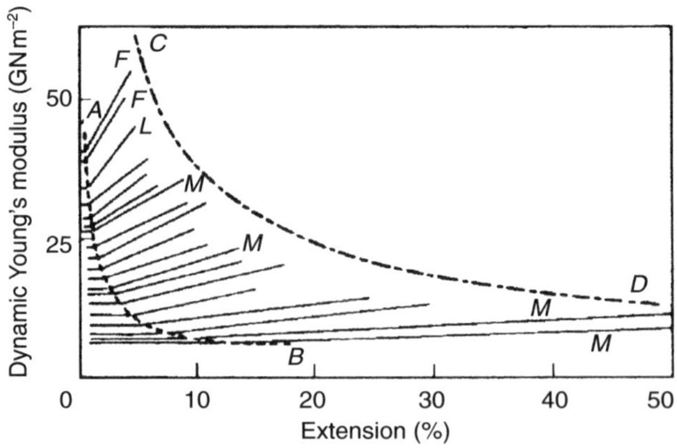

8.21 Dynamic modulus of cellulose fibres.[36] *AB* is locus of critical strain; *CD* is locus break extensions in tensile tests. *F* are saponified acetate; *L* is Lilienfeld rayon; *M* are modal filaments; others are viscose rayon. (From Morton and Hearle.[5])

The last way of showing time dependence is by cyclic loading at different frequencies. At high frequencies the dynamic (or sonic) modulus can be calculated from measurements of the velocity of sound. Figure 8.21 shows results obtained by de Vries[36] for a variety of cellulose fibres. There is an initial constant modulus up to a critical strain and then an increase. The effect of orientation is clearly seen in the rise in moduli from a low orientation modal filament to the highly oriented Fortisan fibre. The dynamic moduli changed little during stress relaxation.

The effect of temperature on the modulus (real part) and the dissipation factor $\tan\delta$ at low frequency is shown in Fig. 8.22 for wet and dry viscose rayon. A drop in modulus and a peak in $\tan\delta$ indicates a transition from a glassy to a rubbery state, caused by hydrogen bonds changing from being held to becoming mobile. Water has a major effect on the transition temperature. In dry fibres the peak is above the possible measurement temperature, which is limited by charring, and in the wet state below that limited by freezing. Meredith[38] found a peak in $\tan\delta$ at 25% rh at 20°C, which was followed by a further increase between 60 and 100% rh.

8.4.5 Directional effects

The study of mechanical properties of fibres is dominated by tensile tests, but properties in other directions should also be considered. Even in a Hookean region of perfect elasticity, the description includes a number of

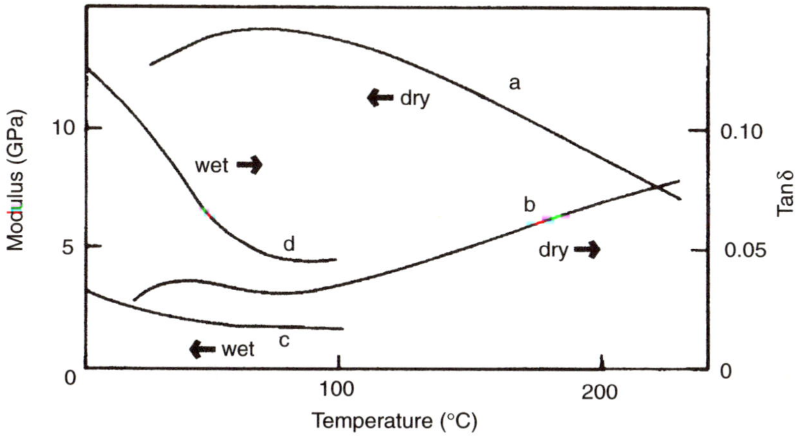

8.22 Dissipation factor tan (a, b) and modulus (c, d), wet and dry, for viscose rayon at about 1 Hz.[37] (From Morton and Hearle.[5])

Table 8.7 Bending and twisting properties (specific rigidities in mN mm^2 tex^{-2}; moduli in GPa)

Fibre	Finlayson[39]		Owen[40] 65% rh, 20°C				
	shape factor	flexural rigidity	flexural rigidity	bending modulus	tensile modulus	torsional rigidity	shear modulus
Viscose rayon	0.74	0.19	0.35	10		0.08–0.083	0.84–1.2
Modal			0.69	20		0.097	1.4
Fortisan	0.83	0.44					
Secondary acetate	0.67	0.08	0.25		4.2	0.064	
Triacetate			0.25		3.8	0.091	

constants in anisotropic fibres. Few measurements are available. Some data on flexural and torsional properties are given in Table 8.7.

Bending produces extension on the outside of the curvature and compression on the inside, so that it is related to tensile properties but also involves axial compression. In the Hookean region, the flexural rigidity B is given by Equation [8.2]:

$$B = (\eta E T^2 / 4\pi\rho) \qquad [8.2]$$

where η is a shape factor, E is the specific modulus in N kg^{-1} m, T is the linear density in kg m^{-1}, and ρ is the density in kg m^{-3}.

Because of the squared term, fibre fineness has a major effect in reducing resistance to bending and hence on the flexibility of fabrics. A specific flexural rigidity, which must be multiplied by the square of linear density to obtain fibre bending resistance, can be defined as $(\eta E/4\pi\rho)$.

Fibre torsion brings in the shear modulus n and another shape factor ε, though at high twist the increase in length of elements near the outside becomes the dominant effect. In the Hookean region of small strains, the torsional rigidity R, defined as the torque to produce one turn per unit length, is given by Equation [8.3]:

$$R = (\varepsilon n T^2/\rho) \qquad [8.3]$$

In isotropic materials the shear modulus is ⅓ of the tensile modulus, but in oriented fibres it is much less. Measurements give ratios of tensile to shear modulus equal to 8.2 in viscose rayon, 28 in a high-tenacity rayon (Tenasco) and 8.1 in acetate. Torsional rigidity falls off rapidly with absorption of water. For both viscose and acetate the value at 65% rh is about ½ that at 0% rh and is down to ⅕ by 90% rh. Breaking twist angles are 35–40° for viscose rayon, 31–33° for high-tenacity viscose, and 40–45° for acetate. Estimates of shear strength are typically about half the tensile strength.

8.4.6 Fracture, fatigue and durability

Viscose rayon and acetate fibres rupture in tension in the form known as granular fracture,[42] which is found in most fibres spun from solution. It is a result of a structure which has discontinuous transverse weaknesses, probably caused by separation during coagulation. The break runs perpendicularly across the fibre with a rough texture, as shown in Fig. 8.23(a)–(c). The smoother surface of the Tricel break in Fig. 8.23(d) may be due to melting or softening as a result of plastic deformation during rupture. Examination of unduly weak viscose rayon fibres was found to be due to the presence of voids, which reduced the load-carrying area, Fig. 8.23(b). Sometimes the break divides between steps, owing to simultaneous rupture with initiation at more than one place. This has been seen in the lyocell fibre Tencel.[41] In another Tencel fibre, the spread of break from initiation at the surface was observed.[41] Twist breaks show similar, but twisted, granular breaks, caused by the elongation of outer segments of the fibre.

Since fibres are subject to repeated shocks of given energy in many uses, a high work of rupture leads to high durability. However, good elastic recovery is also needed, because otherwise the successive inputs lead to a climb up the load–elongation curve to break. Regenerated cellulose fibres do not have as good a combination of these two factors as synthetic fibres like nylon and polyester. Acetate fibres are still poorer.

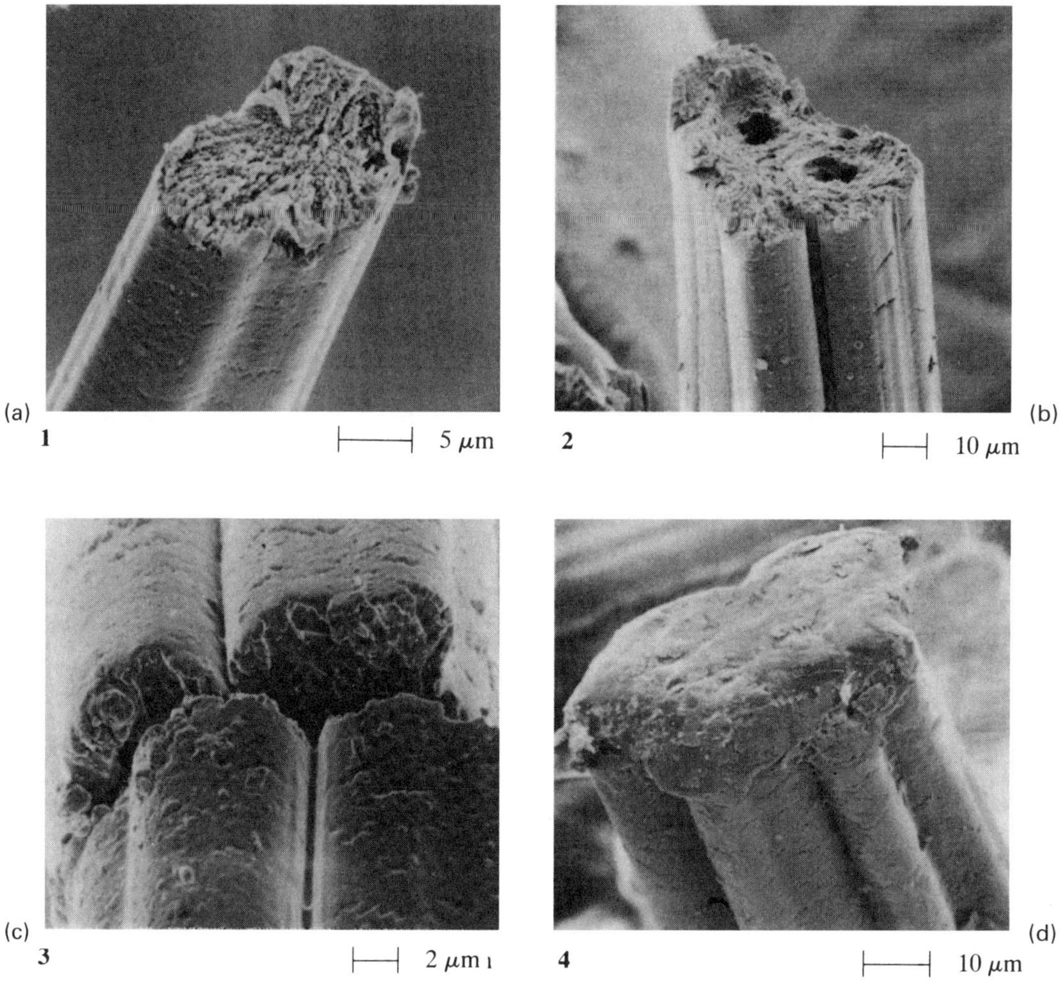

8.23 Scanning electron microscope pictures of tensile breaks of cellulosic fibres: (a) modal fibre (Vincel); (b) weak viscose rayon with holes at point of break; (c) partial break of secondary acetate close to main break, showing cracks on surface; (d) triacetate. (From Hearle et al.[41])

Hearle and Booth[42] showed in the 1960s that cumulative extension fatigue testing, with a fixed stroke but slack removed after each cycle, did not lead to failure for imposed extensions below about 2% in viscose rayon. Above this level, the lack of recovery led to a rapid climb up the load–elongation curve to break at a median value of 79 cycles at 5% stroke and 6 cycles at 10% stroke. Since improved methods of fatigue testing were developed after 1970, there has been little research on manufactured cellulosic fibres. Repeated flexing of viscose rayon over a pin leads to the appearance of kink bands on the inside of the bend, which will weaken the fibre, and leads to some surface wear. However, the breaks are distorted forms of granular fracture.

In use,[41] regular viscose rayon and crimped rayon in carpets show much sharper breaks than are found with most other fibres. However, modal overalls do show the commoner form with multiple splitting at the broken end.

A special feature of lyocell fibres is their tendency to fibrillation, as shown in Fig. 4.7. In some circumstances, this can be a weakness, but provided it is not excessive, it can also be used as a way of increasing water absorption and giving an attractive handle to textile fabrics. The ease of fibrillation can be controlled during the manufacture of lyocell fibres. A high level of fibrillation is beneficial in hydroentangled nonwovens and in paper, as illustrated in Fig. 4.8.

8.4.7 Role of the matrix: setting and recovery

The crystalline regions in cellulose provide a passive framework, which holds the fibres together; the amorphous matrix is the active region which deforms and recovers. Only in the highly oriented, highly crystalline fibres spun from liquid crystals does bond deformation in crystals become dominant. The matrix is also the site for the entry and absorption of water, which, as we have seen, has a major influence on properties. This is very clearly seen in Fig. 8.22. In dry fibres, the molecules in the matrix are firmly bonded together by hydrogen bonds; they only begin to become mobile because of thermal vibrations above 100°C, and are not fully mobile even at 200°C. Under stress, the hydrogen bonds start to break at 1–2% extension and to reform in new places. In wet fibres, there is full mobility at 0°C, and the matrix acts as a rubber. The state of a cellulose fibre depends on how its previous history affects the hydrogen bonding. Fibres can be both set in deformed states and recovered to a reference state.

When fibres are deformed when wet, for example in washing, and then dried, hydrogen bonding will lock the fibres in the deformed state. This is why fabrics are wrinkled after laundering, and why ironing with some steam presses them flat or with creases in the right places.

The lack of recovery from large strains in the dry state is due to the relocation of hydrogen bonds. In the wet state, recovery from a given extension is better, because of the rubbery elasticity of the matrix. Swelling recovery is another manifestation of this behaviour. If a cellulose fibre is mechanically deformed with 'permanent' unrecovered extension in the dry state, then wetting or treatment with steam will cause it to recover to near its original length. Provided there is no external restraint, the wet rubbery structure reverts to its equilibrium, reference state. This is true for individual, free fibres, but the restoring forces are low. In a fabric, the tendency to fibre recovery is not strong enough to overcome frictional interactions between neighbouring fibres.

Crease resistance can be imparted to rayon fabrics by chemical treatments which crosslink the fibres. The crosslinking agents are di- or trifunctional molecules, based on formaldehyde with ureas, melamine or carbamates. They react with the cellulose and link the molecules of the matrix together by covalent bonds. This sets the structure in a given conformation, which, depending on the extent of the treatment, limits the setting of fabrics in wrinkles as a result of mechanical action in wear or wetting and drying in washing. If the crosslinking is activated during garment manufacture, pleats and creases can be permanently set. The negative aspect of such treatments is that they make the fibres brittle, so that they are more easily broken in wear.

Chemical permanent setting has, to some extent, helped cellulose fibres compete with ease-of-care properties of melt-spun synthetics, which can be permanently set by heat treatments. Cellulose triacetate, because of the absence of hydrogen bonding and the lower melting point of the crystals (306°C) can be heat set.

8.4.8 Other physical properties

When dry, the specific heat of rayon is $1.26\,J\,g^{-1}\,K^{-1}$, but this will increase with moisture absorption. In practice, the heat associated with changes in absorbed moisture are important. An increase in relative humidity of 1% will generate about $0.8\,J\,g^{-1}\,K^{-1}$. The thermal conductivity of rayon is about $20\,mJ\,g^{-1}\,mm^{-1}\,K^{-1}$.[43] The melting point of cellulose is above the temperature at which chemical decomposition and charring occur. However, the ends of rayon fibres that have been dipped into a flame do show globular forms that indicate that some melting has occurred.[41]

Table 8.8 gives values of dielectric constant ε and power factor $\cos\phi$, which equals $\sin\delta$ and, for low values, approximates to the dissipation factor $\tan\delta$. These were measured on yarns packed between concentric cones.[44] There are substantial increases as moisture is absorbed. The specific electrical resistance of dry cellulosic fibres is about 10^{11} ohm m, but this falls rapidly and approximately linearly with relative humidity to 10^{5} ohm m at 90% rh. The conductivity is ionic and Hearle[45] showed that resistance was related to dielectric constant through the increased dissociation of oppositely charged ions as dielectric constant increased. The comparatively low

Table 8.8 Dielectric properties;[44] dielectric constant ε and power factor cos φ

Fibre	rh (%)	at 100 Hz		at 100 kHz	
		ε	cos φ	ε	cos φ
Viscose rayon	0	2.1	0.01	2.0	0.01
43% packing	65	8	0.9	3.5	0.05
Secondary acetate	0	1.7	0.005	1.6	0.005
45% packing	65	2.2	0.1	2.0	0.02

electrical resistance means that static electricity is not often generated on cellulose fabrics.

The refractive index and birefringence of cellulose has been referred to above in connection with estimates of orientation. Hermans[19] showed that the refractive index followed the density changes with a small increase from the dry state and then a drop. The value of *n*(parallel) reduced from 1.575 at 0% moisture regain to 1.54 at 20% and *n*(perpendicular) from 1.535 to 1.505.

The surface properties of fibres are highly dependent on the presence and nature of finish. Reported values of coefficient of friction range from 0.14 to 1.02.

References

1 H Boerstoel, *Liquid Crystalline Solutions of Cellulose in Phosphoric Acid for Preparing Cellulose Yarns*, Doctoral Thesis, Rijksuniversiteit Groningen, 1998.
2 J E Ford, (ed), Fibre Data Summaries, *Shirley Institute Pamphlet No. 91*, Manchester, Shirley Institute, 1966.
3 N S Wooding, 'Rayon and acetate fibres', in *Fibre Structure*, eds J W S Hearle and R H Peters, Manchester, The Textile Institute, 1963.
4 W G Harland, 'The degree of polymerization and its distribution in cellulose rayons: Part V', *J Textile Inst*, 1955 **46** T483–T499.
5 W E Morton and J W S Hearle, *Physical Properties of Textile Fibres, 3rd* edition, Manchester, The Textile Institute, 1993.
6 A D French, 'Physical and theoretical methods for determining the supramolecular structure of cellulose', in *Cellulose Chemistry and its Applications*, eds T P Nevell and S H Zeronian, Chichester, Ellis Horwood, 1985 pp 84–137.
7 L M J Kroon-Batenburg, J Kroon and M G Northolt, 'Chain modulus and intramolecular hydrogen bonding in native and regenerated cellulose fibres', *Polym Commun*, 1986 **27** 290–292.
8 J A Howsmon and W A Sisson, in *Cellulose and Cellulose Derivatives, Part I*, 2nd edition, eds E Ott, H M Spurlin and M W Graffin, New York, Interscience, 1954.

9 J W S Hearle, 'Understanding and control of textile fibre structure', *J Appl Polym Sci: Appl Polymer Symp*, 1991 **47** 1–31.
10 C W Bunn, 'Polymer texture' in *Fibres from Synthetic Polymers*, ed. Hill R, Amsterdam, Elsevier, 1953 pp 240–286.
11 J W S Hearle, 'The structural mechanics of fibers', *J Polym Sci*, 1967 **C20** 215–251.
12 J W S Hearle, 'A fringed fibril theory of structure in crystalline polymers', *J Polym Sci*, 1958 **28** 432–435.
13 J W S Hearle, 'Versatile viscose – the first planned fibre', *Skinner's Silk and Rayon Record*, Volume **32**, 1958, Part I New forms, 847 and 849–851, Part II Fashions and furnishings, 961–963 and 965, Part III Industrial uses, 1217–1218 and 1220–1221.
14 H A Coulsey and S B Smith, 'The formation and structure of a new cellulosic fibre', Dornbirn Conference, 1995.
15 Courtaulds Fibres (now Acordis), Lyocell Technical Overview.
16 A R Urquhart and N Eckersall, 'The adsorption of water by rayon', *J Textile Inst*, 1932 **21** T163–T170.
17 A R Urquhart and N Eckersall, 'The moisture relations of cotton, VII A study of hysteresis', *J Textile Inst*, 1932 **21** T499–T510.
18 D K Ashpole, 'The moisture relations of textile fibres at high humidities', *Proc Roy Soc A*, 1952 **212** 112–123.
19 P H Hermans, *Physics and Chemistry of Cellulose Fibres*, Amsterdam, Elsevier, 1949.
20 J M Preston and M V Nimkar, 'Textile fibres: swelling measurement', *J Textile Inst*, 1949 **40** P674–P688.
21 R J E Cumberbirch and C Mack, 'The degree of polymerization and its distribution in cellulose rayons, X. A quantitative theory of the tenacity and breaking extension of regenerated cellulose monofilaments', *J Textile Inst*, 1960 **51** T458–T483.
22 R J E Cumberbirch and C Mack, 'A quantitative theory of the loss of strength of regenerated cellulose filaments on hydrolysis by acids', *J Textile Inst*, 1961 **52** T382–T389.
23 R W Work, 'Cellulose and cellulose acetate yarns: effect of variation in structural order on physical properties', *Textile Res J*, 1949 **19** 381–393.
24 N H Chamberlain and M P Khera, 'Viscose and cuprammonium filaments: radial heterogeneity', *J Textile Inst*, 1952 **43** T123–T157.
25 J W S Hearle, 'Structural mechanics of fibres', *J Polymer Sci* **C**, 1967 No. 20, 215–251.
26 R D Gilbert, in *Handbook of Fiber Science and Technology*, 1993, Volume 3, p 357.
27 T Nishino, K Takano and K Nakama, 'Elastic modulus of the crystalline regions of cellulose polymorphs', *J Polym Sci*, 1995 **B33** 1647–1651.
28 R Meredith, 'Textile fibres: comparison of tensile elasticity', *J Textile Inst*, 1945 **36** T147–T164.
29 J C Guthrie and S Norman, 'Measurement of the elastic recovery of viscose rayon filaments', *J Textile Inst*, 1961 **52** T503–T512.
30 J W S Hearle and H R Plonsker, 'Behavior of fibers in cumulative extension cycling', *J Appl Polym Sci*, 1966 **10** 1949–1971.

31 M T O'Shaughnessy, 'Rayon yarn: creep', *Textile Res J*, 1948 **18** 263–286.
32 R Meredith, 'Relaxation of stress in stretched cellulose fibres', *J Textile Inst*, 1954 **45** T438–T461.
33 J J Press, 'Viscose rayon yarn: flow and recovery properties', *J Appl Phys*, 1943 **14** 224–233.
34 R Meredith, 'The effect of rate of extension on the tensile behaviour of viscose and acetate rayons, silk and nylon', *J Textile Inst*, 1954 **45** T30–T43.
35 H F Schiefer, W D Appel, J F Krasny and G G Richey, 'Impact properties of yarns made from different fibers', *Textile Res J*, 1953 **23** 489–494.
36 H de Vries, 'Regenerated cellulose fibres: dynamic modulus of elasticity in relation to large deformation', *Appl Sci Res*, 1951 **3A** 111–124.
37 S J Van der Meer, *Dynamic–Mechanical Properties and Plastic Deformation of Yarns, Cords and Fabrics*, Doctorate Thesis, Delft University of Technology, 1970.
38 R Meredith, 'Dynamic mechanical properties of textile fibers', *Proceeding 5th International Congress of Rheology* (1968), Tokyo, University of Tokyo Press, 1969, Volume **1** pp 43–60.
39 D Finlayson, 'Yarns for special purposes – effect of filament size', *J Textile Inst*, 1946 **37** P168–P180.
40 J D Owen, 'The application of Searle's Single and Double Pendulum Methods of Single fibre Rigidity Measurements', *J Textile Inst*, 1965 **56** T329–339.
41 J W S Hearle, B Lomas and W D Cooke, *Atlas of Fibre Fracture and Damage to Textiles*, 2nd edition, Cambridge, England, Woodhead Publishing, 1998.
42 J W S Hearle and A J Booth, 'The fatigue behaviour of textile yarns', *Proceeding 4th International Congress on Rheology 1963*, New York, Interscience, 1965 pp 203–225.
43 J Brandrup and E H Immergut (eds), *Polymer Handbook* VI-8, New York, Interscience, 1966.
44 J W S Hearle, 'Capacity, dielectric constant, and power factor of fibre assemblies', *Textile Res J*, 1954 **24** 307–321.
45 J W S Hearle, 'The electrical resistance of textile materials: IV. Theory', *J Textile Inst*, 1953 **44** T177–T198.

9 Applications development

CALVIN WOODINGS

9.1 Artificial silk

The original cellulose nitrate yarns made by Chardonnet looked like silk and were naturally marketed at prices close to those of silk in traditional silk applications. As described in Chapter 1, the desire to emulate the silkworm had been the original driving force behind the technology, and at the end of the 19th century, the successful commercialisation of artificial silk was one of the great symbols of industrial progress. Chardonnet silk was thus assured an early success if only because of its novelty. Despite being less than half the strength of silk and much more prone to shrinkage, it could be converted into decorative trimmings, ribbon and lace, and its astonishing lustre made it ideal for embroidery. Poor wet performance made it more suitable for embellishing home furnishings than for apparel, and natural silk remained largely unchallenged in garments.

Cuprammonium rayon followed cellulose nitrate into the same markets, but having slightly less lustre it did so at a slightly lower price. The first viscose rayon yarns inevitably found applications in similar markets and, despite having been developed to make carbon lamp filaments for the new electric lights, it found a much larger market in braids for the old incandescent gas lighting. Clearly these applications were not really big enough to support a major industry, and the problems of weaving these products into broad, as opposed to narrow, fabrics had to be solved before the market could be expanded significantly.

While the new viscose rayon yarns proved easier to use as weft (across for example a cotton warp), the painstaking work carried out by Samuel Courtauld & Co at Bocking in Essex was a major contributor to both Courtauld's and the artificial fibre industry's rapid growth. In what was one of the early examples of the benefits of captive customers, Courtauld's Bocking weaving mill had to take the entire output of the Courtauld viscose factory in Coventry for the first year of operation. Little if any saleable fabric resulted, but the practical experimentation taught Courtauld's textile technologists how to modify the multitude of processes downstream of fibre making to suit the new yarns. Few if any independent weavers, as Courtauld's competitors who had to sell yarns on the open market discovered, could afford to stay the course.

Towards the end of 1906, the first broad woven viscose rayon fabric from Bocking started to sell and indeed, became fashionable. It was a striped fabric, the stripe being achieved by alternating bands of cotton and rayon across the warp and filling it with a 100% cotton weft. Viscose dyed darker than cotton, and this enhanced the great difference in lustre between the two fibres. A new market was thereby created, and for the next five years the new viscose yarns were used almost exclusively by Courtauld in the ornamentation of cotton fabrics. Not quite the market envisaged by Tetley when he sold the idea of manufacturing artificial silk to the Board of England's leading manufacturers of mourning crepe, but after years of difficult development, a welcome breakthrough. Stripes were in many ways an ideal opening gambit for the new fibres. The bulk fabric properties were largely unaffected by the original narrow stripes, dyeing differences became an effect rather than a problem, and as quality improved wider stripes could carry increased throughput. During the same period, Courtauld gained access, thanks to a curious patent judgment, to the German spin-bath technology that allowed the manufacture of stronger, more consistent yarns (see Chapter 1) and was in an ideal position to increase the use of them.

These new yarns made using the so-called Müller Bath, finally allowed satisfactory quality fabrics to be woven by Courtauld and others with viscose in the weft. A new series of applications resulted. Courtauld branded the new fabrics 'Luvisca' and developed outlets for the new fabrics in shirtings (cotton warp), blouses (silk warp), and, of course, the lining fabric (silk or cotton warps) that remains an important end-use to this day. As yarn quality improved further, viscose replaced the silk in the warp, and by 1914 Courtauld's Halstead Mill was using 100% viscose warps on 200 looms.

The problems of knitting viscose yarns were solved in 1912 by Wardle and Davenport of Leek, who had been Courtauld's first ever external customer, responsible for opening up the braid market to viscose. Silk hosiery had been an important target market since the early attempts to break out of the narrow fabrics niche, but most stocking manufacturers were fundamentally opposed to trying it, despite yarns being offered free of charge to encourage developments. Wardle and Davenport's eventual success owed much to a new knitting machine acquired from Wildt & Co in Leicester, said to be suitable for the new artificial silk yarns. This proved to be the case; others followed Wardle and Davenport's lead, and hosiery became one of the main outlets for viscose until nylon replaced it in the 1940s.

Table 9.1 T.V.C.: distribution of yarn sales by buyers expressed as sales by value, 1916–1918

Trade of buyers	Percentage of sale, by value		
	1916	1917	1918
Hosiery	52.7	40.8	40.3
Cotton	10.4	12.7	11.5
Silk	3.3	3.7	4.2
Woollens	0.9	0.2	..
Jobbers	20.1	31.2	32.7
Knitters	4.5	2.4	2.6
Plush	1.8	3.1	0.8
Tapestry	0.9	1.2	1.1
Underwear	0.6	0.4	0.5
Braid	2.2	1.8	3.5
Embroidery	1.9	2.3	2.8
Webbing	0.6	0.1	..
Lace	0.1	..	..
Exports	..	0.1	..
Total	100.0	100.0	100.0

T.V.C. is The Viscose Company, the US subsidiary of Courtauld.
Source: D.C. Coleman, *Courtaulds: an Economic and Social History*, Clarendon Press, Oxford, 1969.

In the USA, knitted hosiery was the first volume outlet for viscose yarn from the American Viscose Corporation's Marcus Hook (Pennsylvania) plant when it commenced production at the end of 1910. It is not clear whether the US knitting machines were superior to those in Europe, or whether the absence of 'captive' weavers prevented the development of woven fabrics until the viscose yarns were more dependable. By the outbreak of war in 1914, hosiery was the biggest single market for viscose on both sides of the Atlantic (see Table 9.1).

9.2 Artificial cotton

The period between the wars saw a dramatic expansion of rayon usage fuelled primarily by the expiry of the basic patents on the fundamental filament yarn technology. Despite the retarding effects of the Great Depression, every year for 20 years world rayon capacity (including acetate rayon) increased by 60 000 tonnes on average.

This growth was fuelled mainly by the development of a whole range of new applications for what used to be the yarn process waste, staple fibre.

Between 1905 and 1910, the Coventry viscose plant produced thousands of tonnes of waste most of it looking not unlike dirty cotton wool, and most of it dumped on vacant land around the plant. As early as 1907 Tetley proposed extracting some value from this waste and had his chemists briefly investigate ways of making it suitable for the wool and cotton spinning systems. That this development was largely ignored speaks volumes about the preoccupation at that time with making a success of the technically more difficult continuous filament route in the then much more attractive silk market. So it fell to a Frenchman, Augustin Pellerin to patent the first staple process[1] and the first commercial process and applications were later developed in Germany.

The success of this apparently bizarre step of adding value by improving the quality of waste is best illustrated by the fact that at the end of this 20-year period, half of world viscose production capacity was devoted to the waste product that had been largely exterminated two decades earlier. Further incongruity, of course, arose from the need to chop up an eminently weavable and knittable yarn in order to make a different type of yarn by a different type of spinning process. It is hardly surprising that the early proposals to exploit this route did not command the attention of those struggling to improve the quality of the filament route output.

However, in 1930, when the need to expand the application base for viscose into much larger, growth-sustaining markets was becoming evident, the logic for staple becomes clear. Seven million tonnes per year of the natural staple fibres, cotton and wool, were entering the textile market through a route closed to viscose, at a time when viscose filament sales proudly passed the 200000 tonnes per year mark (see Fig. 9.1).

Filament yarns had proved to be very different, in many respects, to yarns spun on the cotton or woollen systems. Consumers proved reluctant to give up the aesthetic advantages of such yarns just because the new filament route was technically superior. Filament cellulose was leaner, weaker, and much softer and slipperier than staple cellulose as exemplified by spun cotton. Furthermore, cotton, being a partially collapsed tubular fibre was bulkier and warmer than the solid viscose fibres.* These were all good reasons why viscose would have to be converted on the cotton system to gain access to broader apparel markets.

It was during World War I, especially in Germany where cotton shortages were more acutely felt because of a successful Allied blockade of the

*The early attempts to use viscose filament in applications such as shirtings were probably only successful in the way that nylon filament was successful in shirtings in the 1960s. It was new and exciting and clever marketing people could persuade technically aware customers to buy it – for a time.

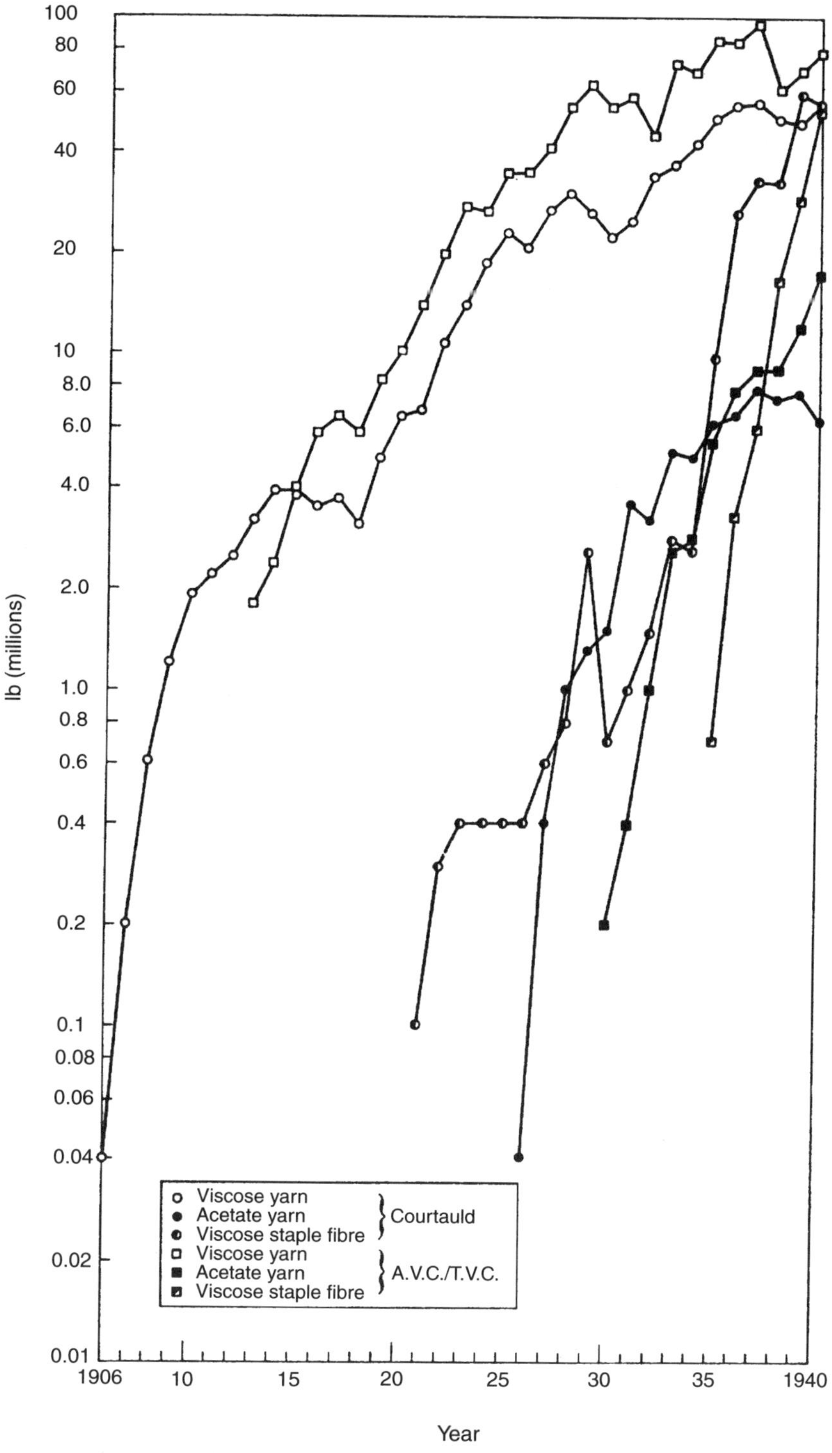

9.1 The introduction of viscose staple (and acetate yarn) by Courtauld's UK and US plants maintains total growth as viscose filament yarn growth rates decrease in the 1930s. (Reproduced from *Courtaulds: An Economic and Social History*, Clarendon Press, 1969.)

ports, that viscose yarn waste was deliberately cut up, cleaned and baled so that it could be blended with cotton and spun into yarns on the cotton system. The German government encouraged further developments by ordering 3000 tonnes from VGF, and the new staple was used instead of cotton in a variety of military textiles, and some apparel.

While this certainly accelerated the technical development of staple fibres and the processes to manufacture them, it did little for the market acceptance of what were, in reality, low quality cotton substitute fabrics with wartime economy connotations.

In the UK, Courtauld's interests in staple were confined to increasing the value of filament waste by making it more suitable for the wool combers. In 1921 they installed the first staple machine in Coventry, launched staple fibre under the 'Fibro' brand, and marketed at a price and quality that made it impossible to compete with the mainstream natural fibres. In this particular incarnation, Courtauld's staple fibre sales (excluding cut-filament waste) were to remain below 400 tonnes per year until 1928.

The Italian company, Snia developed the first efficient continuous precision cutter for large acid tows, launched their product as 'Sniafil' in Italy in 1925, and announced an intention to build a plant in Britain to make it. Snia (Società di Navigazione Italo-America), founded in 1917 as a trading company, acquired the Société Italienne de la Viscose and became Snia Viscosa in 1922 at the start of an Italian rayon boom. By the end of 1925, they controlled 60% of Italian rayon production, were the second largest rayon producer in the world, and thanks to aggressive pricing, were, after Courtauld, the leading supplier to the UK and USA markets. Courtauld had to react and began to produce, at the Aber Works* in Flint, a more economical filament yarn to meet this competition. In 1926 the Italian boom ended, Snia Viscosa weakened, and in 1927 Snia's founder, Riccardo Gualino, sold a controlling interest in the company to VGF and Courtauld jointly.

The Courtauld board, exposed in the course of the deal to the success of the Sniafil process, decided in 1928 to increase emphasis on staple fibre production, and concentrated its production at their more economical Aber Works. Success for Courtauld 'Fibro', however, still had to wait until a 'Sniafil' machine could be started up at Aber a few years later. Courtauld's first dedicated staple plants commenced production in 1936 at Greenfield near Flint in North Wales and at Nitro (an old cellulose nitrate explosives plant acquired by Courtauld in 1921 to process cotton linters) in West Virginia in 1937.

* Originally VGF's plant in the UK, and known as the British Glanzstoff Manufacturing Co., it was bought by Courtauld in 1917, with the encouragement of a UK government suspicious of 'enemy ownership'.

Despite the depression in world economies during the 1930s, the expansion of rayon staple production in Germany, Italy, Japan, Britain and the USA was phenomenal, especially after 1935. Over the 10-year period, sales of staple went from 3000 to 585 000 tonnes per year, overtaking 40 years development of filament (including acetate) which reached 542 000 tonnes per year in 1940.*

9.3 From speciality to commodity

If the development of staple fibre represented the ultimate move down-market for an artificial silk producer, there were many other changes needed to maintain the growth of filament sales. Costs had to be reduced and qualities of yarn had to be developed to suit a broader range of markets.

Cost reductions came initially from the economies of increasing production scale but by the late 1920s further progress was driven by falling market prices and technical solutions were required. The original labour-intensive hank-washing process was still being used to clean up the acid 'cakes' of yarn from the Topham boxes, but automatic 'cake-washing' machines were under development and were brought into use slowly† in the early 1930s. Process chemicals, especially the washing liquors, were routinely discarded, but as improved acid and caustic soda recovery processes were developed they were gradually adopted.

As the end-uses expanded beyond silk replacement, the harsh metallic lustre of the yarn proved disadvantageous and dull or matt fibres had to be developed. A French approach to the problem, patented in 1921[2] was to trap air in the yarn to give it a bulky wool-like texture and appearance. Dull yarn resulted naturally from the gas bubbles in the yarn, but the process was inherently harder to control than the bubble-free process. 'Oil-dulling' was invented by Walter Glover[3] of Courtauld who in 1926 patented the addition of petroleum jelly to the viscose. This development allowed Courtauld to establish the 'Dulesco' brand in hosiery where its opaque appearance met the fashion of the day, and its self-lubricating nature proved ideal for the knitting process. The American, James Singmaster developed an improved dulling method using titanium dioxide[4] in 1929. Courtauld eventually licensed the Singmaster patents from American Tubize, first in the USA (1936) and then in Europe where they had earlier regarded the Singmaster patent as invalid. (There is no record of how the abrasive titania-containing yarns fared on the knitting machines.)

* Source: CIRFS, Brussels.

† 40% of Courtauld's output in 1935 was cake washed, and this had risen to 70% by 1939.

The expansion of both filament and staple viscose rayon production, leading as it did to lower prices and a much broader range of textile applications, did much to improve the range of clothing available to working men and women. In retrospect it is unlikely that natural fibres alone could have met the demands created by increasing populations.

9.4 Industrial yarns

Courtauld's early 1920s attempts to use viscose yarn as a substitute for cotton reinforcement in Dunlop's pneumatic tyres failed because the strength of the filament yarns could not match the strength of spun cotton. Much research and development on the strength problem followed in most viscose companies, the first commercial fruit being the process,* developed in the late 1920s by Lilienfeld. VGF and a new British company The Nuera Artficial Silk Company, licensed it and set out to manufacture tyre yarns (1925). Courtauld acquired the Lilienfeld rights from Nuera in 1927 and commenced small-scale production both in the UK, and, as part of a collaborative venture with VGF, in a new factory in Cologne. They also sold the rights, in 1934, to their US subsidiary, AVC, as a defence against a new viscose tyre yarn emerging from DuPont.

However by then, Courtauld UK had dropped the Lilienfeld approach in favour of what was to become the most enduring solution to the strength problem. This came from the work of John Givens and Leslie Rose[5] who patented the use of hot water or steam to heat the incompletely regenerated yarn while stretching it. This gave strong yarns with much better resistance to abrasion than the Lilienfeld yarns. Cotton shortages and the military importance of tyres accelerated the pace of this line of development, and in 1939, about 20 years after the tyre yarn market had first been targeted by Courtauld research, large scale production commenced in the new factory at Preston in Lancashire. By 1945, 5000 tonnes per year of tyre yarn were being produced by Courtauld's alone. Ten years later this had increased by a factor of five. By the 1960s a significant part of the world's production of continuous filament viscose yarn went for tyre cord and other industrial applications (see Fig. 9.2).

The tyre yarn manufacturing process proved capable of improvement on several occasions as the demands of the tyre industry increased. The process which had first produced yarns with strengths above 2.5 g den^{-1} in the late 1930s was, by the end of the 1960s capable of producing high modulus industrial yarn at 10 g den^{-1}.

* Patented by Lilienfeld in June 1925. The patent revealed a way to make 5 g $denier^{-1}$ viscose fibres by spinning into strong sulphuric acid (~70%) and stretching at high tension before washing.

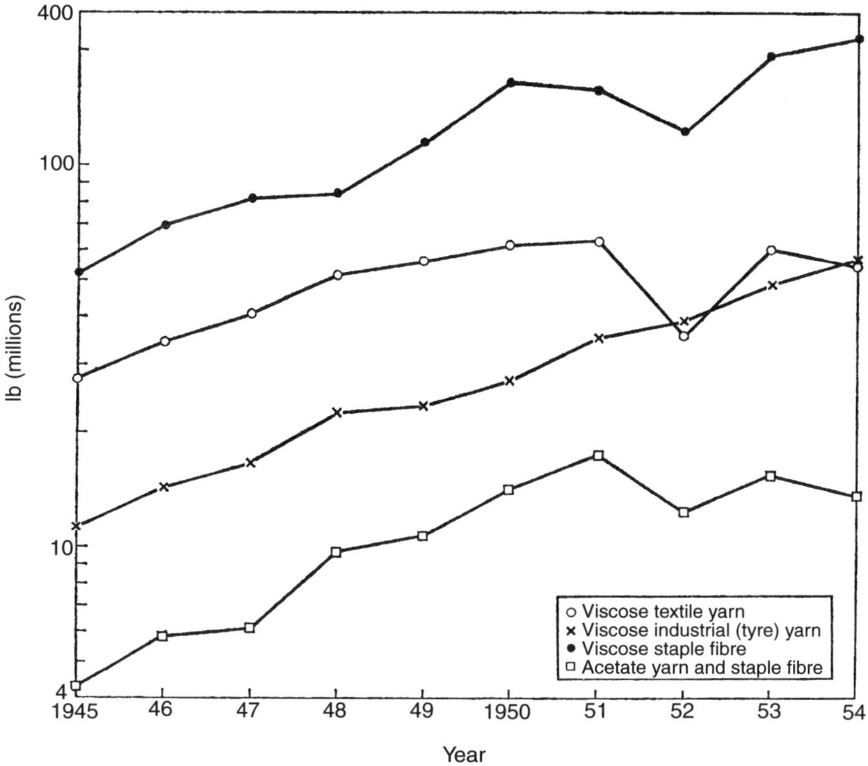

9.2 Courtauld's UK production 1945–1954: The growing importance of staple and tyre yarn developments. (Reproduced from *Courtaulds: An Economic and Social History,* Clarendon Press, 1969.)

After the development of hot stretching, further strength gains were achieved by adding amines to the spinbath. In 1947, Norman Cox of DuPont patented[6] the use of monoamines that have between 4 and 10 carbon atoms as 'modifiers', allowing the production of stronger yarns by permitting the spinning of unripe viscose. Extensive research on the effects of modifiers followed, and a further doubling of strength came from the use of mixed modifiers added to the viscose (developed by Mitchell *et al.*[7–9] for Rayonier). (see Fig. 9.3)

The development of continuous spinning and washing processes allowed the more economical production of yarns both for apparel and industrial use. These compact machines allowed the dried and finished yarn to be wound on bobbins within minutes of spinning. For industrial yarns they allowed higher tenacities to be achieved partly through their ability to strain dry the yarn. For apparel yarns, their inability to produce a yarn completely free of strain (and hence shrinkage potential) meant that the old cake

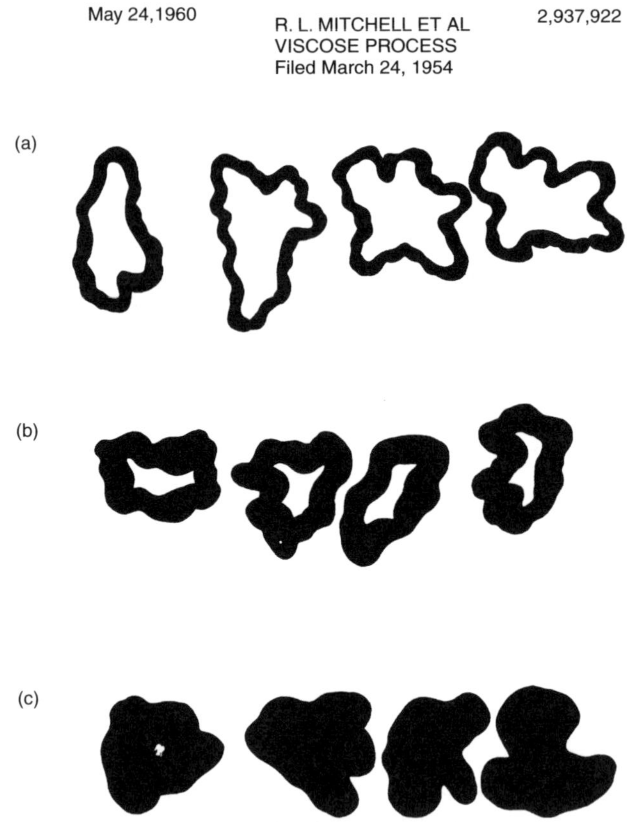

9.3 Development of stronger all-skin tyre yarns with modifiers: (a) textile yarn; (b) an early tyre yarn; (c) all-skin tyre yarn made by adding dimethylamine or dimethylformamide to the viscose and/or spinbath.[7]

process was often preferred for the highest quality materials. While the early continuous processes simply formed a 'warp' of all the yarns leaving the top godets of the spinning machines and then washed and dried this warp over rollers or drums, the invention of the self-advancing reel by the Industrial Rayon Corporation of America (c. 1937: see Fig. 9.4(a)–(c) and Fig. 9.5) provided the most elegant and compact of continuous processes. The reel design was licensed by Courtauld and others and machines based on it produced most of the world's viscose tyre yarn from the 1950s onwards.

The demand for such elegant cellulosics was however diminishing as the new synthetic industrial yarns came down in price. Furthermore the drive for improved modulus and tenacity had highlighted what appeared to be a fundamental weakness of regenerated cellulose technology: the more

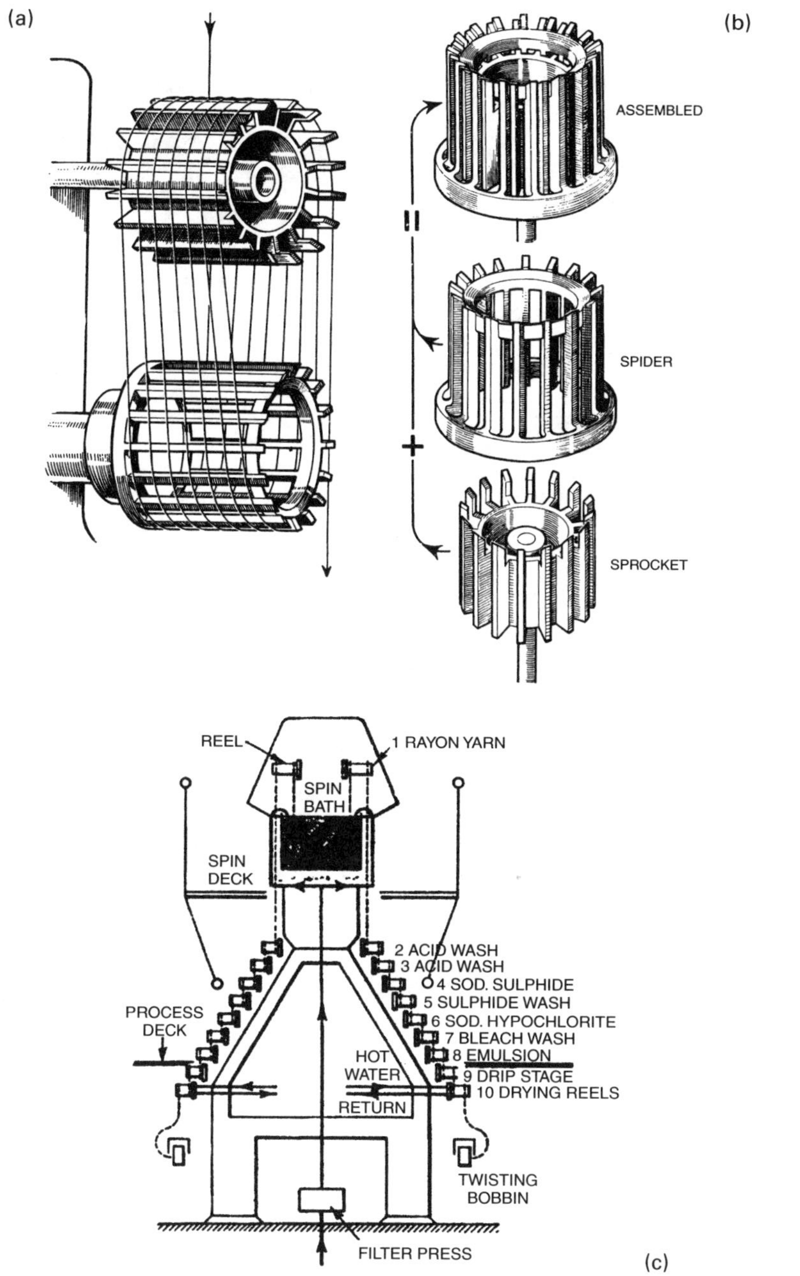

9.4 The development of a continuous spinning process with self-advancing reels to convey the yarn at high speed through the washing liquors was necessary to make the improved tyre yarns. (a) and (b) illustrate how the self-advancing reel is constructed. (a) Two cages (a sprocket above and a spider below) set eccentrically advance the yarn. (b) The sprocket fits into the spider on an eccentric axis to form the complete thread advancing reel. (c) A machine cross-section diagram illustrating the washing stages each taking place on a single self-advancing reel. (Reproduced by courtesy of Courtaulds.)

9.5 Viscose tyre yarn spinning-continuous process on self-advancing reels. (Reproduced by courtesy of Courtaulds.)

oriented and crystalline the cellulose became, the more the fibres fibrillated. Fibrillation brought with it poor lateral strength and hence poor abrasion resistance.

9.5 Modified staple fibres

9.5.1 For wool markets

If the filament process transferred easily to the new staple machines to make a product that could be used in blend with cotton without too much loss of cotton-like character, the same was not the case in wool markets.

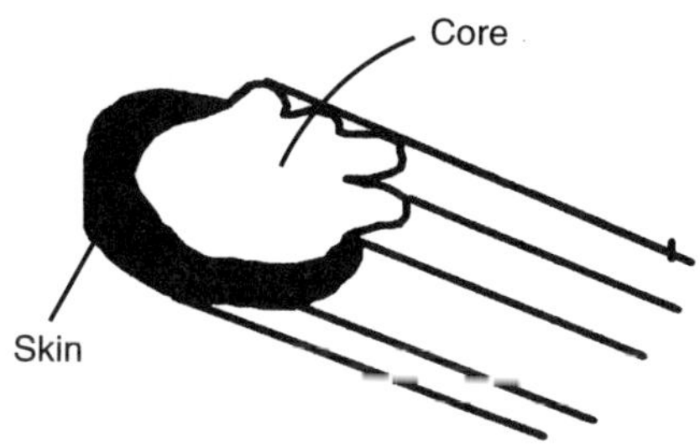

9.6 Chemical crimp arises from the core bursting through the skin.

For success in wool outlets, where altogether bulkier, warmer fabrics were expected, coarser crimped staple fibres had to be developed.

Acetate rayon or the new synthetic fibres were easily crimped by mechanical devices that formed them into waves or spirals and set them into the new shape using heat. Viscose on the other hand had to be regenerated in such a way that the filaments became effectively bicomponent, one half having a more oriented structure and therefore shrinking more than the other half in subsequent washing and drying. This was achieved by using viscose and spinbath combinations that caused the rapid formation of a thick skin, which shrunk so fast that the filament burst open exposing fresh liquid viscose to the spinbath just before the filaments emerged onto the godet (Fig. 9.6). The difference in structure of the early and late fixed halves of the fibre could be amplified in stretching and as soon as the bundle of filaments were free to separate after cutting and sluicing (see Chapter 3) the wavy form developed and was finally set in place by the hot acid treatments on the wash machine.

Even when crimp is fully developed it is easy to pull out (low energy) and difficult to translate into noticeably bulkier woven and knitted fabrics. It does, however, improve the absorbency and the cohesion of the staple (important in spun-yarn and nonwoven making) and gives a subtly different texture to woven fabrics.

While crimped viscose fibres could be made at any denier, the finer crimped fibres were easily straightened in cotton-type yarn manufacture and they added little to the performance of fabrics. However, coarser fibres could be made with more durable crimp and these found applications as wool diluents in carpets, especially in the new tufted carpets that emerged in the late 1950s.

By the mid-1960s staple fibre use in dress fabrics was declining but crimped fibre for tufted carpets had become the single largest application for the cellulosic fibre. It was significantly cheaper and had a better texture than the new acrylic and nylon carpet fibres but was not as hard wearing. Some improvements in durability were achieved by applying tyre yarn process conditions to the staple machines, but the chemical conditions required for the highest durability were incompatible with crimp generation and vice-versa. As nylon, especially the bulked continuous filament version came down in price, the cellulosic fibre found it hard to compete.

9.5.2 For cotton replacement

To increase the marketability of viscose fibres in cotton outlets, two distinct lines of development were undertaken. These were to minimise the two most obvious deficiencies of viscose fibres when compared with the chemically identical cotton fibre in woven fabric form. Viscose fabrics were always leaner, limper, more prone to shrinkage and had a sleazier wet texture than identically constructed cotton fabrics.

The first approach was to boost the strength, modulus and stiffness of the fibre so that its durability and shrinkage could closely match that of cotton, and the second was to alter the shape of the fibre to give bulkier, more cotton-like textures than could be achieved by the chemical crimping process mentioned above. In an ideal world it would have been possible to combine both of these approaches in a single fibre with the strength and bulk properties of cotton. In reality, the chemical constraints of the viscose process means that the simultaneous achievement of high bulk and high modulus has yet to be achieved even in the laboratory.

High wet modulus (HWM) rayons of two types were developed. *Modal fibres*, which were staple versions of the tougher tyre-yarn fibres, were introduced for use in industrial textiles, and for blending with the rapidly growing synthetics. This class was later subdivided into the low, intermediate and high strength variants (LWM, IWM and HWM) as producers tuned the process to market niches. Polynosic fibres with even better wet stability and higher wet modulus were introduced to blend with and substitute for the better grades of cotton.

In fact, the polynosic fibres were developed first, and used radically different process technology from other forms of viscose. Whereas all the developments in viscose to this point appear as a more or less continuous evolution from the first use of the Müller spinbath, the polynosic process developed by Tachikawa[10] in 1952 was revolutionary. He abandoned the zinc and modifier route to higher strength in favour of a very short immersion in a cold low-acid low-salt zinc-free spinbath which did little more than coagulate the viscose. To make the viscose spinnable in such a bath and to get the desired strong high modulus fibres, he had to use cellulose with a high degree of polymerisation (minimal mercerising of the alkali-cellulose), xanthated to a high degree (in fact with an excess of carbon disulphide) and dissolved at an almost uneconomically low concentration in unusually dilute alkali to give a very viscous solution.

His process allowed filaments of cellulose xanthate to be drawn from the bath and stretched to three times their spun length before being regenerated in hot dilute acid. The high stretch oriented the cellulose molecules to a very high degree, giving the resulting fibres a high dry strength, an unusually high wet-to-dry strength ratio, a very high modulus and a characteristically high resistance to caustic soda. This latter point was regarded as important to allow the viscose fibre to be blended with cotton prior to the mercerising (caustic) treatment, regular viscose being almost completely destroyed by this treatment.

Polynosic staple fibres were produced in Japan, England, France and the USA during the 1960s but failed to gain the hoped-for share of the cotton-spinning market. One problem was the fact that they were more expensive than standard viscose, another related to the fact that being highly oriented, they fibrillated on wet-processing and could not easily be dyed to clean bright colours. Yet another problem arose from the almost perfectly circular cross-sectional shape: they made even leaner yarns than irregularly shaped standard viscose fibres. All producers therefore sought ways of reducing the orientation and perhaps surprisingly, with the exception of the Japanese, reverted to the older staple version of the high zinc tyre yarn process to make the above mentioned modal range of fibres.

Toyobo's 'Tufcel' provided an excellent example of how a modern polynosic fibre process, probably the most difficult viscose process to run efficiently, operated.[11] On-line process control allows only four persons per shift to make 10000 tonnes per year of a variety of special fibres including flame retardant, deep-dyeing (two types), activated carbon fibre, and super fine (0.5 denier). Alk-cell and mixing-soda quality are maintained by pressings soda centrifugation, filtration and dialysis to remove 90% of the hemi-cellulose. (Ion exchange membranes are used to give 50 times the life and twice the efficiency of the old dialyser bags used in tyre yarn production.) Dissolution of the 500 DP xanthate is augmented by crumb grinders on the churn outlets and by in-line 'homomixers', which together reduce the dissolution time from three hours to one. Spinnerets for the finer deniers have 40 μm holes, and these are protected by automatic backflush filters removing gels down to 15 μm diameter.

Avtex went further than most in trying to make cotton-like modal fibres from a variant of the tyre yarn process. Using a process developed by ITT Rayonier, they attempted to address the leanness problem, and made a lobed-section, crimped high wet modulus fibre. Rayonier also licensed it to Saeteri, Sniace, Sniafi and Snia Viscosa. Only Avtex introduced the process commercially as Avril Prima® and for its launch ITT Rayonier mounted what was probably the first and last TV advertising campaign in support of a rayon fibre. Avtex also developed a novel way of modifying the cross-section of their standard 'Avril' modal fibre by deforming the incompletely regenerated fibres on the godet. The simple mechanical forces of pressure and shear allowed the production of fibres with a

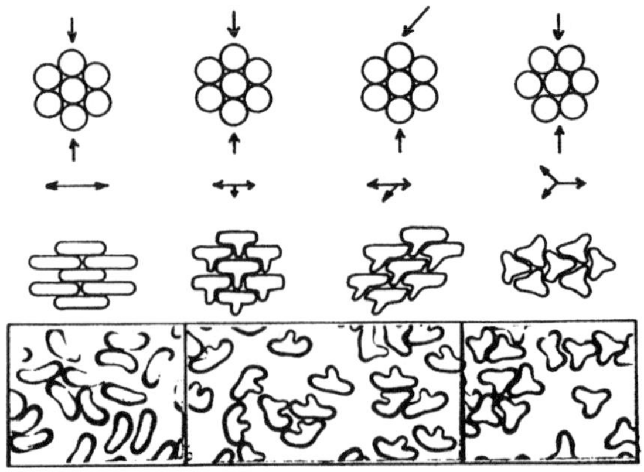

9.7 Bulky high modulus fibres are very hard to make using viscose technology, but this approach deformed the fibres while still in the gel state. Diagram shows flat and three-lobed rayon cross-sections. (Reproduced from chapter by J Dyer and C Daul in *Fibre Chemistry*, ed. M Lewin and E M Pearce, Marcel Dekker, 1985, Volume IV *Handbook of Fibre Science and Technology*.)

roughly trilobal shape and reasonably high modulus (see Fig. 9.7). When dulled with titania pigment, these fibres were said to give a handle and appearance very similar to cotton. They did not, however, have the same high modulus.

As mentioned above, the really high moduli were appropriate for cotton blends, and to start with, the intermediate wet-modulus rayon's targeted polyester blends. However, with the exception of some industrial end-uses such as sewing threads and hose reinforcement, the intermediate wet modulus rayons were easily replaced by cheaper regular rayon simply by increasing the proportion of polyester in the blend.

Courtaulds therefore approached the problem of attacking the rapidly growing polyester/cotton market by getting the bulk right and sacrificing the high modulus. The resulting inflated viscose fibres, introduced in the late 1970s were in fact a spin-off of earlier attempts to replace cotton in a completely different market sector, speciality papermaking. Inflating the fibre allows simultaneous increases of bulk and absorbency but at the expense of fibre strength and processability. Flat fibres are self-bonding and only usable in wet lay or papermaking processes (see Table 9.2).

In the 1960s and 1970s, viscose producers in Europe, the USA and Japan made various attempts to make rayon behave more like cotton in speciality papers. Avtex (USA) and Courtaulds (UK) commercialised inflated/collapsed papermaking fibres under the codenames 'RD101' (Avtex) and 'PM1' (Courtaulds). Kurashiki Rayon (Japan) also entered the market.

Table 9.2 Inflated viscose rayons and their properties

Cross section	Name	Water imbibition (%)	Dry tenacity (g den^{-1})	Dry extension (%)	Comments
	Standard rayon	90–100	1.9–2.5	18–30	Uninflated rayon for comparison
	—	110	—	—	Not commercial
	Viloft	120–140	2.2–2.5	13–15	High bulk cotton-like textile fibre
	PM2 fibre	150–160	1.8–2.0	20–25	Self-bonding fibre for wet laid nonwovens
	PM1 fibre	160–180	1.4–1.8	20–25	Self-bonding fibre for quality papers
	Super inflated fibre	190–350	1.0–1.4	30–50	Highly absorbent, opaque and bulky fibre for nonwovens

Courtaulds fibre met the technical specifications to replace cotton in banknote papers and the rayon technology enabled some useful new security features to be incorporated. However, also for security reasons, the currency manufacturers required total exclusivity for such a fibre and because Courtaulds had identified a major new market for the same fibre in disposable diapers, restriction of sales of this fibre type to the smaller currency market was not acceptable.

Inflated fibres (Fig. 9.8) were made by injecting a gas or gas-forming compound such as sodium carbonate into the viscose.[12] When the regeneration conditions had been adjusted to form an impervious skin on the fibre as it emerged from the spinneret, the gas was retained in the forming fibre and the resulting bubbles would coalesce into a tube. To make fibres suitable for papermaking, the tube had to be sufficiently thin walled to allow it to collapse across a diameter in washing and drying. The resulting smooth-surfaced flat filaments were highly self-bonding simply because they were deformable enough to come into intimate contact with one another as they dried, and when in contact hydrogen bonding locked them together. As with

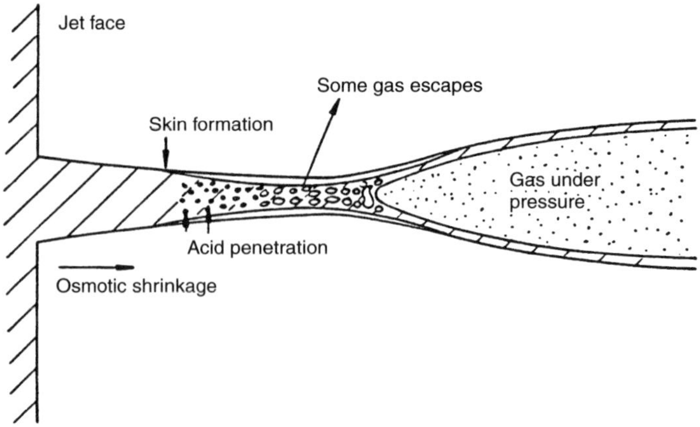

9.8 Inflation levels can be controlled by adjusting the stationary bubble diameter, which in turn depends on the rates of gas evolution and gas escape through the forming filament.

cotton and woodpulp, wet strength could only be achieved by adding resins such as melamine–formaldehyde, but without these resins the inflated rayon papers had an even greater tendency to disintegrate in water.

It was this 'flushability' that attracted the attention of the embryonic disposable diaper makers at a time when the consumers were reluctant to dispose of soiled diapers in the dustbin. The flat fibres could be made into an absorbent cover tissue that could be reinforced to just the right degree to allow coherence in use yet disintegrate in the much higher turbulence of the toilet bowl.

Courtaulds (UK) were alone in commercialising two other inflated fibres targeting cotton markets. The first, an uncollapsed tube (see Fig. 9.9), was not self-bonding, but had the bulk and handle of cotton. While the initial justification for developing and launching this fibre was indeed cotton replacement, the marketing and pricing policy positioned it as a premium product with unique aesthetics for high fashion applications. As with all fashion applications, success can be short-lived and dependent on the size and creativity of the marketing effort. It ultimately found worthwhile and continuing application in thermal underwear until the only factory capable of making it, Greenfield Works in the UK, was closed in 1984. Courtaulds then developed a solid flat cross-section fibre (see Fig. 9.10) that had similar bulk and could be sold under the same brandname ('Viloft').

The other inflated fibre arose from the uneven collapse of a very thin-walled tubular fibre and had a very high absorbency. This fibre briefly enjoyed large and profitable sales as a replacement for cotton in tampons in the 1980s under the brandname 'SI Fibre' (see Fig. 9.11). This fibre was

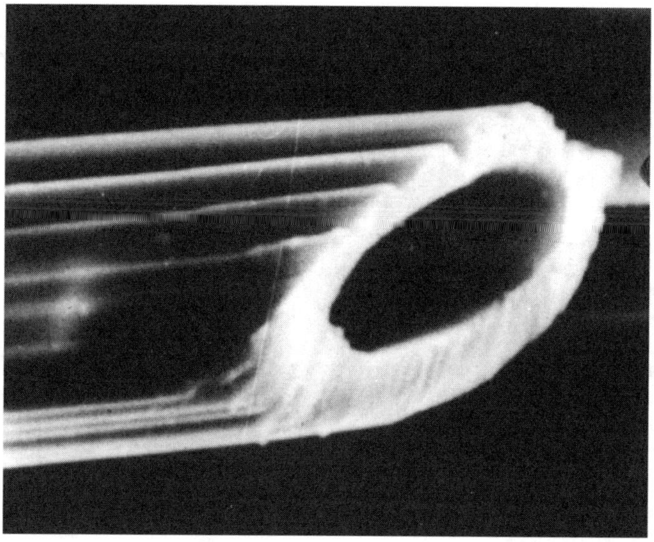

9.9 Courtaulds Viloft®: the first commercial inflated rayon staple. (Reproduced by courtesy of Acordis.)

also adversely affected by the decision to close the versatile Greenfield Works, and after a short period of production in France and Canada, it too was replaced by a solid Y-shaped fibre, 'Galaxy' (see Fig. 9.12). Galaxy, made on ordinary staple machines from Y-shaped spinneret holes of a size impossible to make a decade earlier, provided many of the unique attributes of the collapsed tube fibre while being easier to produce and process. In Japan, the Daiwabo company still manufactures a range of inflated fibres for special paper and hygienic applications.

At the time of writing, both Courtaulds' unusually shaped cotton replacement fibres remain in production. The development of applications for these and their inflated precursors allowed Courtaulds in particular to begin to appreciate the merits of non-traditional, non-textile markets for their rayon fibres. Collectively known as the Unspun or Nonwoven markets for fibre, these markets, unheard of in the first half of the century, became increasingly large users of rayon fibre during its third quarter. Attempts by regenerated cellulose fibre producers to develop further products to suit this new range of applications was a natural consequence.

9.5.3 Flame-retardant rayons

Rayon, like cotton, will burn if ignited, and this characteristic has prevented the fibre being used alone and untreated in nightwear and some workwear applications.

9.10 Solid flat fibres are easier to make and give similar bulk to the hollow Viloft®. (Reproduced by courtesy of Acordis.)

Flame retardancy can be obtained by adding flame-retardant (FR) chemicals to make up about 20% of the fibre weight.[13] The first commercial products used tris(2,3-dibromo-1-propyl) phosphate, but the P—O bonds made these susceptible to hydrolysis by strong alkali, and the bromine increased the rate of photodegradation. The chemical was found to cause cancer in laboratory tests and, in the late 1970s, fell from favour.[14] Propoxyphosphazine (Ethyl Corp) retardants were later used in Avtex's 'PFR' fibre, and a bis(5,5-dimethyl-2-thiono-1,3,2-dioxaphosphorinanyl) oxide powder (Sandoz) was the basis of later European FR fibre developments. Alloys with inorganic salts such as silicates or aluminates are possible, the salts

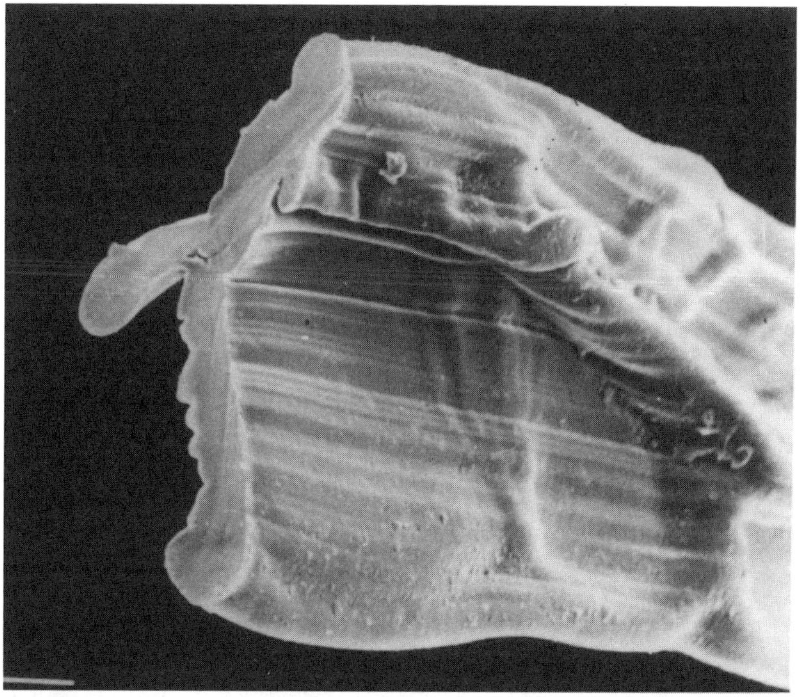

(a)

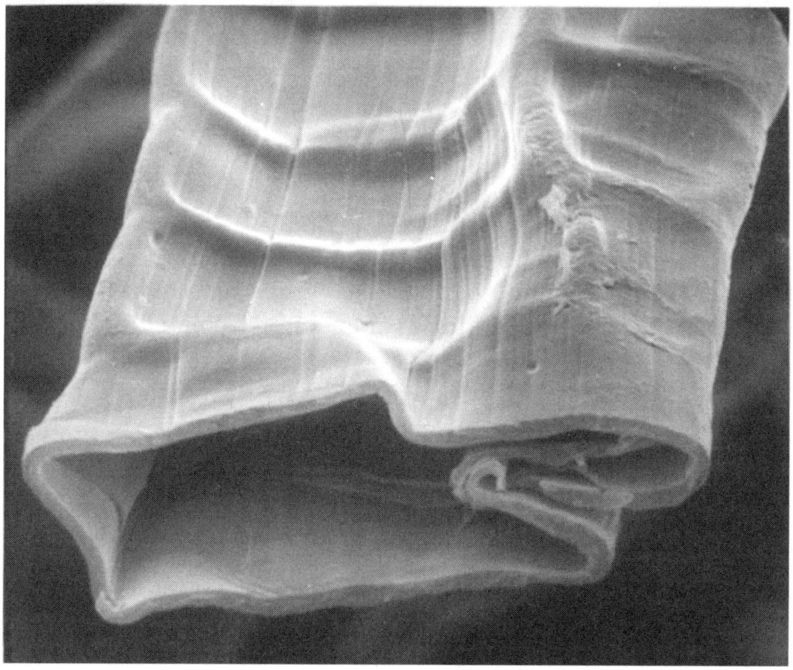

(b)

9.11 (a) Dry 'SI Fibre'®: superinflated fibres give maximum bulk, absorbency and opacity but at the expense of processability.
(b) Wet 'SI Fibre'®: some of the high absorbency of the superinflated fibres arises from the collapsed tube opening up on wetting.
(Reproduced by courtesy of Acordis.)

9.12 Galaxy®: the same shape and bulk as dry SI fibre® but easier to process and lower water imbibition.

being converted to fibrous polyacids when the cellulose is burnt off.[15] This latter approach seems to be the basis of the 'Visil' flame-retardant fibre introduced by the Finnish viscose fibre manufacturer, Kemira Oy Saeteri.[16]

9.5.4 Alloy rayons

It is possible to produce a wide variety of different effects by adding materials to the viscose dope. The resulting fibres become mixtures or *alloys* of cellulose and the other material. The two most important types of alloy arise when superabsorbent or flame retardant fibres are made.

American Enka[17,18] and Avtex[19] both produced superabsorbent alloy rayons by adding sodium polyacrylate, or copolymers of acrylic and methacrylic acids, or sodium carboxymethyl cellulose to the viscose. The resulting alloys contained up to about 20% of the water-soluble polymer giving water imbibitions up to double those of the unalloyed rayons. They performed particularly well in tampons (see below) where the presence of the slippery polymer at the fibre surface encouraged wet-expansion of the compressed plug. Their use in this, the only real market which developed, declined after the 'toxic shock syndrome' outbreak[20,21] in the early 1980s. Other polymers which have formed the basis of absorbent alloys are starch,[22] sodium alginate,[23] polyethylene oxide,[24] polyvinyl pyrrolidone[25] and sodium polyacrylamido-2-methyl-2-propane sulphonic acid.[26]

Alloys of cellulose with up to 50% of synthetic polymers (polyethylene, polyvinylchloride, polystyrene, polytetrafluoroethylene), have also been made, but have never found commercial applications.

In fact any material which can survive the chemistry of the viscose process and can be obtained in particle sizes of less than 5 μm can be alloyed with viscose. An X-ray detectable yarn loaded with barium sulphate (30% cellulose/70% barium sulphate) was an extreme but profitable example of the type, developed and commercialised for medical applications.

9.6 Nonwoven applications

By the end of the 1960s the apparel and home furnishings markets for rayon were highly cyclical and gradually declining, rayon had relatively easily achieved a dominant position in the fast growing nonwovens industry. Since the mid-1970s significantly more effort has been spent on developing new applications for regenerated cellulose fibres in nonwovens than in conventional textiles. A more detailed look at applications development in this sector is therefore appropriate.

9.6.1 Dry-laid nonwovens

The first stage of traditional cotton or woollen textile manufacture involved opening and carding fibres to parallelise them prior to condensing them into slivers for drawing and spinning into yarns. The early cotton cards produced thin, 40 inch wide webs of fibres with just enough coherence to be handled by the sliver-making machinery.

In the late 1930s, the first lightweight* nonwoven fabrics were made by taking this card web and gluing the fibres together with natural latex emulsions or with phenol– or urea–formaldehyde resins. Cotton proved less satisfactory than rayon in these early absorbent disposables because unbleached cotton had a waxy hydrophobic surface layer, and absorbent cotton, made by bleaching, was both too expensive and too difficult to card

*Wool has been felted into what could be regarded as the first nonwovens for millennia. Heavyweight needlepunched 'felts' made from jute and sisal have been made since the 1890s in both the UK and USA. Wet-laid nonwovens from long natural cellulosic fibres were first produced by Dexter Corp. in 1934 in the USA.

into thin even webs. Rayon, especially second-grade staple fibre that would make uneven yarns or unevenly dyed fabrics thus became the fibre of choice for the embryonic nonwoven industry. The so-called 'dry-laid, latex-bonded rayon' nonwovens, were of course far too weak to replace conventional textiles in all but the flimsiest of applications, but they were stronger, more durable and more textile-like than many papers. As cheap new materials with unusual properties they allowed the development of a new range of 'disposable' consumer products for an apparent gap between markets normally served by textiles and those served by paper.

Single-use nappies, sanitary towels, incontinence pads, wipes, and medical swabs were among the early products made possible by the new disposable latex-bonded rayon nonwovens. For wiping applications the absorbent latex-bonded rayon formed the entire product (e.g 'J-Cloth' from Johnson and Johnson in the UK and 'Handiwipes' from Colgate in the USA). As the market grew and consumers demanded higher performance, these disposable wipes could be made to survive 10 washes, further increasing their cost effectiveness compared with paper, while maintaining some hygiene advantages over dishcloths. Interlinings, those out-of-sight textiles used to give stiffness and body to garments could also be made from the new nonwovens, but other 'durable' products required different system of bonding.

Needlepunching and stitchbonding allowed thicker card webs to be consolidated without recourse to latex and such products had a significantly better softness/strength balance that the albeit much lighter latex-bonded varieties. These nonwovens required coarser rayon fibres but achieved significant success in a variety of home-furnishing applications, especially curtains and blinds (for stitchbonded), blankets (for needlepunched) and floor/wall coverings (i.e. carpet tiles, also for needlepunched).

Unbonded absorbent rayon wadding became a common diluent of, or substitute for, the more expensive bleached cotton in sanitary and medical applications. In these and other sensitive wadding applications such as tampons, the rayon fibres had to be produced to meet the purity requirements laid down by the various national pharmacopoeias. Nevertheless some early successes in these more critical applications led to the development of new varieties of rayon staple having better whiteness retention on sterilising and higher absorbency (see above).

Tampons had been developed in the 1930s and by the early 1960s were an established but still growing market for bleached cotton, especially the shorter and hence cheaper varieties based on linters or comber waste. Replacing cotton in such a conservative yet technically sophisticated end-use proved difficult, but occasional shortages of the desired grades of cotton led to 'windows of opportunity' that allowed special versions of crimped viscose to break in. Once established, despite offering no perceivable benefits for the user over cotton, the new rayon fibres were preferred by the tampon producers for their consistency in both processability and price. For the fibre producer, the tampon application offered a large market for special fibres with a modest price premium over regular rayon. At the risk of stating the obvious, the market also required its fibres to be inherently absorbent, biodegradable, and to have a long history of safe use in contact with skin; in other words it was relatively invulnerable to competition from the synthetic fibres.

There were, of course, times when high demand for regular rayon in the more cyclical textile markets made the tampon and other nonwoven markets appear unattractive but over the longer term the very non-cyclicality and recession-proof nature of the nonwoven sector meant most rayon producers were willing to commit a significant proportion of their output to it.

All the US staple producers and several European and Japanese producers also committed many years of serious development effort to producing new fibres specifically for the market. In addition to the inflated and modified cross-section fibres mentioned above, highly absorbent rayons were manufactured by adding water-soluble polymers to the viscose dope prior to spinning. Two varieties were commercialised in tampons, one using sodium polyacrylate as the additive and the other using carboxymethyl cellulose, both at around the 15% level on fibre. These inclusions* increased the fibre water imbibition by about 50%, but perhaps more importantly they made the fibres more resilient and slippery when wet, a feature that allowed good recovery (and hence absorbency) from the highly compressed dry state of the tampon.

9.6.2 Wet-laid nonwovens

Wet-laid nonwovens were manufactured using papermaking machinery adapted to make webs containing synthetic fibres and to bond them using either latex or (if the webs had a synthetic component) thermal means. Less versatile and often more costly to operate than the slower dry-laid route based on textile cards, wet-lay technology was also limited to fibres short enough to be handled in aqueous dispersion: in practise less than 2.5 cm long and typically less than 1 cm in length for 1.7 dtex fibres. Nevertheless rayon fibres, which are very similar to the wood and cotton fibres normally

*Before their use in these alloy rayon fibres, both carboxymethyl cellulose and sodium polyacrylate had, in their crosslinked powder form, been sold as superabsorbents for use in diapers and sanitary products. After they had proved satisfactory in tampons, processes were developed, especially in Courtaulds UK, to convert them into highly absorbent staple fibres.

used in papermaking proved well suited to the technique, provided they could be precision-cut to the short lengths needed.

Unlike the related speciality papermaking sector (see above) wet-laid nonwovens did not require fibres to be self-bonding and hence regular rayons simply cut to the right length were usable. Wet-lay technology was thus not only able to enter some of the markets already penetrated by dry-lay technology, but also to take rayon into new highly technical markets such as tea bags, vacuum cleaner bags, milk filters, and food casings (i.e. salami skins). Wet-laid nonwovens have, however, failed to live up to their early promise and totally failed to justify the late 1960s hype about their prospects in 'paper dresses'. The rayon producers have been blamed at least in part for the disappointing wet-laid market growth. Because during the crucial early years no rayon producer would commit full-scale manufacturing facilities to allow short-cut fibre to be produced at the same low price as staple-fibre for dry-laid nonwovens, the wet-laid industry was unable to compete in the large and fast growing coverstock* and wipes markets. The problem for the rayon producers was that precision short-cut fibre was hard to make on staple fibre machines and the wet-laid market was too small to allow several suppliers to install the large-scale machinery needed to bring the costs down.

9.6.3 Spun-laid nonwovens

Spun-laid nonwoven processes developed in the late 1950s by Freudenberg (Germany) and DuPont (USA) proved much more significant than wet-laid processes. These processes converted synthetic polymers directly into fabrics on a single large machine, the entire fibre and 'textile' manufacturing processes being compressed to the ultimate conceivable degree in both time and space. Polymer chips were fed into an extruder, fibres were spun vertically downwards, oriented in high-pressure airjets, randomly laid on a conveyor and the resulting web bonded and reeled up within seconds of the fibres being spun.

The relevance of a synthetic fibre nonwovens technology to this chapter probably needs some explanation. The direct relevance is that at the time of writing there are two producers of spun-laid regenerated cellulose nonwovens in the world, both in Japan, and the origins of these are dealt with below. The indirect relevance is that the inability of regenerated cellulose to maintain its leading position in the nonwovens market through the last quarter century of dramatic growth can be traced to a progressive deterio-

* Coverstock is the very lightweight nonwoven used to cover the absorbent wadding of disposable diapers, sanitary towels and incontinence pads.

ration in its cost/performance ratio. Producing nonwovens on-line at the rayon plant was a natural evolution from staple fibre production technology to improve the cost/performance of a cellulosic nonwoven.

9.6.3.1 Spun-laid regenerated cellulose nonwovens

Several cellulosic fibre producers did however attempt to improve this performance/cost ratio by making nonwovens themselves.

9.6.3.2 Viscose process based

In Europe Courtaulds, Rhone-Poulenc and Enka Glanzstoff AG (now, like Courtaulds, part of Acordis) developed spun-laid rayon nonwovens on a pilot scale in the late 1960s and early 1970s.

In Japan, Kanebo and Daiwabo researched similar techniques, Asahi worked with viscose and cuprammonium pilot lines, but Mitsubishi Rayon developed a process based on hydroxymethyl cellulose xanthate in which the webs were point bonded in a hot calender before regeneration. Also in Japan Kosabura Miura spun viscose vertically downwards from oscillating spinnerets onto a conveyor, and oversprayed the liquid filaments with acid. The Tachikawa Research Institute developed a polynosic viscose spun-laid process.

Courtaulds introduced the first spun-laid regenerated cellulose nonwoven at the first International Disposables Exhibition in the USA in 1971, but closed the development two years later for reasons associated with the undesirability of competing with its then rapidly growing rayon staple business in nonwovens, and the small size of the available market at that time.

This early cellulosic filament nonwoven was not a true spunbond. It was a wet-laid continuous filament web made from a fully washed viscose tow, laid on-line on an inclined wire former, dried and print-bonded with latex. It nevertheless made an excellent wiping cloth, demonstrably outperforming the market leader (at that time 'J-Cloth' from Johnson and Johnson's Chicopee Division.) However, also at that time the European market for nonwoven wipes of all types was at best 10000 tonnes, and the minimum economic process scale-up would have produced 25% of this.

In 1974, Asahi introduced 'Bemliese', the cuprammonium version of Courtaulds' process (see Fig. 9.13). With cuprammonium the continuous filaments could be laid in the incompletely regenerated state to give a true spunbond of pure cellulose. Because the cupro-based process[27] was significantly more costly to operate than the Courtaulds' viscose-based approach, they targeted higher value medical nonwovens and made much capital in Japan of the fact that they dissolved cotton linters rather than wood pulp.

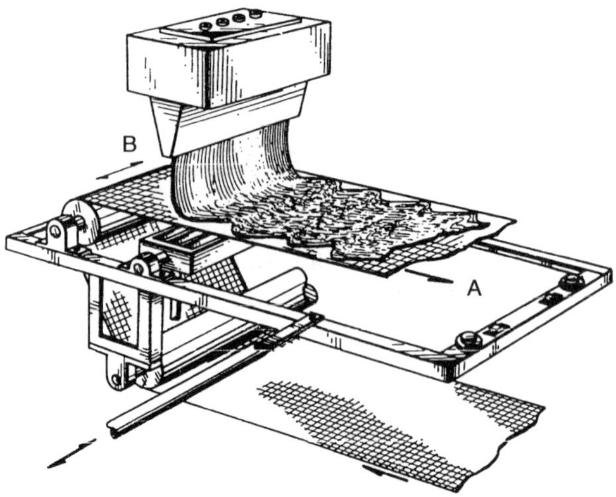

9.13 Asahi's Bemliese® cuprammonium rayon spun-laid process. Filaments are spun vertically downwards onto the wash conveyor and overfed to create lateral interactions prior to washing, bonding and drying as a nonwoven fabric.

This allowed them to describe the resulting nonwoven as made from cotton, just like the woven products they planned to replace.

In 1976, the Mitsubishi Rayon Company introduced nonwovens from their methyol cellulose process[28] as 'TCF' (Textiles Continuously Formed) (see Fig. 9.14). This was made on-line in a viscose plant, and while it looked and felt similar to the Bemliese fabrics it was in fact wet-laid from 10 mm to 15 mm fibres prior to washing and bonding using a hot calender. Today, both of these Japanese processes use the latest hydroentanglement bonding techniques to arrive at softer materials than true spunbonding allows.

Also in 1976, Courtaulds and their in-house nonwoven producer Bonded Fibre Fabrics (now the BFF Nonwovens division of Lamont Holdings) decided to reopen the development of spun-laid viscose. A second large pilot plant was built in the Coventry Research Laboratories, this time along lines more similar to the Asahi approach, with tow-laying preceding washing. This gave the benefit of higher web uniformity from the perfectly parallel filaments laid straight from the special spinning heads, and the ability to use different bonding methods, before or after washing, or both.

Unlike Asahi and Mitsubishi, the Courtaulds route involved either spinning regular fibres and viscose bonding them immediately after laying, or aperturing and hydroentangling the acid tow web using a very early in-house version of the hydroentanglement technology which began to grow

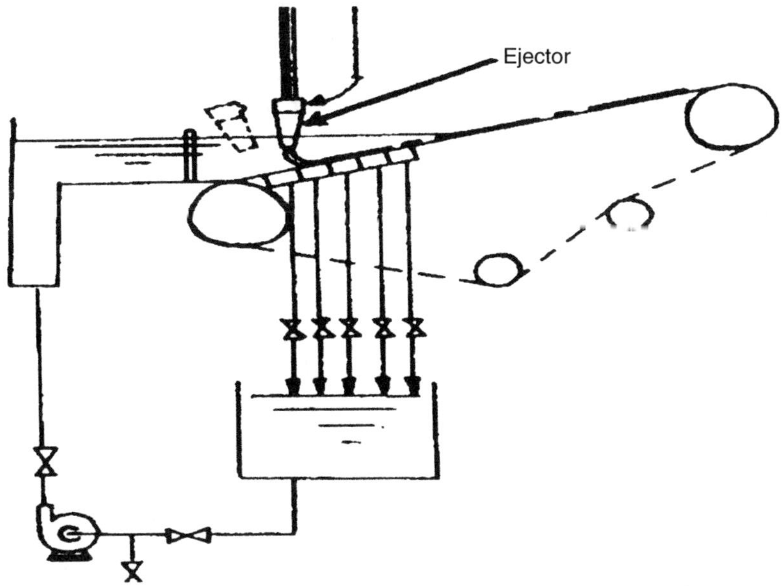

9.14 Mitsubishi Rayon's TCF® spun-laid process uses a conventional wet-lay former and introduces hydroxymethyl cellulose fibres 'ejected' straight from the spinning machine into the headbox.

so rapidly 10 years later. Surgical swabs and binder-free, lint-free wipes were among the main target markets. The development met its technical objectives but plans to install a large commercial line in a UK viscose plant were shelved.*

At Courtaulds the spun-laid viscose equipment had also produced calcium alginate nonwoven, partly to investigate medical applications of alginate and partly because alginate was an easy spinning proposition which allowed the design of the laying section to be finalised. By the time spun-laid viscose was shelved, these alginate spun-laid nonwovens were in demand as advanced wound dressings. The spun-laid pilot plant was therefore converted to alginate and became the foundation of the current Speciality Fibres business of Acordis. The reason for this apparent digression into alginate nonwovens is to note that their development led to an understanding of gel dressings for advanced wound care, which in turn led to the development of an advanced regenerated cellulose nonwoven made by

* In 1982 Courtaulds took the view that the new hydroentangled nonwoven process would allow the manufacture of all-cellulose nonwovens from regular staple fibre, and the new lyocell process would be much more amenable to spunbond manufacture, if ever the market grew enough to justify it.

carboxymethylation of viscose and latterly, lyocell fibres (see Chapter 6). However, one of the first commercially produced carboxymethyl cellulose nonwovens was made by Asahi, and the product, based on carboxymethylation of their Bemliese cuprammonium rayon spunbond was marketed as 'Super AB' superabsorbent nonwoven in the 1980s.

9.6.4 Lyocell nonwovens

Lyocell makes excellent nonwovens, especially in those processes that allow its superior aesthetics to shine through, like needlepunching and hydroentanglement. Its high strength is of little intrinsic value in disposables, but it enables the nonwoven producer to reduce basis weight while meeting strength targets. Its freedom from shrinkage and high wet stability allows higher area yields in hydroentanglement processes, and its high modulus prevents it from collapsing in the wet to the same extent as viscose rayon. Fibrillation, the development of surface microfibres on wet abrasion or in high-pressure entanglement, adds an additional dimension for the nonwoven developer. Unfortunately, while to date it has established itself in several profitable niches, its premium positioning as a fashion apparel fibre has so far prevented its use in mainstream disposables. The Appendix contains a full set of technical data on lyocell in nonwovens.

9.6.5 Spun-laid lyocell nonwovens?

Most fibre-forming polymers or polymer solutions can also be converted into continuous yarns, films, sponges or indeed nonwoven fabrics. Lyocell dope is no exception and many of the characteristics of the lyocell process make it a better basis for spun-laid nonwovens than the viscose process ever was. Technically speaking, the challenges are not great. Economically and commercially speaking, they are enormous.

In the nonwovens market the leading products are nearly always those with the lowest cost, and justification of spun-laid cellulosics on added-value alone has failed several times already. The ultimate in economy arises from inherently low cost raw materials converted on state of the art machinery at the largest possible scale. The nonwoven industry enjoys the economies of (say) polypropylene (PP) because PP is a by-product of the world's largest industry, energy. Viscose rayon, a premium product of the timber industry, requires the most costly grades of wood pulp. Lyocell is currently similar, but its simple production process has the so far unexplored potential to use cheaper pulps. It also has the potential to achieve very high levels of sales in textiles, and hence the economies of scale that may ultimately attract the major nonwoven converters.

9.6.6 From commodity to speciality

Prior to 1960, regenerated cellulosic fibres enjoyed 50 years of rapid expansion. Since then, synthetics have grown to dominate the market. Cotton, for centuries the most important of all fibres is taking second place to the combined weight of synthetics and viscose rayon. But rayon now appears relegated to little more than a niche in a global fibre market driven by the ready availability of cheap fossil fuels and the demand for commodity textiles and nonwovens.

Nonwoven production was founded on the ready availability of low-cost viscose rayon fibres and these continued to dominate the industry until the mid-1970s. Since then the reducing cost of synthetics, coupled with their easy conversion into binder-free spun-laid and melt-blown fabrics, caused a steady decline in rayon's nonwoven market share.

Is the relative decline in use of cellulose, both natural and regenerated, in nonwovens and textiles just another example of the last stage of the inevitable growth/maturity/decline life cycle of most markets? Or is there any suggestion that, on a longer time scale, the biopolymers will prove to be a serious rival to the synthetics?

This section reviews how rayon arrived at its current position in the nonwovens market, records the technology-based attempts to reverse the decline, and examines possible futures.

9.6.6.1 Current positioning of cellulosic fibres

Figure 9.15 illustrates how fibres based on fossil fuels have replaced fibres based on biopolymers over the 20th century according to the CIRFS statistics on world fibre usage in all markets.

Roughly half of the 45 million tonnes of fibre consumed annually in the world are now made from synthetic polymers. The only perturbations in an impressively smooth growth curve appear in the 1973–1974 period, the 1978–1984 period and the 1990–1991 period and are explained later.

On a more local level, fibre usage in the European nonwovens industry is illustrated in Figs. 9.16 and 9.17.

While the tonnage of viscose rayon sold has held remarkably constant for 30 or more years, rayon has not participated at all in the massive growth of the industry and its market share is now one tenth of the 1970 figure.

The enormous expansion of the synthetics in the 1960s and 1970s put the viscose rayon industry on the defensive. Despite having taken almost 25% of a 14 million tonne cellulosic fibre market without ever promoting rayon strongly against 'King Cotton', and despite having the potential to double this share, the viscose producers felt the future of manufactured fibres to be synthetic.

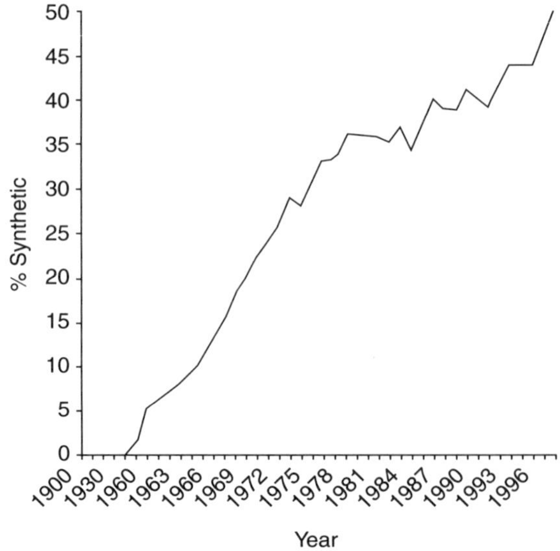

9.15 Synthetics versus polymers in the 20th century.

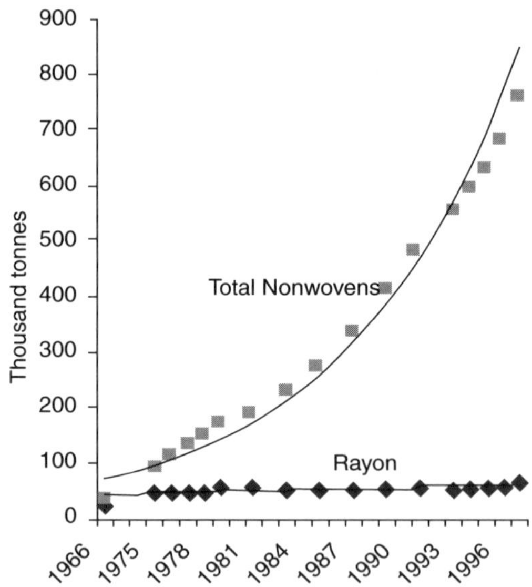

9.16 Rayon in European nonwovens (EDANA).

With hindsight, the viscose industry entered an end-game strategy at this time. Profits that had been substantial during the upside of the textile cycle, were spent on synthetic fibres or in diversification ventures rather than in marketing rayon against cotton or in building efficient new plants.

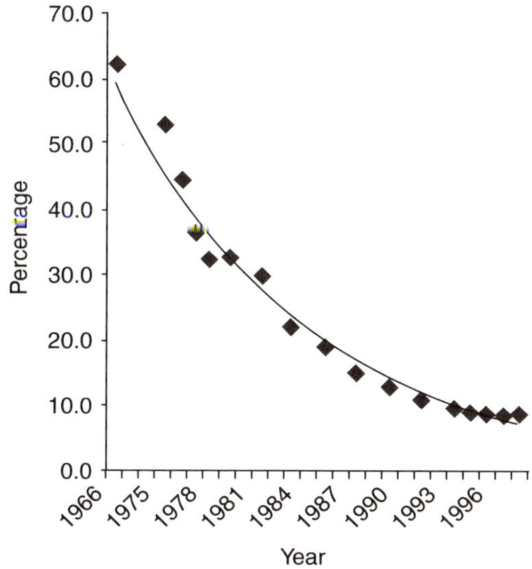

9.17 Rayon share of EDANA tonnage.

What rebuilding and expansion there was became largely confined to replicating existing production technology in parts of the world where cheap labour allowed more competitive manufacturing costs. A progressive tightening of the regulations governing the release of gaseous and liquid effluents from the viscose process compounded the problem and adversely affected the older plants. Effluent control projects and 'modernisation' took whatever funds were available. During the downside of the textile cycle, the higher cost viscose capacity was simply closed down. By the end of the 1970s nobody expected to see a new viscose plant being built anywhere.

In the absence of substantial reinvestment in new plants, the repositioning of viscose rayon became inevitable, and from 1985, the fibre was successfully transmogrified from a commodity to a premium-priced speciality fibre for the more lucrative niches in apparel and nonwoven markets. The reasons for the decline are obviously related to relative fibre price and performance, but a more detailed analysis is needed to decide if a simple extrapolation of the graphs correctly identifies the most likely future.

9.6.6.2 Price

The graph shown in Fig. 9.18 brings together fibre, oil and pulp prices from a variety of sources to throw some light on the competitive positioning of

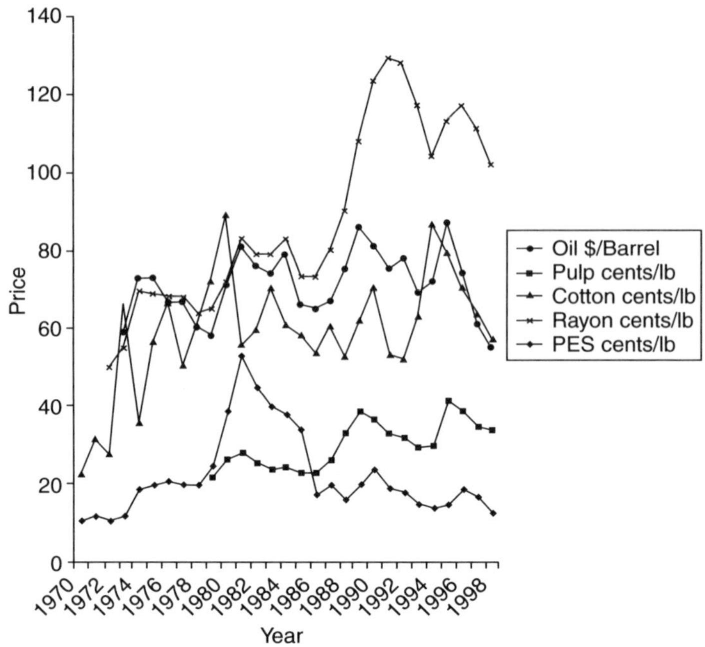

9.18 US price indications from various sources.

the key fibres in the period from 1970. They are not all on the same basis and the more readily available US figures are used in preference to less complete and more volatile European data.

They nevertheless illustrate the following:

- A doubling of oil price was caused by the oil embargo during the 1973 Yom Kippur War.
- The oil price controls during the Iranian revolution/Iran–Iraq war and aftermath (1978–1986) caused a massive disturbance in oil price.
- A return to 'normal' oil pricing was interrupted by the Gulf War in 1990.
- Cotton price increased from the 20 cents/lb to the 60 cents/lb region during the 1970s, opening the door to polyester.
- Ignoring the political perturbations in oil price, the pulp/oil price ratio is increasing.
- A quadrupling of oil price (1972–1982) hardly affected the relative price of polyester.
- Polyester and cotton prices appear to be linked.
- Rayon prices remained broadly in line with polyester and cotton until 1986 when they moved rapidly ahead of both the competitive fibres and of reasonable price expectations based on pulp prices.

- By 1990 the repositioning of rayon at prices 50–100% higher than cotton was complete.

9.6.6.3 *Performance*

The aesthetic, absorbency and comfort advantages still enjoyed by cellulosic fibres has slowed synthetic penetration of the apparel sector, and has extended their life at premium prices in the hygiene sectors of the nonwovens industry. Synthetics are used for their low-cost, thermal bondability, resilience, dryness, and durability characteristics. In the absence of fibres combining all these properties, synthetic/cellulosic blends have been a most popular combination. Ratios varying between 35% cellulosic and 35% synthetic depending on the market positioning of the fabric and the relative prices of the fibres have been typical.

Within the cellulosic part of a nonwoven fabric, however, rayon now has to compete with wood pulp and to a much lesser extent, cotton.

9.6.6.4 *Key markets*

While the versatility of rayon has ensured its continued use in a wide range of absorbent disposables, its fortunes appear to have been dominated by events in a few key markets. Development aimed at responding to the needs of these markets extended the boundaries of rayon technology and identified ways of significantly altering its performance. These are worth reviewing, but it goes without saying that none were capable of reversing the market share decline.

The first major market share loss occurred in coverstock, a market where the skin-friendly absorbent nature of cellulose was felt to be a major advantage.

During the late 1960s when disposable nappies came in two pieces (reusable plastic pant with rectangular absorbent pad), latex bonded rayon was the cover of choice. At this time 'flushability' was becoming a key development issue. The rectangular inserts with their heat-sealed latex-bonded rayon covers were too stable to be disposed of in the toilet even after tearing in half. New wet-laid nonwovens made from the specially developed self-bonding collapsed-tube rayon fibres had no wet-strength at all and dispersed easily in flowing water. However, when treated with the standard wet-strength agents used in the paper industry it became strong enough in use and remained disintegratable in toilet turbulence. Rayon producers in Europe, Japan and the USA developed such fibres and a small market developed in the USA. The introduction of the more convenient one-piece happy pushed mothers' concerns about flushability into the background.

Latex-bonded 100% rayon continued as the leading cover on one-piece nappies, but in 1974 coverstocks containing 50% polyester were market tested for the first time. Consumers found they could not really spot the difference from the 100% rayon fibre versions, so in a second test the latex bonded rayon/polyester blend was put through a point embossing calender to give it a textured surface. This time the mothers could express a preference for the patterned over the plain and a 'unique' new product was born. This first use of synthetic fibre in what had been regarded as an absorbent fibre 'fortress' appeared to be driven by nothing more than the concern over the escalating price of rayon. Technically, however, the success of polyester was put down to its easy embossability.

In the course of the introduction of polyester to coverstock, the nappy industry discovered a major new marketing angle to support a move up to 100% polyester. Coverstocks containing polyester were found to be drier to the touch than the rayon versions. This was initially explained as a consequence of the fact that the synthetics did not absorb water, and so the rayon industry was asked to develop hydrophobic embossable fibres to stay in the game.

Embossability, achievable through alloying rayon with polyethylene emulsions, proved difficult to scale up, but hydrophobic rayons made simply by using hydrophobic finishes, were nevertheless commercialised in both the USA and Europe.

One other feature of the new synthetic coverstocks was proving to be at least as important to surface dryness as their hydrophobicity. Resilience when wet, coupled with much greater dry-bulk associated with their low-collapse in latex bonding allowed them to provide a greater mechanical barrier to urine wetback than a water-plasticised viscose fibre ever could.

Solutions to cellulose's wet collapse problem have been many and various. Dry crosslinking technology had been used on-line in rayon plants in the 1960s to improve the resiliency of fibres used in carpets. Wet crosslinking had been possible since the early 1970s but was rejected as compromising the chemical simplicity of rayon. Hollow and multilimbed fibres gave benefits that allowed them to become major products, but only in markets where the premium prices allowed them to be considered.

Despite their advantages, 100% polyester latex-bonded coverstocks had a short life span. Concerns over latex chemistry (e.g. formaldehyde) led to reformulation of the binders, but the progress in making thermal-bonded polypropylene nonwovens allowed the nappy producers to move swiftly on to this even cheaper, even cleaner technology.

Several attempts were made to develop a thermally bonded rayon nonwoven: Acetylation of the surface of rayon and the use of a solvent bonding process perfected for stabilising acetate in cigarette filters was one obvious approach. Like the surface grafting of thermoplastic materials, this was felt to add more cost than value. Alloying with polyethylene, a technique developed to improve embossability, gave insufficient strength in calendering.

Through-air bonding of 70/30 rayon/bicomponent fibres to lock the rayon into a high volume configuration allowed the manufacture of coverstocks that had attractive softness and a good balance of strikethrough and surface dryness. The former was achieved without recourse to the high levels of surface finish necessary for polypropylene (PP). However the rapid expansion of point-thermal bonding on wide high-speed calender lines allowed 100% PP coverstock to be made at prices that could not be matched on any less dedicated hardware using more costly fibres. The rayon industry quickly decided that this was a wave of nonwoven expansion that held no opportunity.

The rayon tonnage that was lost in the coverstock market was largely regained in new latex-bonded fabric softeners for use in tumble driers. These were lightweight latex-bonded rayon nonwovens produced from rather coarser fibres than had been possible in coverstock. They were made on purpose-built wide lines at speeds well in excess of those possible on coverstock lines a few years earlier. Rayon gave better thermal stability than polypropylene and more strength than latex-bonded polyester. As this market matured, the fact that rayon absorbed too much softener was identified as a technical disadvantage because it allowed the sheets to be used in more than one drying cycle. Technical solutions were, of course, possible, and one involved harnessing the technical advantages of the new lyocell fibre (high strength, less absorbent rayon which could be converted into very light nonwovens – see Chapter 4 on lyocell). However spun-laid polyester emerged as a lower cost alternative.

With fabric softeners as with coverstock, the absorbency of rayon appeared to have changed from a fundamental advantage at the outset, to a fundamental disadvantage as the converters' experience in the marketplace grew. Rayon's absorbency advantage has been less transient in the wiping and hygiene markets, and it is here that ingress of cheaper synthetics has been resisted best. This has been aided by the rapid growth of the hydroentanglement (HE) bonding method that allowed the true character of pure cellulose to be realised in major nonwovens markets for the first time. The silky softness of rayon had, until the development of HE, always been masked by the need to use external bonding agents. Here again though, the poor wet resilience of cellulose has been disadvantageous, and many wipe producers now use up to 50% of synthetic fibres to prevent wet-collapse.

Coming right up to date, the growth of HE bonding has led to a turnaround in rayon's fortunes and in 1998 and 1999 tonnages consumed in nonwovens have reached their highest ever levels in Europe. World HE capacity now exceeds 350 000 tonnes and an estimated 40% of this remains to be utilised. One of the major new products made possible by HE bonding of rayon is the ultra soft baby wipe as exemplified by the European 'Pampers' product from Procter & Gamble. Many other companies are now imitating this and it may be just a matter of time before similar products appear in the US market.

References

1 A Pellerin, French Patent 410 776 (1907).
2 Lazare Drut, British Patent 143 253 (Aug 15th 1921).
3 W H Glover and G S Heaven for Courtaulds, British Patent 273 386 (June 29, 1926).
4 J A Singmaster, US Patent 1 875 894 (Sept 6, 1932).
5 J H Givens for Courtauld, British Patent 467 500 (1935).
6 N L Cox for DuPont, US Patent 2 535 045 (Dec 26, 1950).
7 R L Mitchell, J W Berry and W H Wadman for Rayonier Inc, US Patent 2 937 922 (May 24th 1960).
8 R L Mitchell, J W Berry and W H Wadman for Rayonier Inc, US Patent 2 942 931 (June 28th 1960).
9 R L Mitchell, J W Berry and W H Wadman for Rayonier Inc, US Patent 3 018 158 (Jan 23rd 1962).
10 S Tachikawa, US Patent 2 592 355 (8th April 1952).
11 T Nagata, *New Process Developments for Polynosic Fibres*, Proceedings of the 28th Dornbirn International Man Made Fibres Conference, Sept 1989.
12 C R Woodings and A J Bartholomew, *The Manufacture Properties and Uses of Inflated Viscose Fibres*, TAPPI, Nonwovens Symposium, Myrtle Beach, SC, USA, 1985, pp 155–159.
13 B V Hettich, *Textile Research Journal*, 1984 **54**(6) 382–390.
14 S M Suchecki, *Textile Industry*, 1978 **142**(2) 29.
15 B V Hettich, *Regenerated Cellulose Fibres – Technology for the Future*, 50th Annual TRI Meeting, Textile Research Institute, Princeton, USA, 1980.
16 Sh. Heidari, *VISIL: A New Hybrid Fibre*, Chemifasern/Textilindustrie, Vol 41/93, Dec 1991.
17 T C Allen and D B Denning (to Akzona Inc), US Patent 4 066 584 (Jan 3, 1978).
18 T C Allen and D B Denning (to Akzona Inc), US Patent 4 263 244 (Apr 21, 1981).
19 F R Smith (to FMC Corporation), US Patent 3 844 287 (Oct 29, 1974).
20 *T.S.S. – Assessment of Current Information and Future Research Needs*, Institute of Medicine, USA, National Academy Press, Washington 1982.
21 S F Berkley *et al.*, *JAMA*, 1987, **258**(7) (Aug 21).
22 F R Smith (to Avtex Fibers Inc), US Patent reissue 31 380 (Sept 13, 1983).
23 F R Smith (to Avtex Fibers Inc), US Patent 4 063 558 (Dec 20, 1977).
24 F R Smith (to FMC Corporation), US Patent 3 843 378 (Oct 22, 1974).
25 F R Smith (to Avtex Fibers Inc), US Patent 4 179 416 (Dec 18, 1979).
26 T C Allen (to Akzona Inc), US Patent 4 242 242 (Dec 30, 1980).
27 Kaneko *et al.* (to Asahi Kasei Kogyo KK), US Patent 3 833 438 (Aug 30th 1972).
28 Suzuki *et al.* (to Mitsubishi Rayon Co), US Patent 3 718 537 (Feb 27th 1973).

10 Current and future market trends

TIM F N JOHNSON

10.1 Introduction

Cellulosic fibres form part of the artificial fibre industry, a 20th century creation that sits, sometimes uneasily, between three great industries upstream – agriculture, forestry and petrochemicals – and the ancient textile industry downstream of it. It is necessary to define a few terms. 'Textiles' in the general sense can cover the manufacture of fibres, yarns, fabrics and apparel. But in the narrow sense the textile industry makes yarns and fabrics from fibres, for the making-up (or apparel) industries to process into garments. Although this breakdown (fibres, textiles, apparel) is commonly used, it ignores the fact that an appreciable proportion of textiles do not end up as apparel, but as household goods and in industrial and technical uses.

The textile industry (in its widest sense) abuts the plastic film, leather and fur, and the paper industries. There are products in each case that could be categorised in the textile industry, or in one of these three – for instance baler twine, reinforced papers, airlaid wood pulp, some agritextiles.

Some yarns are produced as continuous filament (silk and many synthetic fibres), others are spun by the modern equivalents of the manual cottage industry spinning whorl seen in medieval paintings. But artificial staple fibres are also said to be spun, because spiders (and silkworms) have also been described as spinning their web (or cocoon). In fact we should say that spiders extrude a polypeptide monofilament to produce a structured non-woven web. But such phrases would not inspire poets. Synthetic fibres, however, should be described as extruded rather than spun. The very term 'fibre' is used in different ways – in Asia it usually means staple (and by implication not filament), elsewhere the term will include both products. Staple again can be used generically to cover staple, tow and tops (i.e. to distinguish from filament), or it may be used to mean staple as opposed to tow or tops. (Tow is the precursor to staple, which is produced by either cutting or stretch-breaking tow. Tops are produced from stretch-broken tow for woollen or worsted spinning).

The meaning of these basic terms has to be deduced from their context, a skill usually quickly acquired by newcomers to the industry, if a constant source of confusion to its financial analysts.

The fibre industry buys material in large units of mass (tonnes, bales) and sells in smaller units (kg or pounds), albeit with a linear dimension specified (denier or dtex, the weight in grams of 9000 m and 10 000 m, respectively. A silkworm produces on average 9000 m of filament in each cocoon). The textiles are then sold in a like manner (yarns) or by area (square yards or square metres). Fabric is sometimes specified just in metres, the width being understood by context. Then the making-up industry changes to numbers of pieces (often in dozens, for clarity), before the retail outlet sells on an individual basis. A fibre maker therefore runs a large capital intensive factory, resembling a chemical works, but has to understand the very different environments existing before the final buyer. The industry is still going through remarkable transformations as some of the great names of the past disappear from it – Courtaulds, Akzo, and Hoechst following ICI, and even DuPont having second thoughts about some products. One interesting trend is for firms to be broken up into single fibre business units which then become part of privately owned vertically integrated firms – DuPont/Sabanci/Alpek, Hoechst/Koch/Saba/Texmaco, Enichem/Orlandi, Acordis (via CVC).

Any market analysis has to take all this into account.

10.2 The broad picture

The cellulosic fibre market is part of the total fibre market, which includes natural and synthetic fibres as well. There is no general agreement on the scope of the fibre market – many agencies collect and publish statistics, but include or exclude different fibre types. It is useful to bear in mind the breakdown given in Table 10.1

Given that some US sources do include glass, metal and asbestos fibres, and that total jute consumption is almost 3 million tonnes it is clear that some care needs to be taken with such statements as 'the total fibre market is *x* million tonnes'. Polypropylene is the fastest growing fibre (at the expense in part of cellulosics), but worldwide coverage is still patchy, and there are no reliable time series back to the 1980s or further. Table 10.2 shows the 'big picture' for the worldwide fibre market from 1900 to 2000, with an indication of the total market in 2050.

By the year 2000, cellulosic fibres declined from a peak in absolute tonnage at some 3.5 million tonnes in 1975 (but that was actually at a market share of 2% points lower than in 1950 when they had a 13% share)

Table 10.1 Conventional statistical coverage of fibres

	Usually included:	Usually excluded:
Natural	Cotton, wool	Flax, sisal, kenaf, silk, coir, jute
Synthetic	Polyester, acrylic, nylon, (polypropylene)	Polyethylene, elastanes, aramids (polypropylene)
Cellulosics	Viscose/rayon, acetate, cupro	Lyocell
Others	None	Metal, glass, asbestos

Table 10.2 World fibre market 1900–2050 (million tonnes)

	Natural fibres	Cellulosics	Synthetics	Others[a]	Total
1900	4	0		1.5	5.5
1925	6	0.015		2.5	8.515
1950	7	1.5	0.07	3	11.57
1975	14	3.5	8	6.5	32
2000	20	3	25	7	55
2050					140[b]

[a] Others includes all but cotton, wool, viscose/rayon, acetate, polyester, nylon, acrylic, polypropylene and metal/glass/asbestos. Some of these figures are estimates.
[b] TFNJ estimate.

to about 3 million tonnes, 5% share. The most noteworthy element of this table is the swift rise in the share of synthetics (polyester, nylon, acrylic and polypropylene).

The world population in the years shown in Table 10.2 was (in billions) 1.6, 1.9, 2.5, 4.1, 6.2 and a projected 11. The 10-fold increase in the total market, from 5.5 to 55 million tonnes from 1900 to 2000 compares therefore to a 3.8-fold increase in population, indicating that the increase in economic prosperity per head was of the order of 2.7-fold. The effects of population and economic growth in the projection to 2050 are in similar relative proportion. What the table does not show is the recent (since 1990) decline in market share for natural fibres, both cotton and wool.

Cotton is, of course, chemically a cellulosic fibre, but is never classified with the man-made cellulosics. It is, however, their main competitor, and the future level of production of cotton is crucial to the size of the total cellulosic market. Table 10.3 shows the world production of cotton together

Table 10.3 World cotton production and yields per hectare

	Production (million tonnes)	Area (million ha)	Yield ($kg\,ha^{-1}$)
1930	5.5	33.1	0.20
1940	6.2	33.1	0.20
1950	5.8	30.2	0.25
1960	10.1	32.7	0.31
1970	11.2	32.3	0.38
1980	14.2	33.4	0.43
1990	18.7	32.1	0.56
2000 est	19.7	34	0.58

with the acreage planted, and hence the yield per unit area, from 1930 to 2000.

During the 1990s the actual production varied between 18.0 and 20.4 million tonnes (cotton is subject to the vagaries of the weather, pest attack, other infections), but the area has remained under 34 million hectares and the yield, using a 10-year moving average to smooth out variations, seems to be asymptotically approaching $0.60\,kg\,ha^{-1}$.

The importance of this is that most of the cellulosic fibre produced shares with cotton the property of being hydrophilic (moisture absorbing), which makes it the fibre of choice for garments or household uses which are used in contact with the skin. All current synthetic fibres by contrast are not at all hydrophilic, although ironically, for extreme conditions, athletes and other sports players have discovered that polypropylene (the least hydrophilic) makes the best inner layer as it quickly wicks away large amounts of excess sweat. *[Even babies have discovered the effect: disposable diapers used to have absorbent covers but now use polypropylene nonwovens. Ed.]*

A world population of 6.2 billion using 20 million tonnes of cotton and 2 million tonnes of relevant cellulosic fibres implies a use per head of 3.5 kg of hydrophilic fibre. Almost doubling this population implies the need for about another 20 million tonnes of hydrophilic fibre.

There are a limited number of candidates for this supply:

- Increase of cotton area – unlikely, in fact it is expected to decrease in China and India
- Spread of best practice – seems to have come to a halt
- More irrigation – more likely to be restricted (as in Turkey now)
- Yield increase by genetic modification – the wild card, impossible to forecast

- A new hydrophilic fibre – a polyester variant (possible), or fibres based on lactic acid for example
- Increase in viscose/rayon capacity – the industry expects pollution constraints to prevent this
- Increase in lyocell capacity – possible, but hardly at the scale needed
- A new cellulosic fibre variant – fibrils grown by bacteria???

In all, while it seems unlikely that all of the extra requirement for hydrophilic fibre can be met by new cellulosic fibre capacity, it is probable that the rundown of production over the last two or three decades of the 20th century will be reversed in the early decades of the new century. Even so a review of 150 years of fibre demand can be summed up by observing that while our one billion or so grandparents wore only natural fibres (80% cotton) our 10 billion grandchildren will wear predominantly synthetic fibres (mostly polyester).

10.3 Breakdown by fibre type

World capacity and production data are available for the following types of cellulosic fibre:

- Viscose; staple, textile filament, and high tenacity (HT) filament.
  Some capacity breakdown by regular, modal and polynosic is available, see Table 10.4 below
- Acetate; tow (for cigarette filters in the main), textile filament.
  Some capacity breakdown by di- and triacetate, and cupro, is available

Table 10.4 World cellulosic fibre capacities (kilotonnes per annum)

	1985	1990	1995	2000	2001
Viscose					
Staple	2395	2195	2190	2400	2350
Textile fil	465	495	395	390	390
HT fil	320	260	150	105	105
Acetate					
Tow	415	500	600	700	690
Textile fil	285	275	275	220	215
Cupro	40	40	35	30	30
Lyocell					
Staple	0	c2	c20	c100	c100
Filament	0	0	0.5	(0.5)	(0.5)

Table 10.5 World cellulosic fibre production (kilotonnes per annum)

	1980	1985	1990	1995	1998
Viscose					
Staple	2085	1960	1910	1740	1590
Textile fil	405	390	370	340	300
HT fil	290	275	185	85	80
Acetate					
Tow	335	350	465	555	585
Textile fil	330	240	245	205	155
Cupro	30	30	30	30	25
Lyocell	—	—	—	Not declared	Not declared

- Lyocell; staple (filament only at a pilot plant stage, and staple capacity data only).

Acetate tow capacity has increased throughout and viscose capacity is fairly constant. For the other variants, however, the picture is one of steady decline. This is confirmed in Table 10.5, where even acetate tow growth has begun to falter (1999 preliminary data suggests that some capacity cutbacks will occur).

10.4 Breakdown by main area

Market data for these fibres is difficult to come by on a comprehensive basis, and the following data is based on production data. In the main, there is not a lot of trade at the fibre stage between major blocs, but it can be considerable within blocs. For instance, Turkey is a major textile area but has no cellulosic fibre capacity. Lyocell, being a new fibre, is another exception, and is omitted from Table 10.6.

Interbloc trade in fabrics, however, can be considerable (e.g. in lining fabrics from Asia to Europe), and trade in apparel items can be huge. It is not possible to track trade in apparel items by fibre content, although some claim to do so using vast arrays of factors.

10.4.1 Capacity by the main producers

Table 10.7 shows the 1999 capacity for each fibre for the main producers worldwide, with an indication of fibre variants for each.

Table 10.6 Cellulosic fibre production by bloc (kilotonnes per annum)

	1980	1985	1990	1995	1998
Viscose staple					
W Europe	505	365	365	360	340
E Europe	725	710	530	140	75
Americas	280	220	220	180	125
Asia	565	650	780	1050	1040
Others	10	15	15	10	10
World	2085	1960	1910	1740	1590

The surge in Asia has mainly been in India and Indonesia, with growth also in China. The decline in East Europe is particularly striking.

	1980	1985	1990	1995	1998
Viscose textile filament					
W Europe	75	70	65	70	65
E Europe	150	150	115	75	55
Americas	30	25	25	25	15
Asia	140	135	155	160	160
Others	10	10	10	10	5
World	405	390	370	340	300

The 1999 figures (not yet finalised) showed a continuing severe drop in European production.

	1980	1985	1990	1995	1998
Viscose HT filament					
W Europe	90	80	70	55	50
E Europe	150	155	95	10	15
Americas	25	20	10	5	2
Asia	25	20	10	15	13
Others	0	0	0	0	0
World	290	275	185	85	80

Again a striking fall in East Europe. This product depends on the continuance of unrestricted speed limits on German autobahns.

	1980	1985	1990	1995	1998
Acetate tow					
W Europe	70	80	115	130	135
E Europe	5	10	5	0	5
Americas	220	215	275	305	305
Asia	40	45	70	120	140
Others	0	0	0	0	0
World	335	350	465	555	585

Recent growth has been in Asia, where, however, some progress is being made with acceptable (to smokers that is) cheaper polypropylene filters.

	1980	1985	1990	1995	1998
Acetate and cupro textile filament					
W Europe	70	50	50	45	45
E Europe	55	60	55	10	10
Americas	175	105	110	115	75
Asia	60	55	60	65	50
Others	0	0	0	0	0
World	360	270	275	235	180

1999 was another difficult year, according to preliminary figures.

Table 10.7 Cellulosic fibre; 1999 capacity by main owner (kilotonnes per annum)

Main firm (%)	Viscose staple	Viscose filament	Acetate tow	Acetate/Cu filament	Lyocell textile
Top 10 firms have 75% of world capacity					
China (15)	480	118			
	(many plants, some of poor environmental record)				
Birla (12)	408	43			0.15
	(world no 1 as a company)				
Acordis (10)	169	69	38	33	c80
	(widest spread of products, plants in USA & Europe)				
Lenzing (8)	315				c20
	(plants in Europe, America and Asia, over 30 ktonnes of modal viscose)				
Celanese (8)			224	90	
	(USA, Canada, Belgium & China, 16 ktonnes of filament closed end of 1999)				
Eastman (6)			196	20	
	(USA, trade talk of closures due, but possibly not in fibres)				
Russia (5)	157	44		5	
	(many plants producing hardly any volume)				
RGM (4)	170				
	(plants in Finland and Indonesia, now spun off by APRIL)				
FCFC (4)	162				0.4
	(plants in lyocell)				
Rhodia (3)			106		
Other companies					
Daicel (plant in China) (2)			85		
SIV (1)	47	8			
MRC (1)			33	21	
Vicunha (1)	44	5			

Table 10.7 (cont.)

Main firm (%)	Viscose staple	Viscose filament	Acetate tow	Acetate/Cu filament	Lyocell textile
Asahi (1)		16		29 (cupro)	
Toho (1)	41 (polynosic)				
Chemapol (1)	38				
Fuji (1)	35 (polynosic)				
Banesto (1)	32				
Kohjin (1)	31				
Jilin (1)		31			
Daiwabo (1)	30				
Svenska (1)	26				
Bosnia (1)	26				
Gruppe (1)		24			
Omikenshi (1)	22				
Kabool (1)	20				
Firms with under 0.5% of world capacity each					
Toyobo	18				
Ukraine		18			
Sunkyong			10	8	
Belarus		18			
Goenka		16			
Martis	15				
Lithuania				15	
Serbia	10	3			
Impressa	12				
Teijin				12	
G Veneto		11			
Kuraray		11			
Inacsa				10	
Hindaya	7	4			
LSB		10			

Table 10.7 (cont.)

Main firm (%)	Viscose staple	Viscose filament	Acetate tow	Acetate/Cu filament	Lyocell textile
Yunica				9	
Etma		7			
Shriram		7			
Abranov				7	
Cydsa		6			
Snia		6			
MISR	5				
Bulgaria		5			
CI	5				
Locals		5			
Beaulieu		5			
Travancore		4			
Poland		4			
Slovakia		4			
Uzbekistan				3	
Bemberg				3 (cupro)	
CMMF		3			
ESCO		3			
Sirsilk				2	
Industrias		2			
CTMTC		1			
Possible new lyocell entrants?					
TITK					0.11
Hanil					0.01
World total, 3888 kilotonnes per annum (100)					

10.5 Capacity and production trends, 1980–2010

Although there is overcapacity in cellulosic fibre, it is not nearly as severe as for polyester. However, its impact will be more deeply felt because there is not the underlying growth in demand to eventually absorb it. Long term growth in viscose/lyocell staple should eventually return, and possibly could be dramatic after 2010. For filament, however, the long decline looks set to continue, although it appears that cellulosic will always be the material of choice for linings, cost apart. Thus polyester will take the commodity clothing lining business, leaving acetate and viscose in higher quality garments, where they might be joined by filament lyocell. The fashion business will provide periodic booms for these fibres. Acetate tow seems likely to be less attractive, as smoking continues to be discouraged by authorities around the world, and polypropylene provides low cost competition. HT filament demand has become a small niche associated with performance sports cars and the continuance of unrestricted speeds on German motorways. *[There are some signs of an upturn in the fortunes for tyre-yarn associated with the development of run-flat tyres for which rayon seems ideally suited. Ed.]*

10.6 Trends in markets by end-use

This section is heavily dependant on West European data (see Table 10.8), where a great deal of effort has been put into market analysis by end-use. Somewhat coarser data is available for the USA, and a little for Japan. The European data has probably been maintained for longer than the continually eroded resources for market research could cope, and has now been cut back in scope. This section compares the years of 1990 and 1996, and attempts to draw broad conclusions about the implications worldwide. More recently industrialising countries, such as China, are and will be able to leapfrog European experience – for instance they will start with heat

Table 10.8 West European fibres by spinning route

	Cellulosic	Polyester	Acrylic	Cotton	Others
Cotton spinning (ktonnes)					
1990	186	199	120	1070	25
1996	134	154	92	1042	24
(%)					
1990	12	12	8	67	2
1996	9	11	6	72	2
Woollen spinning (ktonnes)					
1990	25	48	56	47	361
1996	22	45	60	42	380
(%)					
1990	4.7	9	10	9	67
1996	4.0	8	11	8	69
Worsted spinning (ktonnes)					
1990	13	32	221	2	231
1996	11	37	188	1	254
(%)					
1990	2.6	6	44	0	46
1996	2.2	8	38	0	52
All spinning (ktonnes)					
1990	224	279	397	1119	617
1996	167	236	340	1085	658
(%)					
1990	8.5	11	15	42	23
1996	6.7	9	14	44	26

Table 10.9 West European unspun fibre market

	Cellulosic	Polyester	Acrylic	Cotton	Others
For apparel (ktonnes)					
1990	na	na	na	na	na
1996	5	49	4	2	11
(%)					
1990	na	na	na	na	na
1996	7	69	6	3	15
For household uses (ktonnes)					
1990	na	na	na	na	na
1996	21	172	4	17	160
(%)					
1990	na	na	na	na	na
1996	5.6	46	1	5	43
Other uses (ktonnes)					
1990	na	na	na	na	na
1996	146	93	4	61	190
(%)					
1990	na	na	na	na	na
1996	30	19	1	12	38
All unspun (ktonnes)					
1990	160	241	31	46	284
1996	172	314	12	80	361
(%)					
1990	21	32	4	6	37
1996	18	33	1	9	38

bonded polypropylene nonwovens without the preliminary latex bonded viscose products used in Europe in the 1960s and 1970s.

During the period 1990 to 1996 the total spinning market declined by almost 6% (cotton spinning by 10%) and the use of cellulosic (only viscose in spun yarn) fibre by 25% (28% in cotton spinning). Wool is, of course, the main component in 'others' in woollen and worsted spinning. Open end spinning accounted for 27% of the viscose spun in 1996, compared to 16% for polyester, 9% for acrylic, 14% for cotton.

Table 10.9 gives data for the market for unspun fibre for 1990 and 1996.

During this period the total unspun market, in contrast to the spinning market, grew by over 23% and the use of cellulosic (both viscose and acetate tow) fibre by 7.5%, although its share dropped 3%, to 18%. Acetate tow is in the 'other uses' section above, and polypropylene is of course the main component in the 'others' column. Unfortunately a change in methodology by the industry prevents the 1990 data being displayed by broad end-use sector in the same format as for 1996.

Table 10.10 gives data for the market structure for synthetic staple fibre for the same years.

Table 10.10 West European synthetic staple fibre market (spun and unspun)

	Cellulosic	Polyester	Acrylic	Total	Decline
Production (ktonnes)					
1990	404	470	664	2405	
1996	473	459	681	2596	8%
(%)					
1990	17	20	28	100	
1996	18	18	26	100	
Imports (ktonnes)					
1990	256	377	247	1393	
1996	254	426	214	1423	–2%
(%)					
1990	18	27	18	100	
1996	18	30	15	100	
Exports (ktonnes)					
1990	269	313	505	1581	
1996	399	351	566	1550	2%
(%)					
1990	17	20	32	100	
1996	26	23	37	100	
Net imports (ktonnes)					
1990	–13	65	–258	–188	
1996	–145	75	–352	–127	33%
Consumption (ktonnes)					
1990	390	535	406	2217	
1996	328	534	329	2469	–11
(%)					
1990	18	24	18	100	
1996	13	22	13	100	

During this period the total market for synthetic staple in Europe declined by 11%, with local production falling by 8% and net imports rising by a third. Although cellulosic production is shown growing strongly, this is because of the inclusion of Finland and Sweden in the figures in 1996, by which time the European Community had been enlarged (Austria had been incorporated for some time). For consumption data (in this and the above tables) this change makes little difference because of the tiny textile industries in Scandinavia. The fibres omitted (derivable *in toto* by difference) are nylon and polypropylene, neither of which compete with cellulosic staple, in the main (the latter does compete with acetate cigarette filters, in China).

Table 10.11 shows data for the market structure for yarn (both spun and synthetic filament).

During the period 1990–1996 the total market for yarn in Europe grew by 2.4% (3841–3922 ktonnes), with local production rising by 1.3%

Table 10.11 West European yarn market (spun and filament)

	Cellulosic filament	Synthetic filament	Cellulosic staple	Synthetic staple	Cotton
Production (ktonnes)					
1990	177	1081	215	843	865
1996	161	1155	159	741	942
(%)					
1990	5.0	30	6.1	24	24
1996	4.5	32	4.4	21	26
Imports (ktonnes)					
1990	135	819	83	196	475
1996	127	1056	89	329	536
(%)					
1990	7.6	46	4.7	11	27
1996	5.9	49	4.1	11	25
Exports (ktonnes)					
1990	140	791	68	187	230
1996	135	940	73	221	352
(%)					
1990	9.4	53	4.6	13	15
1996	7.4	52	4.0	12	19
Net imports (ktonnes)					
1990	–5	28	15	9	246
1996	–8	116	16	18	184
Consumption (ktonnes)					
1990	172	1109	230	852	1110
1996	153	1271	176	758	1126
(%)					
1990	4.5	29	6.0	22	29
1996	3.9	32	4.5	19	29

(3554–3600 ktonnes) and net imports rising by 17% (277–322 ktonnes). Cellulosic filament consumption has fallen from 172 to 153 ktonnes, about a single percentage point loss of share. In 1999, as this book was being prepared, cellulosic filament suffered further dramatic losses in its share worldwide, in the face of competition from cheap polyester following the 1997 Far East financial crisis. Cellulosic staple lost a similar share, falling from 230 to 176 ktonnes, but was not to be further hit by the after effects of the financial crisis in the Far East.

The fibre omitted is wool, which does not compete to any significant extent with cellulosics. In all this end-use data it should be remembered that, as is normal in fibre statistics of this nature, some natural fibres are omitted – mainly jute, sisal, kenaf, coir, flax (linen), ramie, silk and the like.

Table 10.12 gives data for the yarn market by end-use.

Table 10.12 West European yarn market (spun and filament)

	Cellulosic filament	Synthetic filament	Cellulosic staple	Synthetic staple	Cotton
WEAVING					
Weaving, linings (ktonnes) (total market decline 12%)					
1990	48	35	6	2	4
1996	38	35	4	2	3
(%)					
1990	51	37	6	2	4
1996	47	43	5	2	4
Weaving, other apparel (ktonnes) (total market decline 11%)					
1990	27	111	102	167	448
1996	33	105	79	139	368
(%)					
1990	2.6	11	9.8	16	43
1996	3.5	11	8.6	15	40
Weaving, household (ktonnes) (total market decline 8%)					
1990	10	66	56	143	364
1996	8	77	41	127	331
(%)					
1990	1.5	10	8.6	22	55
1996	1.3	13	6.7	21	54
Weaving, industrial (ktonnes) (total market growth 7%)					
1990	51	159	11	27	91
1996	43	207	9	31	71
(%)					
1990	15	47	3.3	8	27
1996	12	57	2.5	9	20
All weaving (ktonnes) (total market decline 7%)					
1990	136	370	175	338	906
1996	122	424	133	299	774
(%)					
1990	6.4	17	8.2	16	43
1996	6.2	22	6.7	15	39
KNITTING					
Warp knitting (ktonnes) (total market growth 8%)					
1990	11	134	4	19	22
1996	11	153	4	24	13
(%)					
1990	5.5	70	2.3	10	12
1996	5.1	74	2.1	12	6
Weft knitting, apparel (ktonnes) (total market growth 5%)					
1990	7	206	24	267	284
1996	15	225	26	256	285
(%)					
1990	0.8	23	2.7	30	32
1996	1.6	24	2.8	28	31
Weft knitting, household (ktonnes) (total market growth 52%)					
1990	0	7	1	3	6
1996	0	15	3	4	5
(%)					
1990	0	39	6.3	18	34
1996	0	56	9.4	14	18

Table 10.12 (cont.)

	Cellulosic filament	Synthetic filament	Cellulosic staple	Synthetic staple	Cotton
Weft knitting, industrial (ktonnes) (total market growth 28%)					
1990	0	9	2	3	14
1996	0	14	3	6	13
(%)					
1990	0	32	8	12	49
1996	0	38	9	18	35
All knitting (ktonnes) (total market growth 6%)					
1990	18	355	32	292	326
1996	25	406	36	290	316
(%)					
1990	1.6	32	2.8	26	29
1996	2.1	34	3.0	24	27
ALL OTHER PROCESSES					
Narrow fabrics (ktonnes) (total market growth 11%)					
1990	3	25	2	3	13
1996	3	33	3	4	9
(%)					
1990	7.2	55	3.5	7	27
1996	5.7	65	5.5	7	17
Carpets (ktonnes) (total market growth 6%)					
1990	0	242	2	172	9
1996	0	304	1	150	9
(%)					
1990	0	48	0.3	34	2
1996	0	57	0.3	28	2
Hand knitting, household (ktonnes) (total market decline 49%)					
1990	1	0	1	37	6
1996	1	2	0	13	5
(%)					
1990	1.6	1	1.1	65	10
1996	1.7	6	1.0	44	16
Miscellaneous (ktonnes) (total market decline 13%)					
1990	3	83	6	26	29
1996	1	80	3	15	28
(%)					
1990	1.8	56	4.3	18	20
1996	1.0	62	2.2	12	21
All other processes (ktonnes) (total market decline 23%)					
1990	4	83	7	63	34
1996	2	82	3	28	32
(%)					
1990	1.8	41	3.4	31	17
1996	1.1	52	2.0	18	20
All yarn markets (ktonnes) (total market decline 2.4%)					
1990	161	1076	218	869	1287
1996	152	1249	176	771	1140
(%)					
1990	4.0	27	5.4	22	32
1996	3.9	32	4.5	20	29

Table 10.12 illustrates well the niches where cellulosic fibres thrive – linings (50% cellulosic), non-melting reinforcement fabrics (10–15% cellulosic), warp knitting and narrow fabrics (5–7% cellulosic). Overall cellulosic fibre has about 4% of the European market for yarns. The overall decline is in part caused by the general decline in upstream activity in Europe as the retail sector switches to imports from Asia, and in part by the continuing gain by synthetics, especially polyester.

How relevant is this Western European market breakdown to the markets in other areas? It is probable that in most areas there will be a similar pattern, if only because in the main textile producing areas much of the downstream industry is producing for West Europe and North America. East Europe is a special case: their fibre portfolio was frozen some decades ago as the lack of investment in civilian projects caused them to remain overcommitted to cellulosic (basically a pre-World War II version of the fibre) and later to nylon. The polyester revolution that swept Asia from the 1970s just passed them by.

Appendix A
Lyocell end-use development datasheets

The following datasheets are presented with the kind permission of Acordis Fibres.

But for changing the fibre name from 'Courtaulds Lyocell' to 'Acordis Lyocell' or 'Tencel' they are the original documents prepared by Mr Manny Coulon, Ms Pam Johnson, Mr Robert Morley, and Mr Calvin Woodings for launching lyocell into the technical textiles, nonwovens, and special paper markets between 1996 and 1998.

They cover the following properties of lyocell in the end-uses mentioned. Technical textiles:

1 Tensile and tear strength
2 Dimensional stability
3 Heat ageing
4 Abrasion resistance
5 Drape and fluidity
6 Moisture
7 Fibrillation

Nonwovens:

8 Web bonding – Hydroentanglement
9 Web bonding – Needling
10 Web bonding – Latex
11 Wet-laying
12 Air-laying
13 Wipes
14 Medical swabs and gauzes
15 Absorbency
16 Chemical and UV resistance
17 Biodegradation

Special papers:

18 100% Acordis Lyocell papers
19 Blends
20 Filtration

N.B. The ‘Acordis Lyocell’ brand has been dropped. Acordis’s lyocell fibre is now branded Tencel® in all markets.

Tensile and Tear Strength

Whether TENCEL® fibre is wet or dry, its strength exceeds that of cotton and is in fact stronger than all other cellulosic man-made staple fibres. Its compatibility with polyester means that blends with TENCEL® form yarns of similar strength to 100% polyester. Resulting fabric strengths can be exceptional, particularly in respect of tear resistance.

The dry tenacity of TENCEL® is up to 44 cN/tex (5 g/den), considerably greater than other cellulose fibres. Even when wet, this highly moisture absorbent fibre will retain up to 80% of its strength. Of course, when dried, the original fibre strength is recovered.

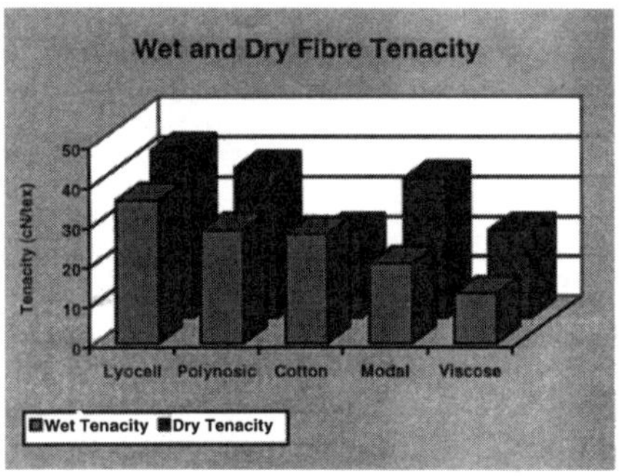

This high fibre strength translates into high tensile strength in the spun yarn. In finished fabrics, TENCEL® can easily give a 25% tensile strength advantage over cotton.

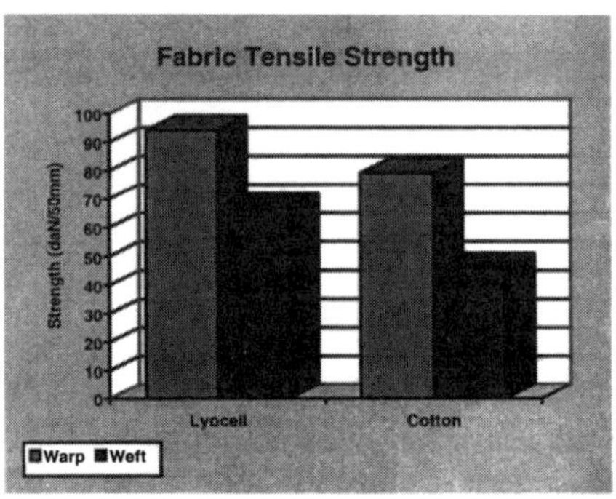

The tear strength performance is even more impressive: tear strengths can double that expected for a similar cotton fabric, depending upon the fabric construction used.

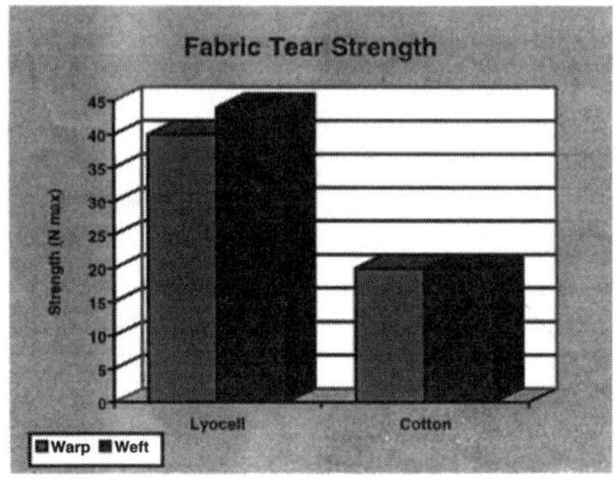

Blends with polyester

TENCEL® blends particularly well with polyester. The stress / strain curves for both fibres are compatible, such that yarns of high strength are possible.

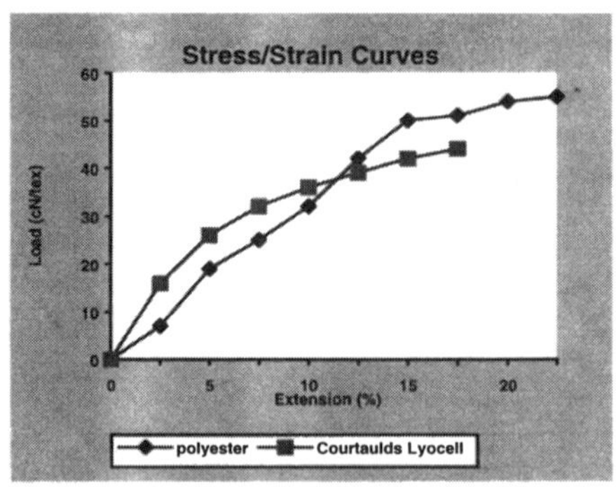

In combination with polyester, yarn strengths, and therefore fabric strengths, are comparable to 100% polyester at all blend levels with TENCEL®.

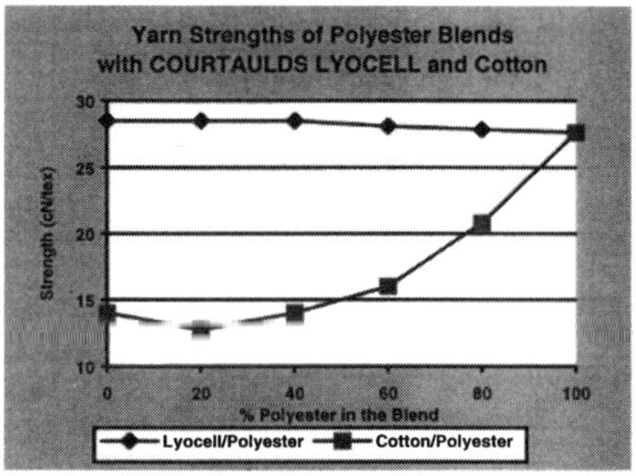

This means that the proportions of each fibre used in a polyester blend with TENCEL® can be adjusted to meet other performance requirements whilst still maintaining optimum strength. For instance, high compatibility with polyester allows freedom to balance the abrasion resistance offered by the polyester with the absorbency and comfort brought by the TENCEL®, and without compromising the fabric tensile properties.

Polyester blends with cotton give inferior strength. Even with 30% polyester in blend with cotton, yarn strength is no greater than for 100% cotton. This is because the stress/strain curve of cotton is a poor match to polyester, whilst that of TENCEL® is highly compatible with polyester.

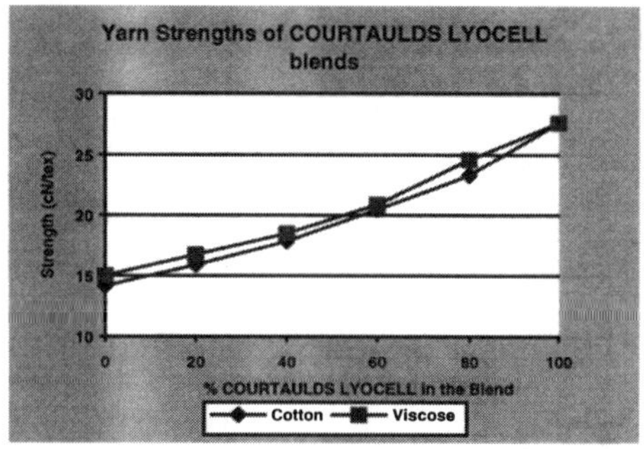

A blend of TENCEL® with cotton or viscose also yields improvements in yarn strength. Strength increases are achieved in almost direct proportion to the percentage of TENCEL® fibre added to the blend.

TENCEL®, either as 100%, or in blend with polyester produces strong yarns which translate into excellent fabric tensile and tear strength. The absorbency of TENCEL® results in fabrics which are comfortable to wear and perform well in demanding applications.

Dimensional Stability of Woven Fabrics

Woven fabrics produced from TENCEL® fibre have excellent stability both in processing and in use. Resin free fabrics give lower shrinkage in laundering than comparable cotton fabrics and do not show the progressive shrinkage problems often associated with cotton and other cellulosic fabrics.

Wet Stability - The superior wet stability of TENCEL® is due to its high wet modulus and, to a lesser extent, its low linear swelling in water. Fabrics from TENCEL® can produce smaller losses in area during wet processing as well as low laundering shrinkages in use. Combined processing and shrinkage losses can be better than half that for cotton and other cellulosics. TENCEL® fabrics can achieve stable dimensions rapidly when wet, and still offer lower residual shrinkage.

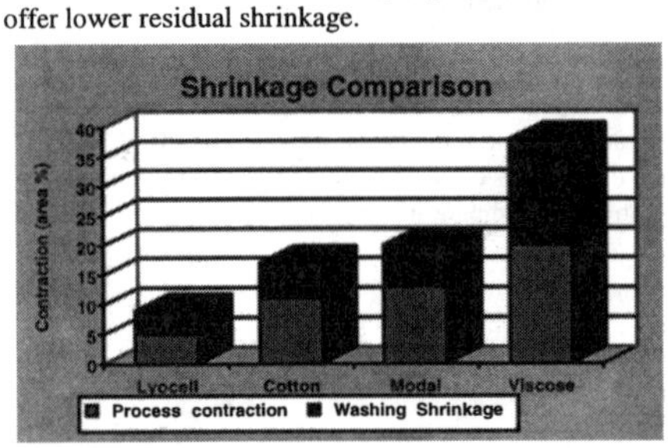

Such dimensional stability is not unusual for TENCEL®, even without the use of resins if fabrics are relaxed correctly in finishing.

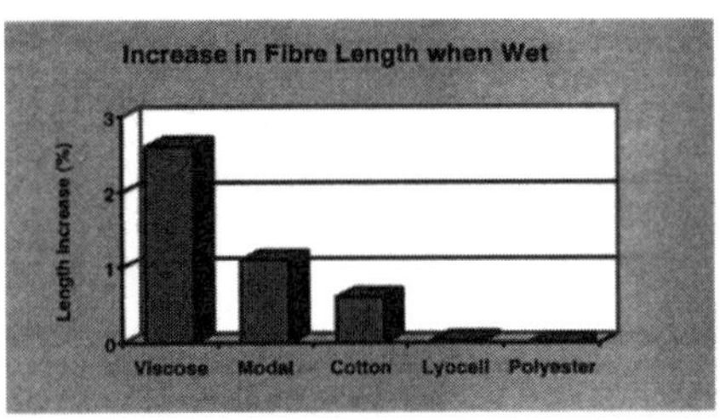

The basis for the inherent stability of TENCEL® is that its fibres do not change in length when wet. All other cellulosic fibres grow in length when wet, which means that for every wet/dry cycle there is potential for progressive shrinkage as the fibres alternately extend in length and then contract.

If required, the good dimensional stability of TENCEL® fabrics can be enhanced still further by the use of suitable resins or by blending with a synthetic fibre such as polyester or nylon.

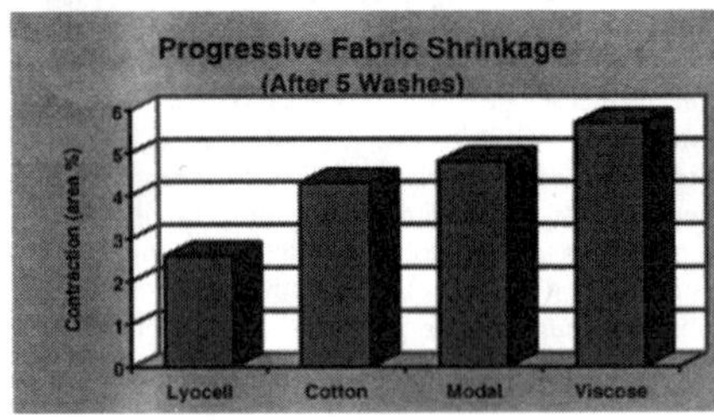

TENCEL® fibre is stiffer than other cellulosics when wet. Fibres with low wet stiffness will easily distort during wet processing which can lead to shrinkage problems. Low wet modulus is one cause of such shrinkage problems with viscose and is also a contributing factor in the progressive shrinkage of cotton. Progressive shrinkage is accepted as a characteristic feature of cotton but is typically negligible in fabrics made from TENCEL®.

Fibre Tensile Moduli Compared to Other Cellulosics

	TENCEL®	Polynosic	Modal	Cotton	Viscose
Dry cN/tex	1113	-	717	500	371
Wet cN/tex	265	210	125	150	53

Dry Stiffness - TENCEL® fibre has a dry tensile modulus comparable to polyester and is resistant to stretching. This high fibre stiffness means that appropriately finished fabrics are stable under load and will resist distortion better than nylon.

Fibre Tensile Moduli Compared to Synthetics

	TENCEL®	HT polyester	HT nylon
Dry (cN/tex)	1113	1014	500
Dry (GPa)	16	14	6

Heat Ageing

In addition to having a higher initial strength, TENCEL® outperforms cotton in its ability to retain fibre strength when subjected to elevated temperatures. In suitably engineered structures, TENCEL® can withstand higher temperatures over longer periods than cotton or viscose.

TENCEL® fibres do not melt and are stable below 150°C. Above 170°C the fibres start to lose strength gradually, beginning to decompose more rapidly by 300°C. At 420°C the fibres will ignite. Under controlled conditions they can be oxidised and carbonised.

Cellulosic fibres are not well known for their thermal resistance or thermal stability. However, TENCEL® exhibits better resistance to dry heat at 150°C than cotton and viscose fibres.

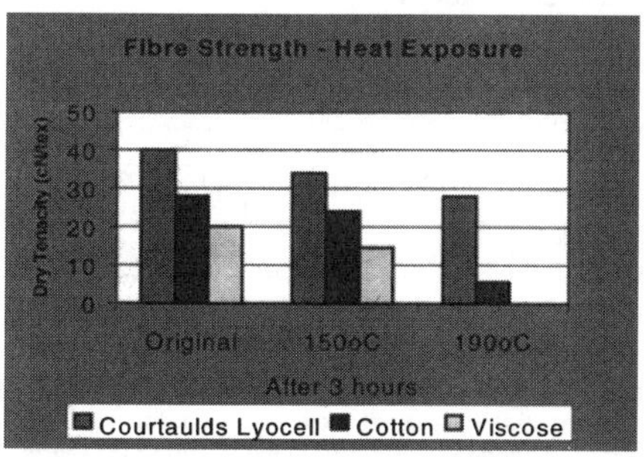

The strength benefit of TENCEL® is also apparent after three hours exposure to 190°C. The strength of viscose and cotton fibres are reduced by over two thirds. In comparison, TENCEL® loses less than one third.

After 80 hours exposure at 150°C, TENCEL® remains as strong as the unexposed cotton fibre.

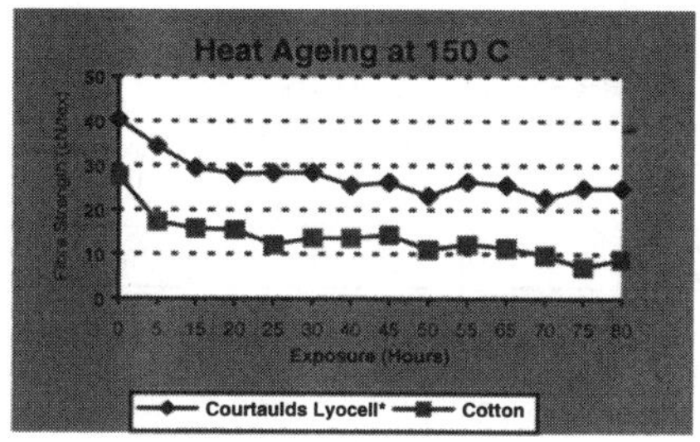

The strength advantage of TENCEL® was maintained throughout the test, which was extended to over 1000 hours.

TENCEL® discolours significantly less than viscose and cotton fibres, when exposed to temperatures above 120°C for extended periods of time.

Abrasion Resistance

In line with other cellulosics, the abrasion performance of TENCEL® is sufficient for the general needs of apparel. For more demanding uses in industrial and technical applications, enhanced abrasion resistance may be required. Practical options include blending with low proportions of synthetic fibre or application of special finishes or resins. Such fabrics can retain many of the desirable features of 100% TENCEL®

20% Blends with polyester or nylon

Blends with synthetic fibres can provide a considerable improvement in the abrasion performance of TENCEL® fabrics. Substitution of 20% polyester doubles the Martindale abrasion resistance to 35,000 cycles for a typical 200g/m^2 fabric. Blending with 20% nylon in a similar fabric construction has raised the Martindale abrasion level to more than 100,000 cycles.

	100% TENCEL®	80:20 blend TENCEL® / polyester	80:20 blend TENCEL® / nylon
Martindale cycles	15,000	35,000	100,000+

Blending with 20% of a synthetic fibre produces fabrics which retain much of the original character and physical performance benefits of 100% TENCEL® fabrics.

Synthetic blend levels of 20% and below are not considered to present a hot melt hazard in next-to-skin applications provided that fibre blends are used rather than yarn on yarn mixing.

TENCEL® fabrics that also contain synthetic fibres can be cross dyed to solid shades if required. However, TENCEL® fabrics with up to 20% polyester or nylon can give attractive marl effects where only the lyocell fibre has been dyed.

10% Blends with nylon

Using even lower blending levels of synthetic fibre, significant performance benefits are still possible. With only 10% nylon in an intimate blend, Martindale abrasion performance can be raised to over 60,000 cycles for a 200g/m^2 fabric.

	100% TENCEL®	90:10 blend TENCEL® / nylon	80:20 blend TENCEL® / nylon
Martindale cycles	15,000	66,000	100,000+

Use of Fabric Finishes

Abrasion resistance can also be improved by the use of suitable fabric finishes.

For example, the application of an Axis treatment to TENCEL® fabric results in a small but worthwhile improvement in measured abrasion performance.

	100% TENCEL® (without resin)	100% TENCEL® (Axis finish)	100% TENCEL® (Perapret HVN)
Martindale cycles	15,000	20,000	29,000

Abrasion performance can be increased still further by the use of a resin binder. By applying 100g/l Perapret HVN, the abrasion resistance of 100% TENCEL® fabric can be raised to 29,000 Martindale cycles (again, for a 200 g/m^2 fabric). Although such finishes tend to stiffen the fabric, the natural drape and fluidity of TENCEL® means that the final handle is still softer than for other cellulosics without a finish.

Drape and Fluidity

Fabrics in TENCEL® are characterised by their unique drape and fluidity. Such tactile properties are a direct result of the space created within the fabric when it is prepared, dyed and finished. The spaces between the warp and weft yarns allow the fabric to drape well and the ease with which the yarns can move relative to one another gives the impression of fluidity in the fabric.

Loomstate fabrics in TENCEL® are similar to other loomstate fabrics in terms of stiffness, due to the presence of size and the frictional forces between the yarns. However, after wet processing, the handle of the fabric changes significantly: the fabric develops better drape and a fluidity that clearly differs from other fabrics. Hence, the fluidity and drape are related to changes that take place during wet processing of the fabric.

What happens during wet processing?

When TENCEL® fibre is wetted it absorbs water and the fibre diameter increases by about 35%, but the fibre length does not change.

Figure 1 - greige fabric

Figure 2 - wet fabric

Figure 3 - dried fabric

When a greige fabric (Figure 1) in TENCEL® is wetted, the swelling of the fibres/yarns causes an increase in the crimp of the yarn but without an increase in the length of the yarn (Figure 2). Therefore, the distance between the extremities of the ends of the yarn in the fabric must decrease to accommodate the increase in yarn crimp. The fabric dimensions therefore decrease, i.e. the fabric shrinks.

When the fabric is dried the fibres/yarns reduce in diameter back to their original size (Figure 3) but the dimensions of the fabric are not recovered, therefore the fabric remains in the "shrunk" state. This means that space is created around each yarn in the fabric as a result of the wet treatment and subsequent drying. The more a TENCEL® fabric is wetted and dried, the more the fibres and yarns will take up the new configuration of the wet fabric. Fibre movement occurs within the yarns as the fibres reduce in diameter during drying. Fabric treatment processes that involve a number of wetting and drying cycles are therefore very effective in creating space between yarns.

The creation of space between the warp and weft yarns increases the bulk of the fabric and allows the yarns to move easily relative to each other. This ease of movement within the fabric produces both the drape characteristics and the sense of fabric fluidity.

Why other fabrics behave differently

- TENCEL® yarns are very lean, so there is little space within the yarn for the fibres to swell into. The yarn diameter therefore increases significantly when it is wetted. The yarns return to their lean state after the wetted fabric has been dried, so space is created around the yarns rather than within the yarns.
- Other fibres and yarns retain their bulk after wetting and drying, and this reduces available space that has been created around the yarns. There is therefore a greater force between the yarns to limit their movement. Some other cellulosic fibres also give lean yarns but they are not robust enough to withstand the wetting and drying processes required for "space creation".
- Synthetic fibres are generally hydrophobic and therefore do not absorb water. Space cannot therefore be created by the swelling and contraction of the fibres as they are wetted and dried.

To generate drape and fluidity

- Fabric must be allowed to shrink during wet processes
- Tensioning of the fabric should be minimised during wet processing and stentering.
- Dry beating and tumble drying the fabric will enhance drape and fluidity.
- The use of lubricants can help to generate drape and fluidity.
- Increasing the bulk of the yarns will not enhance drape and fluidity.

Moisture Properties

In common with other cellulosics, TENCEL® absorbs water readily to give excellent wearer comfort. Moisture absorbency is combined with an inherently high strength capability to provide a combination of properties that is unusual, if not unique in a cellulosic man-made staple fibre. Absorbency may be enhanced further by using the fibrillating capability of TENCEL®.

The moisture absorbency of TENCEL® fibre is high, and this means that fabrics made from it provide the kind of wearer comfort normally associated with natural fibres such as cotton.

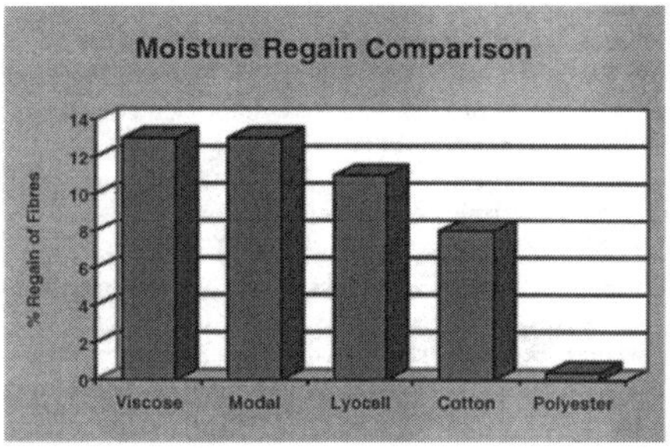

The natural moisture regain of TENCEL® is slightly higher than for cotton and much greater than for synthetic fibres such as polyester. This helps to ensure static free handling of TENCEL® fibres, yarns and fabrics, both in processing and in use.

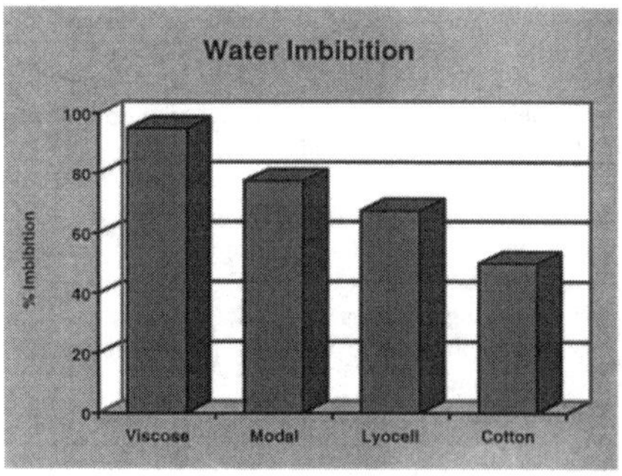

Another measure of water absorbency of a fibre, imbibition, again shows that, TENCEL® can be more absorbent than cotton.
The total moisture absorbency of a fabric made from any cellulosic fibre also will depend on the physical structure, as well as the accessibility of the cellulose. In the case of TENCEL® the physical structure of the final fabric can be modified during wet processing to create surface microfibrils. These fibrils provide an increase in the surface area available for water absorption. This fibrillation capability offers the potential for features normally only available from expensively produced synthetic microfibres.

Surface of fabric without fibrillation

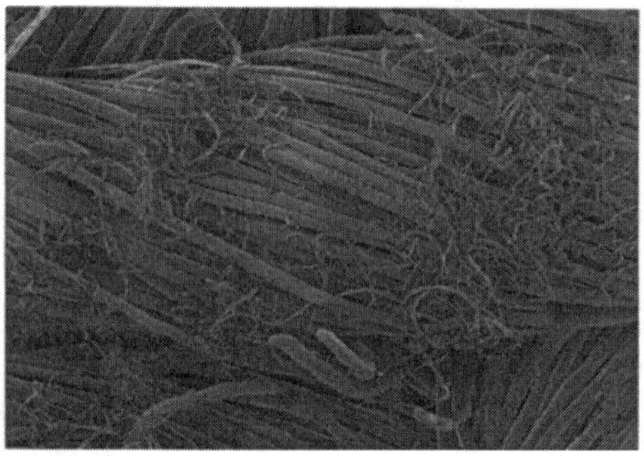

Fibrillated fabric surface

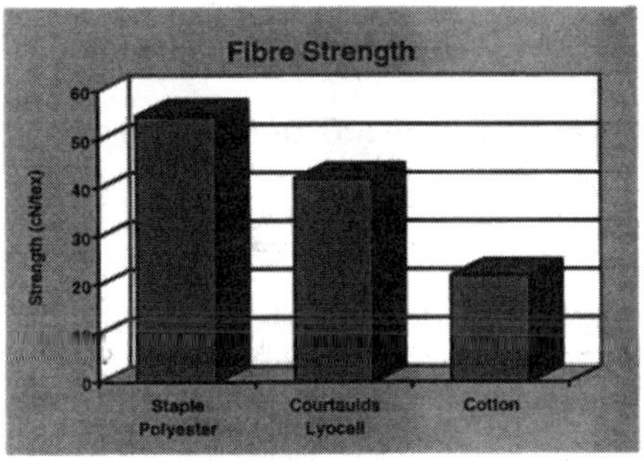

High strength together with good moisture absorbency are an unusual combination in any fibre. TENCEL® provides both in good measure, whilst cotton has a lower strength, and polyester provides the strength but not the absorbency required to give wearer comfort.

TENCEL® retains over 80% of its inherent fibre strength when wet. Full strength is of course recovered when dried again.

The fibre has a high moisture absorbency and yet still has low shrinkage both in processing and in use (see Technical Textiles 6 datasheet). This stability draws from the relatively high fibre stiffness and the fact that the length of a TENCEL® fibre is completely unchanged when wetted or dried.

In comparison, other cellulosic fibres become considerably swollen in length when wet which contributes to potential shrinkage problems in fabrics made from them.

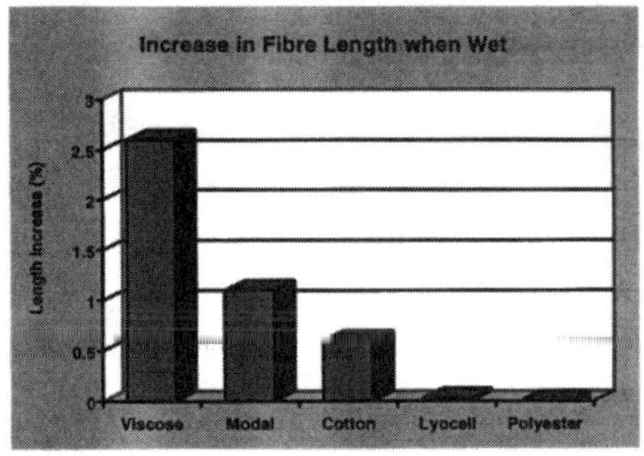

Although the length of a TENCEL® fibre is unaffected by moisture, there is significant swelling in fibre diameter. Indeed, where required, fibre and yarn swelling can be used to close up the spacing between yarns within a fabric in applications such as barrier fabrics, filters and water repellent products.

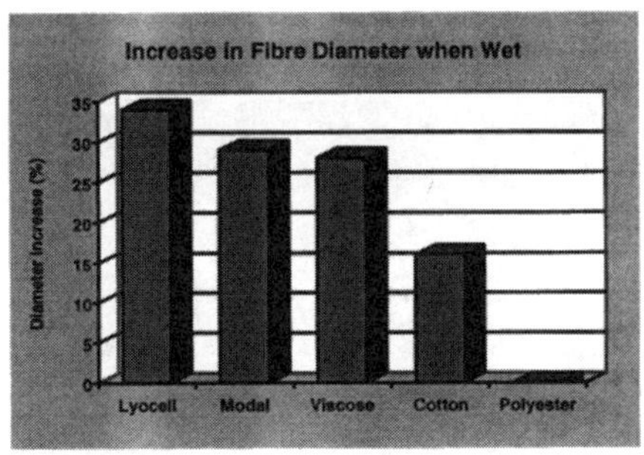

Fibrillation

TENCEL® can be fibrillated to develop interesting aesthetic effects that can also have practical benefits in technical and industrial applications.

Fibrillation is the longitudinal splitting of a single fibre into microfibres of typically less than 1 to 4 microns in diameter. The splitting occurs as a result of wet abrasion against fabric or metal. The fibrils are so fine that they can become almost transparent, giving a white or 'frosty' appearance to finished fabric. In cases of extreme fibrillation, the micro-fibrils can become entangled, giving a pilled appearance.

Using fibrillation

Fibrillation can occur in the wet processing of fabrics and garments made from TENCEL®. The microfibres generated can be used to create a variety of interesting tactile aesthetics. A peach skin effect, which can also withstand repeated domestic washing at 40°C is possible, providing that the fibrillation is developed such that the fibrils cannot become long and entangled. This is key.

Fibrillation can be used in both piece dyeing of fabric and in garment washing/dyeing to produce characteristic soft and drapey aesthetics.

Unfibrillated TENCEL®

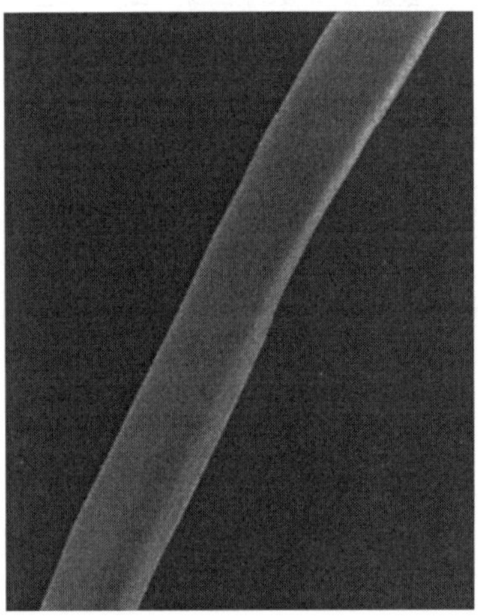

Fibrillated TENCEL® fibre

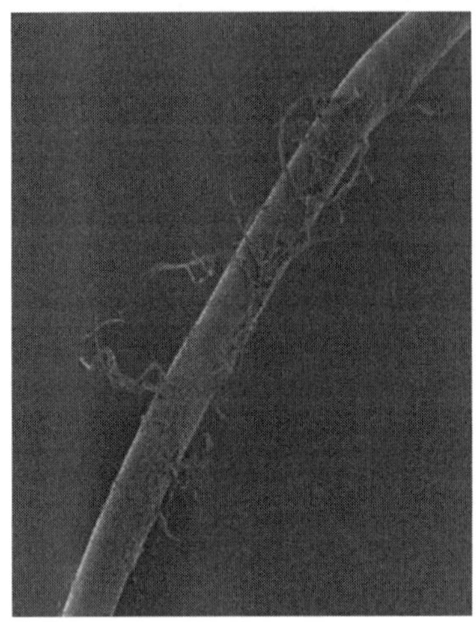

Web Bonding - Hydroentanglement

ACORDIS LYOCELL exhibits good hydroentanglement efficiency, having good aperturing clarity, and providing strong, stable, low linting, soft absorbent fabrics over a wide range of pressures. Hydroentangled ACORDIS LYOCELL fabrics are significantly stronger than those produced from other cellulosic fibres. By using lower decitex ACORDIS LYOCELL, strength, absorbency and handle of hydroentangled fabrics can be further improved

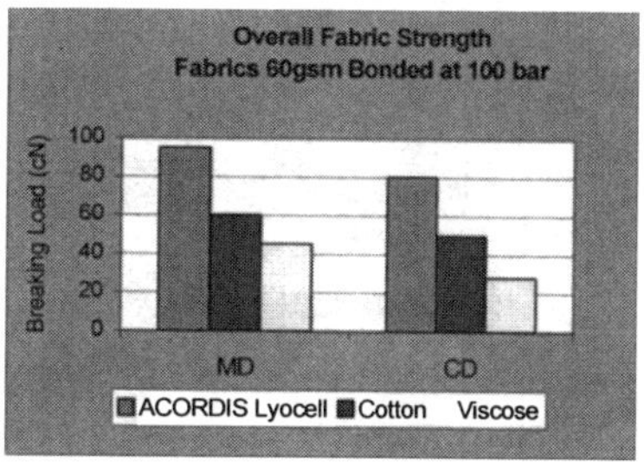

ACORDIS LYOCELL is ideally suited to the hydroentanglement process. The fibre's high wet and dry strength can be translated into high fabric strengths, with the added opportunity to generate fibrillation using this process.

Plain and apertured hydroentangled fabrics can be produced using ACORDIS LYOCELL down to 20gsm. Strong stable fabrics can be made at a range of pressures without the use of binders. When bonded at low pressures, lyocell fabrics are exceptionally soft, have an attractive, silky appearance and particularly good drape characteristics.

ACORDIS LYOCELL fibres can be fibrillated by using bonding pressures over 120 bar.

Fibrillation can alter fabric properties significantly, depending upon the fabric construction - opacity, wicking, wiping performance, strength, and barrier performance can all be improved.

Low levels of fibrillation do not increase fabric linting, as the fibrils remain attached to the fibre.

Fibrillated ACORDIS LYOCELL fibres

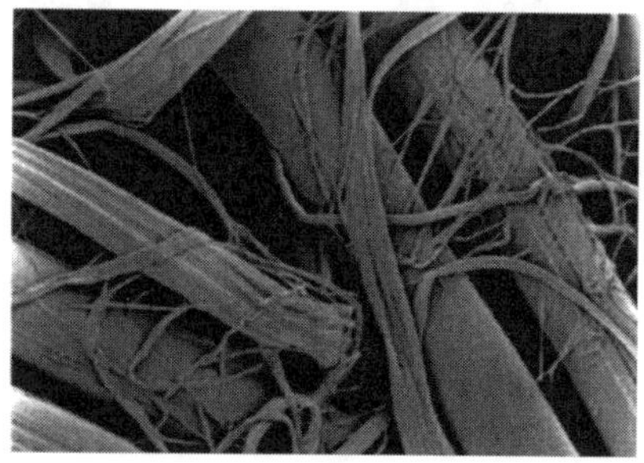

The abrasion and delamination resistance of hydroentangled lyocell fabrics can be improved by blending with nylon or polyester. Superior, high strength, bulky, porous and absorbent coating substrates, having a smooth surface, particularly suitable for use in artificial leathers can be made.

Durable fabrics with a more textile like appearance are also possible by manipulation of the bonding conditions used. High pressure bonding of ACORDIS LYOCELL fibres has produced fabrics of comparable strength to equivalent weight woven cotton.

ACORDIS LYOCELL has been found to have excellent aperturing and embossing clarity compared to other cellulosic fibres used in hydroentanglement, making excellent fabrics for gauzes and wipes.

ACORDIS LYOCELL can be readily used in blends with viscose to improve both wet and dry strengths of hydroentangled nonwovens. Blends with polyester have also produced highly absorbent products.

ACORDIS LYOCELL has similar absorbency to viscose fibre, but this is significantly improved when blended with polyester. This is primarily due to the higher resilience of lyocell fibre, which does not suffer from wet collapse as viscose does.

As with any fibre, some optimisation of fibre type, blend ratio and pressure profile will be required to achieve the desired parameters. Lower decitex variants such as 1.4 can offer improved strength, softness and absorbency over 1.7 decitex fibres, for the same basis weight and bonding conditions.

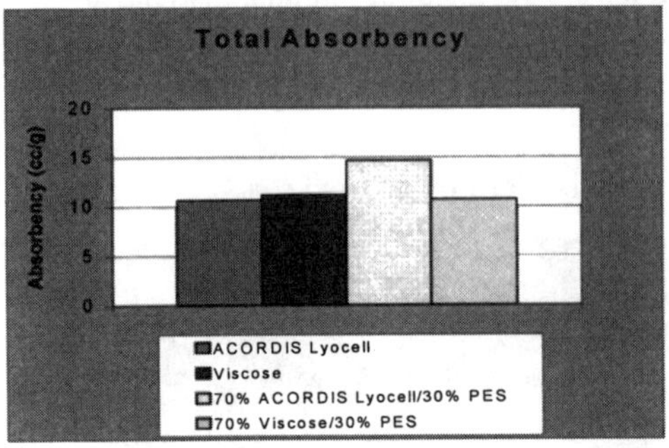

Because of the solvent spinning process, ACORDIS LYOCELL is a remarkably pure form of cellulose. After hydroentanglement ACORDIS LYOCELL has been found to have lower levels of cations and anions than other cellulosic fibres, making it suitable for products such as high performance clean room wipes.

Hydroentangled ACORDIS LYOCELL now has US FDA 510K covering its use in sensitive medical applications, where its strength, absorbency, low linting, precise aperturing and disposability are key features. Unlike its synthetic counterparts, lyocell products are completely biodegradable.

Web Bonding - Needling

ACORDIS LYOCELL fibres give stronger and more stable needlefelts than viscose. Needled ACORDIS LYOCELL webs are more resistant to wet collapse than viscose equivalents and can form more open, bulky structures with a higher absorbent capacity.

At moderate needling densities, ACORDIS LYOCELL needlefelts are markedly stronger and more stable than viscose equivalents. This is illustrated by the tensile and burst strength data below, which particularly highlight the strength retention of ACORDIS LYOCELL when wet.

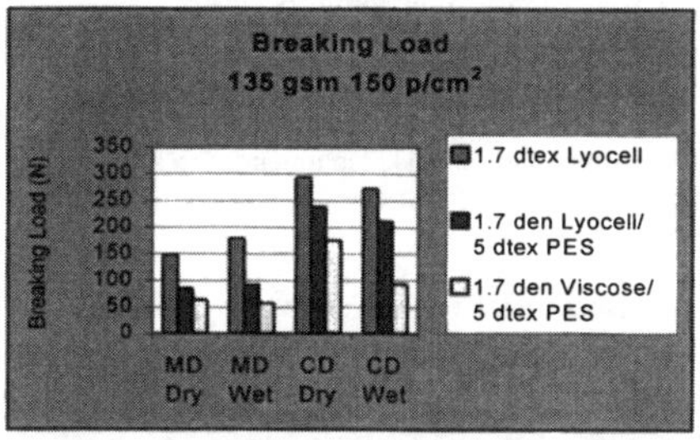

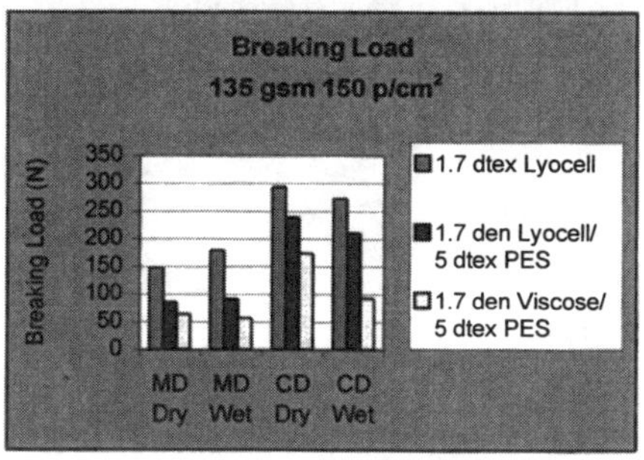

ACORDIS LYOCELL is so strong that it can easily be used at finer counts. In trials, fabrics made from 1.7 dtex ACORDIS LYOCELL were over twice as strong as viscose when dry and three times as strong when wet. ACORDIS LYOCELL can also boost the strength of cellulosic/polyester blends.

Higher strength can be valuable in itself, but also offers the potential to make lighter products or to reduce the level of non-absorbent binding materials. ACORDIS LYOCELL which forms more open, bulky needlefelt structures than comparable viscose fibres can help increase total absorbent capacity.

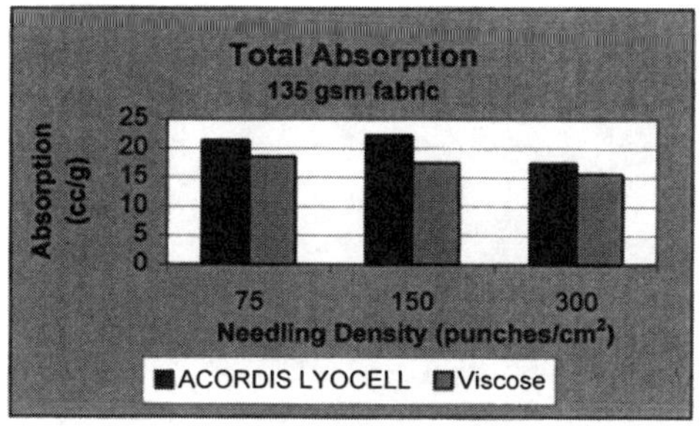

The high wet modulus of ACORDIS LYOCELL also leads to improved wet resiliency. Needlefelts made from 1.7 dtex ACORDIS LYOCELL show only half the wet collapse of equivalent 3.3 dtex rayon structures. This is clearly advantageous in applications where liquid retention is important.

The balance of strength, absorption rate and capacity can be further manipulated for a particular end-use by the choice of appropriate needling parameters.

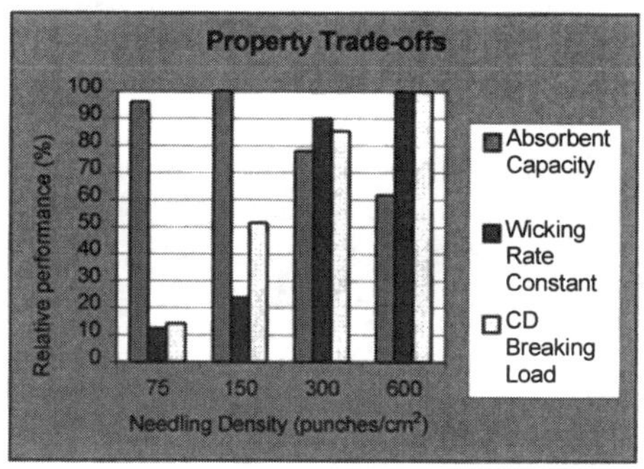

Web Bonding - Latex

Replacing viscose with ACORDIS LYOCELL in latex bonded fabrics doubles dry strength and gives an even greater improvement in wet strength. This opens up opportunities for stronger or lighter products, or reduced binder levels for greater absorbency.

Latex bonded ACORDIS LYOCELL

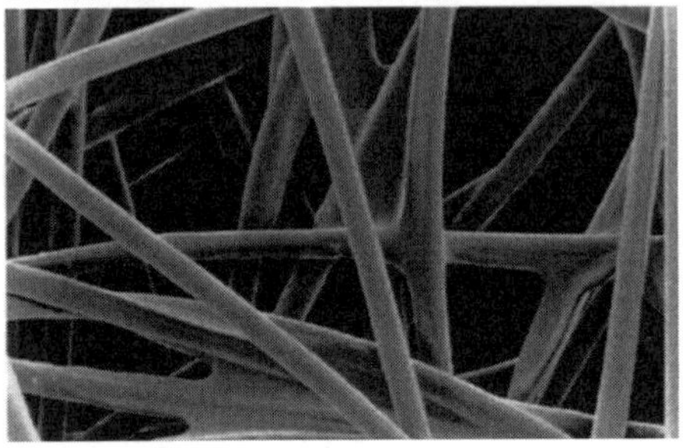

In saturation bonding, ACORDIS LYOCELL gives twice the dry fabric strength of viscose rayon and an even greater improvement in wet strength.

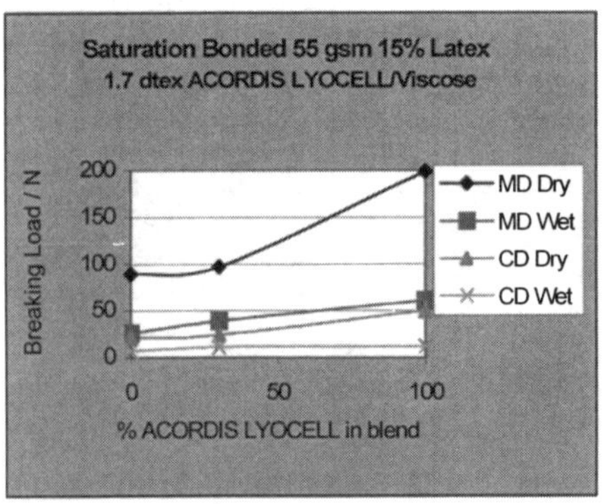

This improved strength offers various opportunities. Fabrics can be made stronger. Alternatively products can be made to current strength specifications, but at lower basis weight or with reduced binder levels. Lower binder levels create the potential for increased absorbency and better fabric aesthetics.

Because ACORDIS LYOCELL has a higher fibre modulus than viscose rayon, the resulting saturation bonded fabrics are stiffer than when viscose is used under the same bonding conditions. This may be beneficial in some end uses, e.g. interlinings.

If greater drape and softness is required, the use of a softer (lower Tg) latex is recommended to achieve the optimum balance of fabric aesthetics and performance.

Similar effects can be achieved in print bonded fabrics, e.g. for household wiping applications. Partially replacing viscose with ACORDIS LYOCELL boosts strength, opening the way for stronger, lighter or more absorbent products.

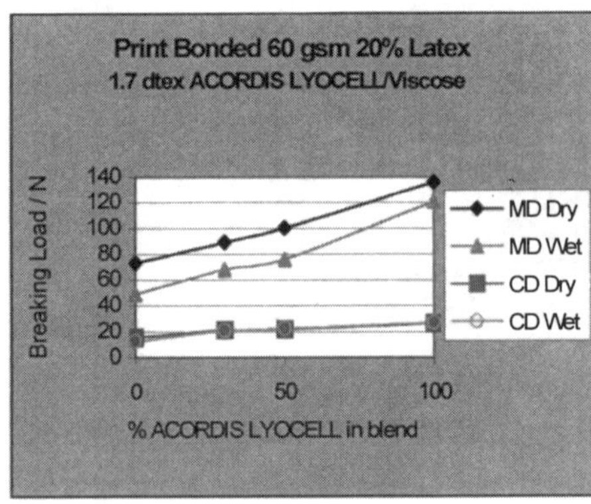

Wet Laying

ACORDIS LYOCELL is produced as a tow, which is parallel and twist-free. In this form it is ideal for precision cutting to short fibre lengths for wet-laid nonwovens, allowing clean, uniform webs to be produced.

ACORDIS LYOCELL can be dispersed easily in water. The fibre has a relatively high modulus, so long lengths can be used for wet-laying, allowing very strong webs to be produced by this route. ACORDIS LYOCELL has been successfully wet-laid up to 16mm, although the fibre does need a high degree of dilution at such long staple lengths.

The circular cross-section of lyocell fibres produces a large amount of inter-fibre contact, resulting in a high wet web cohesion and making transfers of the web prior to bonding much easier.

ACORDIS LYOCELL has a low water imbibition which allows easier drying of wet fabrics whilst the fibre's high wet stability results in reduced fabric shrinkage losses.

The fibre can be fibrillated by vigorous mixing of the slurry in a hydrapulper or similar mixer. Fibrillation adds strength to the web, which can then be processed as a paper. Alternatively, bonding can be carried out by hydroentanglement, where, again, fibrillation can be induced if so desired.

Fibrillation dramatically increases the mean length to diameter ratio, and agglomeration will result if fibres are close to the critical length and concentration.

Hydroentangled wet-laid lyocell webs produce strong fabrics. Further significant strength improvements can be achieved by increasing the fibre length: fabric produced from 12mm lyocell has three times the dry tensile strength of 5mm fabric.

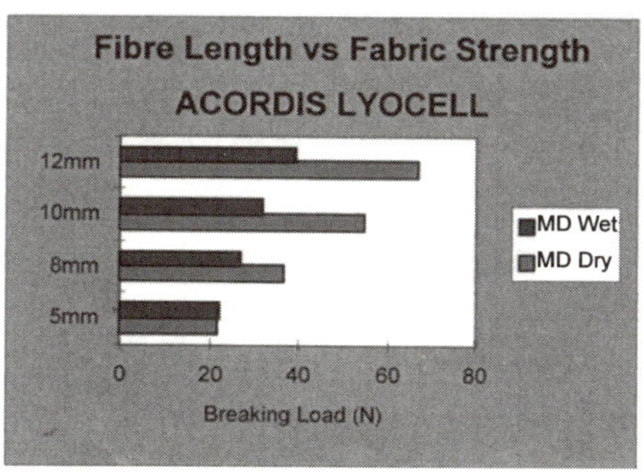

Decreasing fibre from 1.7 to 1.4 dtex, also produces significantly stronger wetlaid nonwovens by increasing the number of fibres within the fabric. The 1.4 dtex wet-laid nonwoven also exhibits significantly higher wet than dry strength.

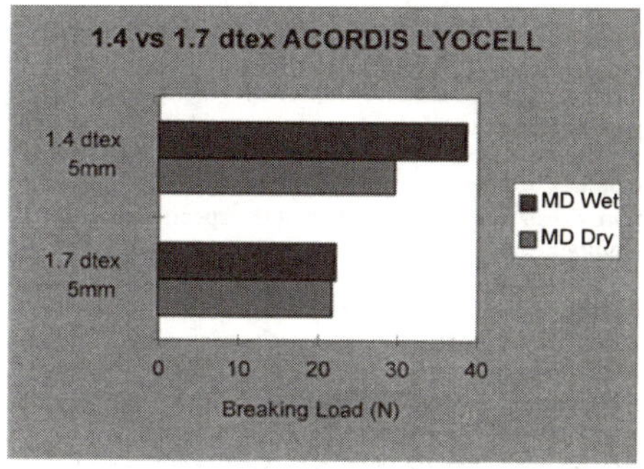

Further strength improvement is possible by using longer length 1.4 dtex fibres.

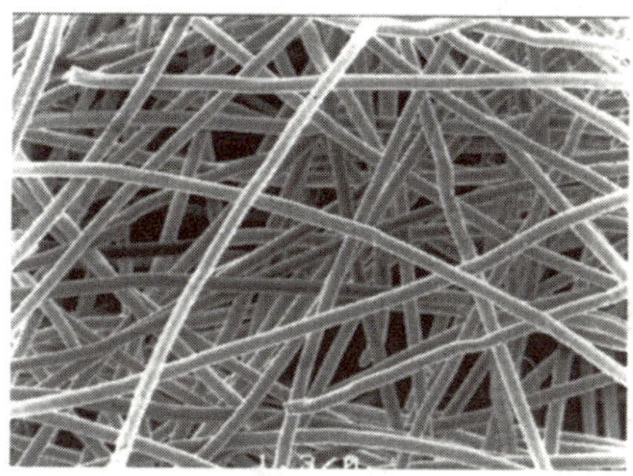

Air Laying

ACORDIS LYOCELL tow can be precision cut to short fibre lengths, either crimped or uncrimped, which are ideal for air-laying. Inclusion of the fibre in air-laid pulp products can improve bulk, softness, absorbent capacity and strength.

The importance of air-laying as a technology is growing significantly, with more sophisticated products and demanding applications being introduced. This extends the opportunity for using new fibres in addition to woodpulp, in order to achieve the desired fabric properties. As ACORDIS LYOCELL is manufactured as a tow product, it can be cut to short lengths suitable for air-laying processes. Additionally, there exists the possibility to use crimped or uncrimped fibre the choice depending upon the process detail and fabric properties required. A range of fibre decitexs is available, and fibre finish can be tailored to suit the specific process requirements.

Results from fabrics produced using flat-bed air-lay technology demonstrate the benefits of using ACORDIS LYOCELL fibre in air-laid structures. In a series of trials 4mm crimped ACORDIS LYOCELL was used as a replacement for pulp in a 80 gsm fabric with 20% bicomponent fibre (included for bonding purposes). The length of the lyocell fibres and excellent strength properties result in a significant strength improvement, especially when wet - where the fabric containing ACORDIS LYOCELL shows a doubling in strength.

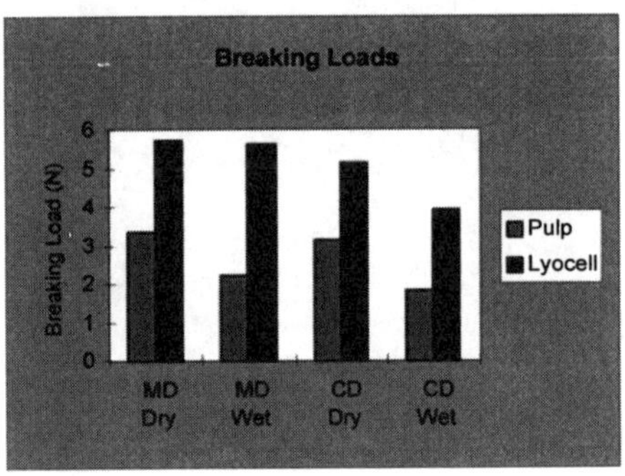

The use of ACORDIS LYOCELL also results in an improvement in bulk properties. Dry results show the fabric including lyocell to be thicker (lower bulk density). Lyocell also shows superior resiliency when wet, indicated by the smaller increase in density compared to pulp.

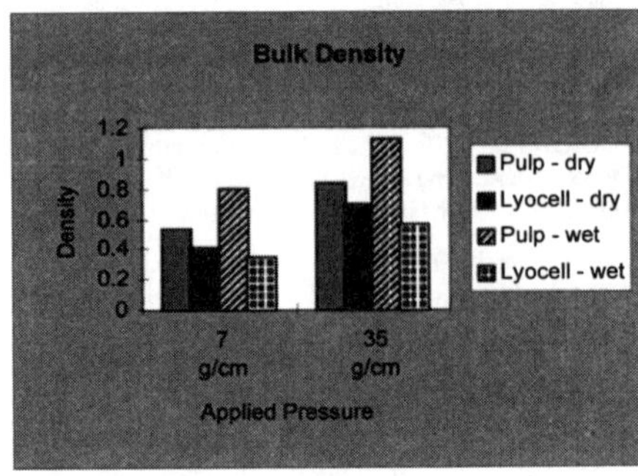

This improvement in bulk and resiliency leads also to an increase in total absorptive capacity. The fabric containing lyocell has a total absorption of 21.9 cc/g - an improvement of around 50%. Panel testing used to evaluate fabric softness shows a significant improvement in fabric hand when lyocell was used. ACORDIS LYOCELL can, therefore, also bring benefits in fabric softness.

8-10mm ACORDIS LYOCELL can be used in cylinder-type air-lay processes. Here, the use of longer fibre lengths gives further improvements in properties.

Wipes

ACORDIS LYOCELL's unique combination of strength, absorbency and fibrillation potential make it an ideal component in products for a variety of wiping applications. Lyocell is already being used commercially in drying cloths, wipes for critical tasks (clean manufacturing environments, print and paint shops), food service wipes and patient washcloths in the medical industry.

Drying Cloths

A commercial drying cloth based on ACORDIS LYOCELL shows faster rates of water absorption than competitive products. Owing to its high rate of demand absorbency and absorbent capacity, this wipe absorbs more surface water than competitive products in a test to simulate wiping.

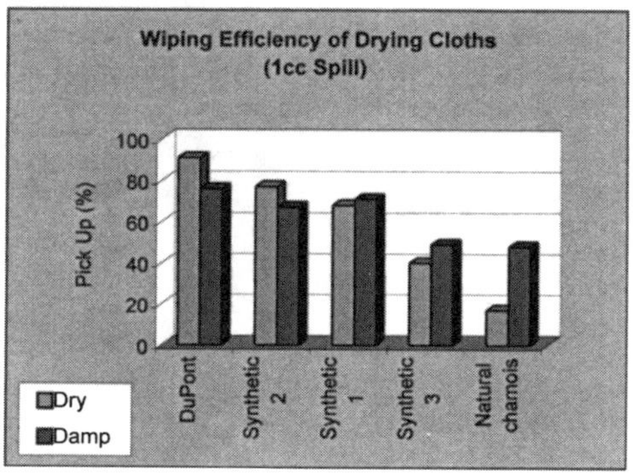

The lyocell based drying cloth is as strong as natural chamois and can be washed repeatedly. Wiping performance actually improves with repeated washing.

ACORDIS LYOCELL

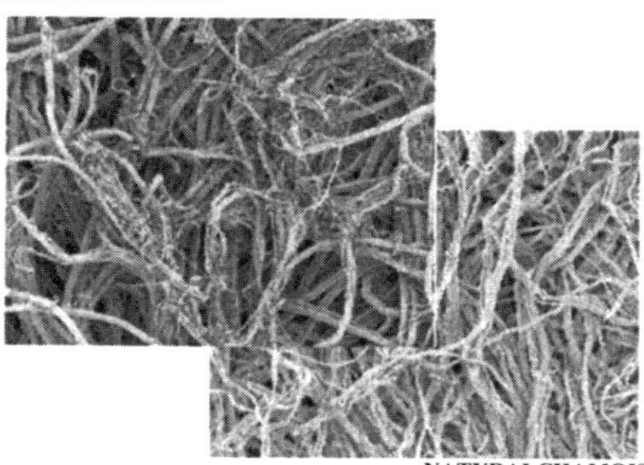

NATURALCHAMOIS

The fibrillated ACORDIS LYOCELL fibres are able to simulate the collagen microfibre structure of natural chamois better than typical man-made drying cloths.

Furthermore, unlike natural chamois, the lyocell-faced cloth does not feel slimy when wet and stays soft even when dry.

Critical Task Wipes

These wipes are used in a variety of applications where low linting fabrics are required, such as clean manufacturing environments and automotive or boat painting and refinishing. The presence of ACORDIS LYOCELL enhances various critical performance features of hydroentangled critical task wipes. The product absorbs liquid faster and has a higher wet tensile strength, but stays soft, preventing the risk of abrasion damage. ACORDIS LYOCELL gives lower linting without the use of binders.

Food Service Wipes

Apertured spunlaced food service wipes made from a blend of ACORDIS LYOCELL and polyester offer several advantages over an industry standard 2 oz/yd^2 wipe. Demand absorbency is 7 times faster and capacity is 60% greater. There is also less residual smearing when wiping ketchup in a wiping simulation test.

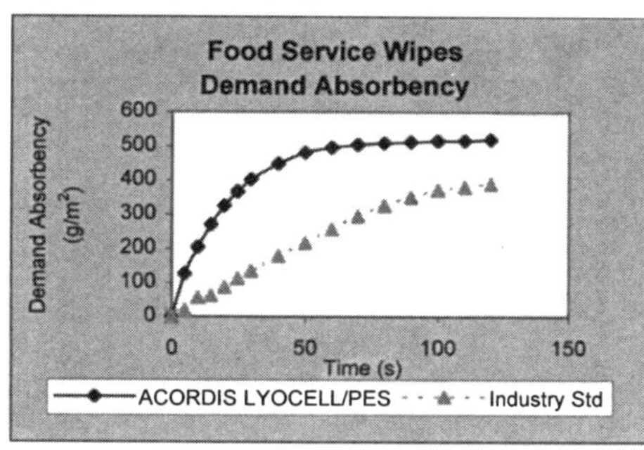

The product has twice the CD wet strength and remains stronger after extended bleaching.

As well as being stronger and more absorbent it is also softer.

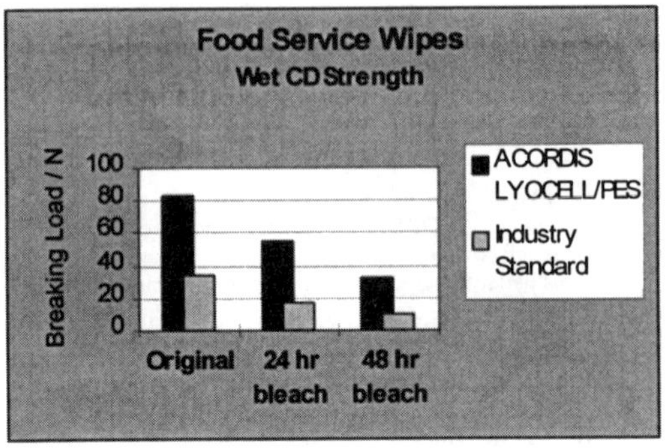

Baby wet wipes
ACORDIS LYOCELL also has the potential to create improved wet wipes, although these are not yet commercial. Lyocell enhances strength and stability, increases wet resiliency and gives improved crease definition. ACORDIS LYOCELL also gives excellent embossing clarity so is ideal for creating novel surface texture effects.

Medical Swabs and Gauzes

ACORDIS LYOCELL exhibits good hydroentanglement efficiency, delivers aperturing clarity, and provides strong, stable, low linting absorbent fabrics ideal for use in medical applications. Hydroentangled ACORDIS LYOCELL fabrics are significantly stronger than those produced from other cellulosic fibres. It meets the necessary standards required of fabrics for medical uses and provides an ideal partner for blends.

Hydroentangled fabrics continue to take an increased share of the swab and gauze market. The performance of ACORDIS LYOCELL is ideally suited to the hydroentanglement process. The fibre's high wet and dry strength can be translated into high fabric strengths.

Plain and apertured hydroentangled fabrics can be produced using ACORDIS LYOCELL down to 20gsm. Strong stable fabrics can be made at a range of pressures without the use of binders. When bonded at low pressures, ACORDIS LYOCELL fabrics are exceptionally soft, have an attractive, silky appearance and particularly good drape characteristics. At higher bonding pressures a firmer handle is obtained which can assist debridement.

Hydroentangled ACORDIS LYOCELL has US FDA 510K approval covering its use in sensitive medical applications. 100% ACORDIS Lyocell products are completely biodegradable.

A cellulosic fibre needs to perform acceptably following sterilisation as a pre-requisite to finding widespread acceptance in nonwoven medical end-uses.

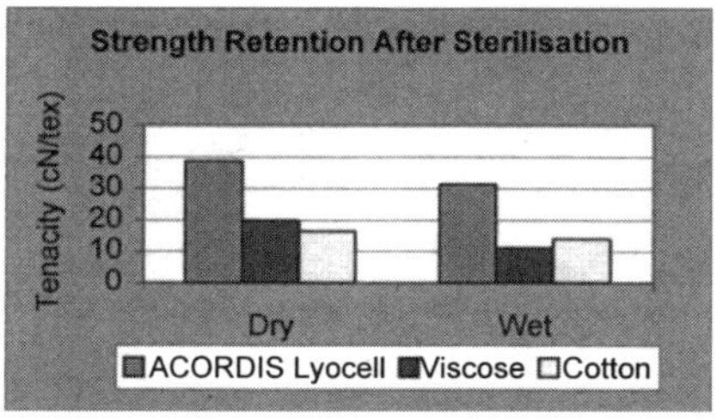

ACORDIS LYOCELL has been found to have excellent aperturing and embossing clarity compared to other cellulosic fibres used in hydroentanglement, making excellent fabrics for gauzes. ACORDIS LYOCELL is supplied in a delustred form to provide a non-reflective surface in use.

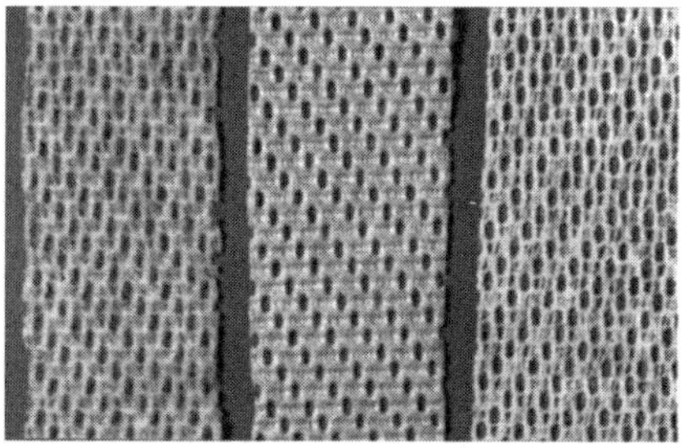

ACORDIS LYOCELL can be readily used in blends with viscose to improve both wet and dry strengths of hydroentangled nonwovens. Blends with polyester have also produced highly absorbent products.

ACORDIS LYOCELL has similar absorbency to viscose fibre, but this is significantly improved when blended with polyester. which is primarily due to the higher resilience of lyocell fibre.

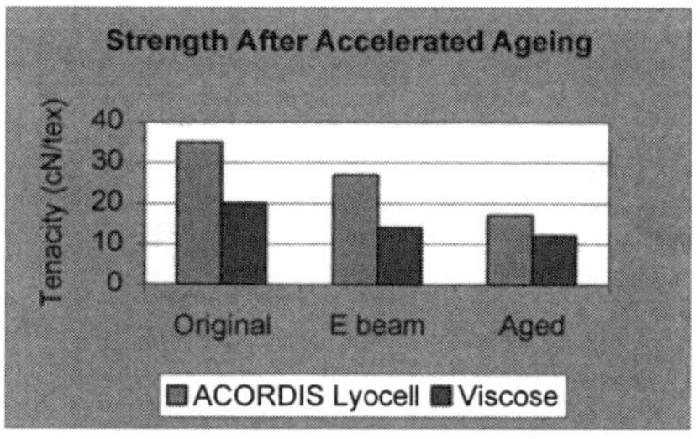

ACORDIS LYOCELL fibres are produced to meet the pharmacopoeia requirements originally devised for woven cotton products. Product development is designed to meet customers specific requirements.

Absorbency

The affinity of ACORDIS LYOCELL for water lends itself to a range of applications in absorbent products. Like viscose rayon, ACORDIS LYOCELL has a highly wettable surface and inherent fibre absorbency. However, ACORDIS LYOCELL exhibits less wet collapse than rayon, thanks to its high wet modulus. This resiliency can boost absorbent capacity as well as maintaining pore integrity for faster wicking.

ACORDIS LYOCELL is hydrophilic and swells in water, making it suitable for a range of applications in absorbent products.

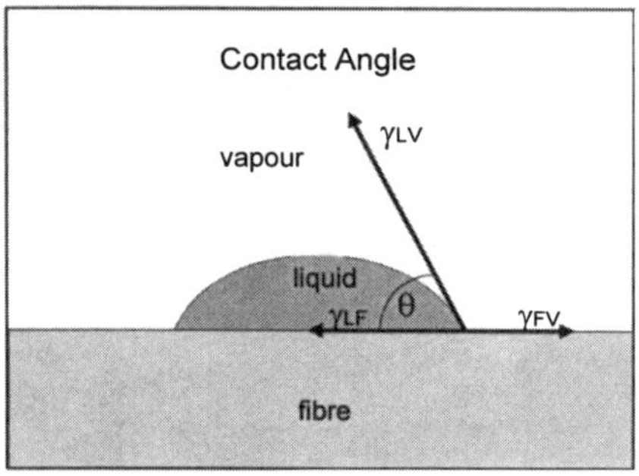

The fibre has a similar fibre/water contact angle to viscose rayon (10-40° depending on surface finish), leading to rapid fibre surface wetting on exposure to water.

After water penetration, the fibre cross sectional area increases by 50% - over twice the swelling of cotton.

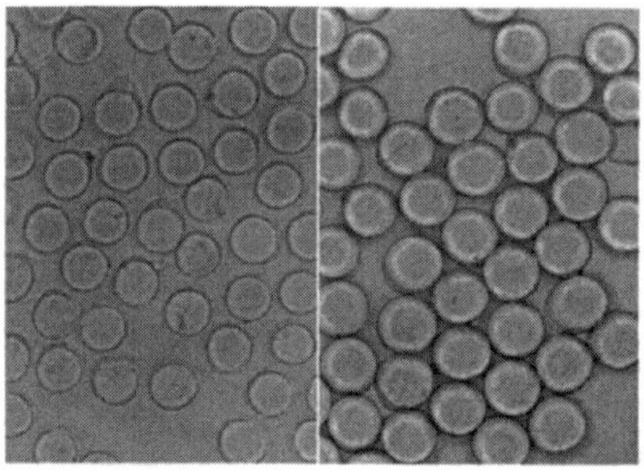

ACORDIS LYOCELL swelling in water

Such swelling enhances liquid transport within and between fibres, which can be particularly important in applications where liquid is moved vertically against the force of gravity.

ACORDIS LYOCELL has a higher water imbibition than cotton, but lower than viscose (ACORDIS LYOCELL 65%, cotton 45%, viscose 95%).

Generally, structures made from viscose lose bulk when wetted, due to the fibre's low wet modulus. This reduces inter fibre pore volume and so lowers the total absorbent capacity. The high wet modulus of ACORDIS LYOCELL makes it resistant to such wet collapse, offering the potential to improve absorbent capacity. For example, ACORDIS LYOCELL gives a higher total absorption than viscose in this series of needlefelts.

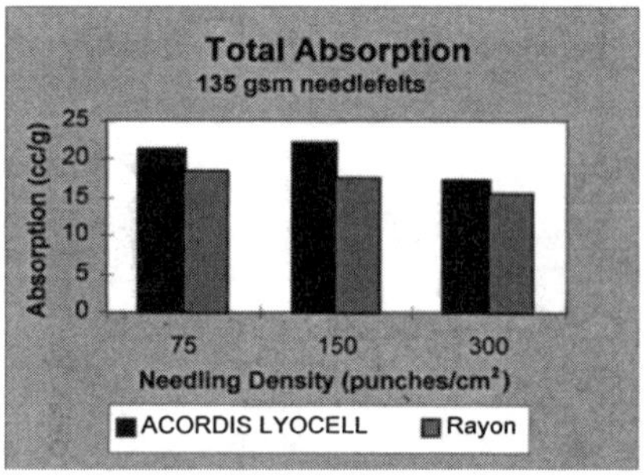

The rate of fluid absorption depends on pore size. Smaller dry pores increase the capillary force which drives fluid uptake, whilst larger wetted pores reduce the viscous drag that slows down fluid transport. The rates of transplanar absorption and wicking in a fabric are therefore highly dependent on fabric construction.

In viscose nonwovens, wet collapse can drastically reduce the size of pores in the wet structure, increasing the drag forces that inhibit flow. By contrast, the resiliency of ACORDIS LYOCELL helps to maintain pore integrity and favours faster wicking rates.

Chemical and UV Resistance

In addition to having a higher initial strength, ACORDIS LYOCELL outperforms other cellulosic fibres in its ability to retain fibre strength and integrity when exposed to mineral acids or UV irradiation.

Chemical Properties

ACORDIS LYOCELL degrades hydrolytically when in contact with hot dilute or concentrated mineral acids. The fibre swells when first exposed to alkalis (maximum at 9% NaOH solution at 25°C) and then ultimately, disintegrate.

Strength Retention

Cellulosic fibres are not usually regarded as being resistant to degradation by acids or alkali. However, ACORDIS LYOCELL not only resists the degradative effects of mineral acids better than both cotton and viscose, but also retains a higher proportion of its already superior wet and dry strengths.

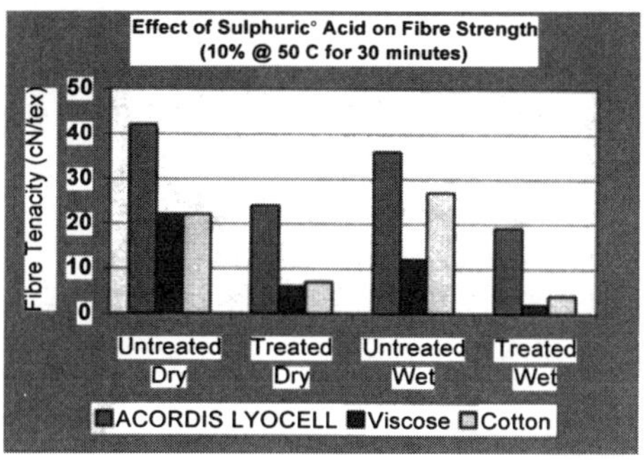

UV Resistance

ACORDIS LYOCELL retains its significant strength advantage over viscose following prolonged exposure to UV irradiation. After 14 days exposure to a Xenon lamp (equivalent to 140 days of direct sunlight) the fibre tenacity is reduced by less than 10%.

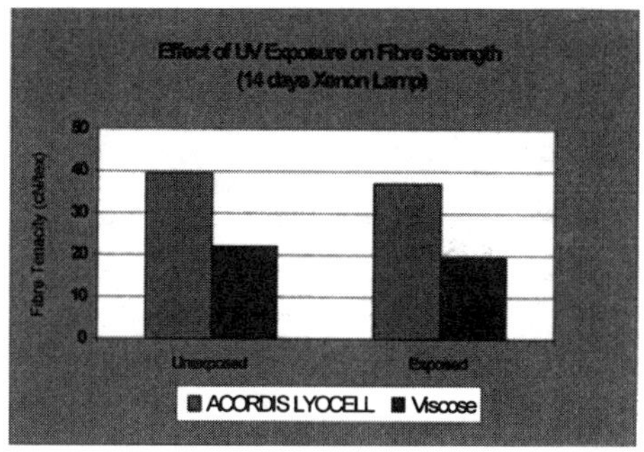

Swelling

Lyocell swells when in contact with either acids or alkali in a comparable way to viscose.

% Fibre Swelling	Sulphuric Acid Concentration		Sodium Hydroxide Concentration	
	0.1%	1.0%	1.0%	10.0%
ACORDIS LYOCELL	46	47	50	1345
Viscose	69	49	101	1059

Biodegradation

Cellulose is the natural polymer that makes up the living cells of all vegetation. It is the most abundant and renewable biopolymer on Earth. Like all cellulosic fibres, products made from ACORDIS LYOCELL are completely biodegradable and can be composted, digested in sewage, landfilled or incinerated.

Biodegradation occurs through the action of enzymes created by living organisms, breaking a product down to carbon dioxide (CO_2) and water (H_2O). Cellulosic fibres commonly used in nonwoven products such as ACORDIS LYOCELL, viscose and cotton are known to be biodegradable, whereas synthetic fibres are not.

Composting

ACORDIS LYOCELL and viscose fibres were found to have degraded completely after 6 weeks in a static aerated compost pile, and cotton fibre had suffered a weight loss of 80%. The synthetic fibres tested, polyester, polypropylene and polyethylene, showed very little signs of degradation.

Sewage Treatment

Sewage treatment plants and septic tanks operate mainly under anaerobic conditions, though some parts of the process are aerobic. The microorganisms present in sewage are accustomed to breaking down cellulosic products such as tissue paper. The resultant natural gas generated can be used to power the sewage treatment works.

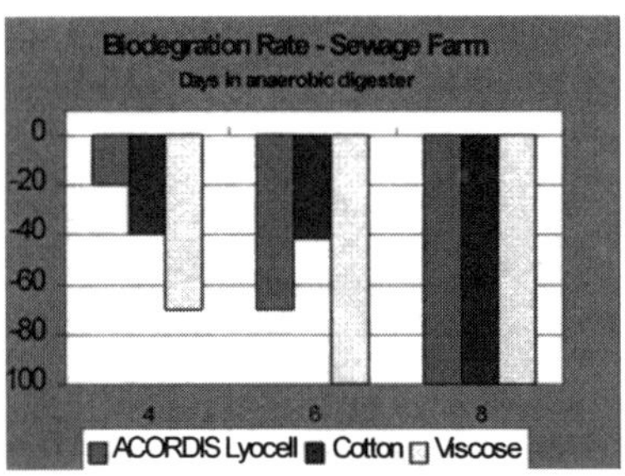

ACORDIS LYOCELL, viscose and cotton fibres degrade completely within 8 days in a typical sewage farm anaerobic digester, where the residence cycle is about 20 days. The synthetic fibres tested show slight reductions in tensile strength after 12 weeks in an anaerobic digester. Similar results would be obtained in septic tanks.

Landfill

Organic matter buried in the ground rots over a period of time by the bacterial process of anaerobic digestion. A landfill site is not easy to define or simulate, as it is somewhat heterogeneous. Soil burial tests (BS 6085 /AATCC 30) are accepted methods of assessing the biodegradability of a product. ACORDIS LYOCELL, viscose and cotton fibres degrade completely within 12 weeks. Synthetic fibres gain weight initially, and only show slight strength and weight loss after 24 weeks burial.

The result of the biodegradation studies carried out to date correlate well with the knowledge we have regarding the structure and chemical resistance of cellulosic and synthetic fibres.

Incineration

Mass incineration plants especially in large conurbations, with facilities for recovering energy from waste, can earn substantial amounts of money from the sale of electricity to power generating companies. Cellulosic fibres, such as ACORDIS LYOCELL burn readily with a heat of combustion of 15 kJ/g. Such plants also have the added advantage of reducing the volume of refuse sent to landfill sites by as much as 90%.

100% ACORDIS LYOCELL Papers

The ability of ACORDIS LYOCELL fibre to split into micro-fibres during wet processing makes it ideal for use in papers. The long fibre length and round sub-micron fibrils can improve paper properties such as tear strength, opacity and air permeability. These properties can be tailored by controlling the amount of fibrillation generated on the fibre.

ACORDIS LYOCELL Low fibrillation

ACORDIS LYOCELL Highly fibrillated

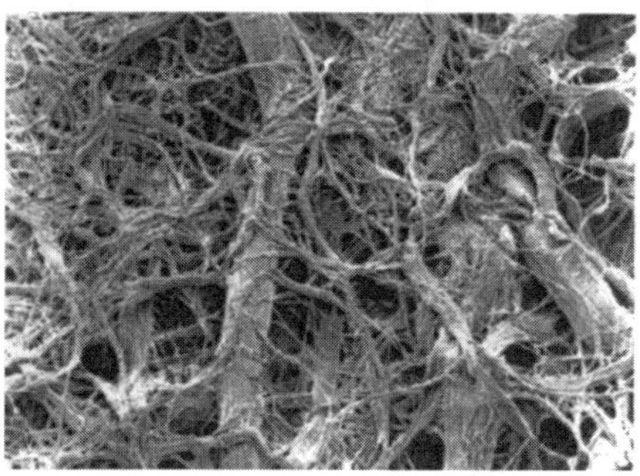

The potential for ACORDIS LYOCELL fibre to fibrillate into micro-fine fibrils makes it ideal for use in special papers. This fibrillation can be achieved using papermaking equipment such as beaters or refiners or by vigorous mixing in a hydrapulper or high shear mixer.

Laboratory beating demonstrates the potential for lyocell fibre to be fibrillated to a range of levels, as indicated by the Canadian Standard Freeness test for drainage rate.

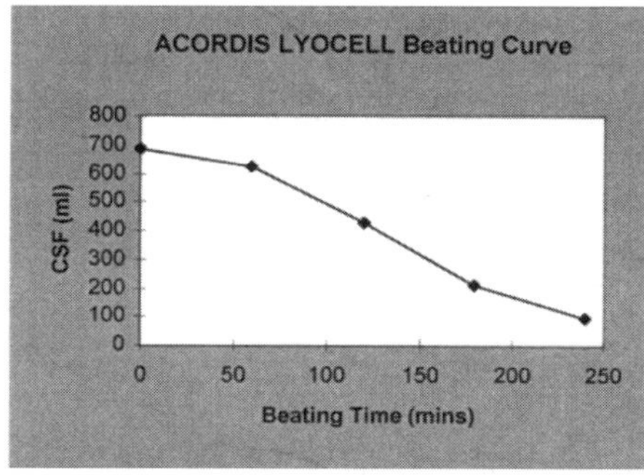

Pilot scale refining work has demonstrated the potential for fibre fibrillation, whilst plant scale trials have confirmed that lyocell can be processed using standard papermaking equipment. ACORDIS LYOCELL has been successfully fibrillated in beaters, refiners and hydrapulpers

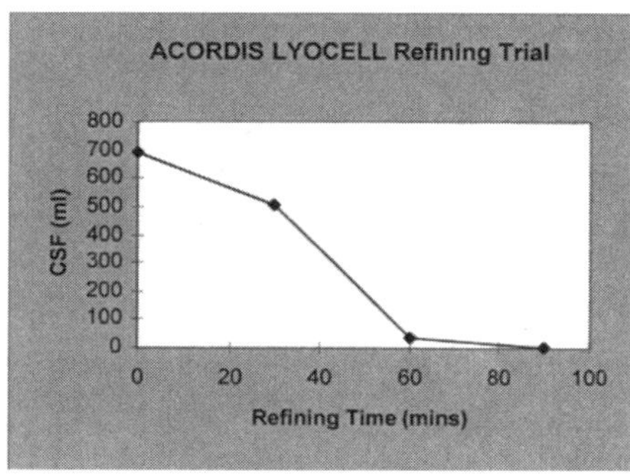

The unique nature of ACORDIS LYOCELL and of the fibrillation generated from it, results in characteristic paper properties: papers with good tensile strength, outstanding tear strength and high opacity can be made.

Lyocell papers also have relatively high air permeability compared to woodpulp papers due to the circular cross-section of the fibres and fibrils.

Typical water leaf paper results show how the generation of fibrillation influences sheet performance and the interesting combination of properties which are generated. As more fibrillation is produced, tensile and tear properties increase. Air resistance increases although it remains low compared to a paper produced from woodpulp. Opacity of the paper increases as refining progresses.

Recommendations for wet strength resins suitable for use with ACORDIS LYOCELL can be supplied. Laboratory work has shown that wet and dry properties can be improved without adversely affecting air permeability.

ACORDIS LYOCELL is suitable for a wide range of special paper types covering end-uses such as filters, battery separators, food casings, map and chart papers, tea bags, cigarette papers, bank-note and security papers.

ACORDIS LYOCELL papers have gained German BGA approval for use in food contact applications.

100% Lyocell, refined at 0.75%	Refining Energy (kWh/tonne)			
consistency, SEL 0.2 Ws/m	0	200	400	600
Freeness (ml)	800	607	93	0
Basis Weight (gsm)	79.3	68.5	66.8	66.4
Thickness (microns)	256	178	151	135
Bulk (cc/g)	3.2	32.6	2.3	2.0
Tear Strength (mN)	584	2237	1823	973
Burst Strength (kPa)	18.8	101	120	136
Tensile Strength (N/15mm)	3.8	15.8	25.3	30.7
Breaking Length (km)	0.3	1.8	2.57	3.2
Stretch (%)	1.9	1.7	2.6	2.2
Double Folds	0	15	69	287

Woodpulp Fibrillation

ACORDIS LYOCELL Fibrillation

Blends

ACORDIS LYOCELL is ideal for use in blend with woodpulp and can enhance specific properties, such as tear, opacity and filler retention. The fibre can also be used as a binder for microglass fibres.

ACORDIS LYOCELL can be used to enhance specific properties of papers when combined with other fibres and it is particularly effective in improving tear strength and opacity as well as in reducing air resistance.

ACORDIS LYOCELL can be used readily in blend with a wide range of other fibres. For example, in laboratory work, addition of lyocell to a typical pulp stock (70% hardwood, 30% softwood) gave improvements in a range of properties.

	Tear Index ($mN.m^2/g$)	Tensile Index (Nm/g)	Burst Index ($kPa. m^2/g$)	Air Resistance (secs)	Opacity (%)
100% Pulp	11.0	52.5	4.7	10.4	71.3
90% Pulp 10% unbeaten ACORDIS LYOCELL	13.2	50.4	4.35	4.01	73.2
90% Pulp 10% 230 CSF ACORDIS LYOCELL	13.8	60.4	4.0	9.8	74.8

The improvement in properties which can be obtained by inclusion of ACORDIS LYOCELL makes it a highly suitable alternative to other reinforcing fibres.

For example, short cut polyester is commonly used for tear strength enhancement.

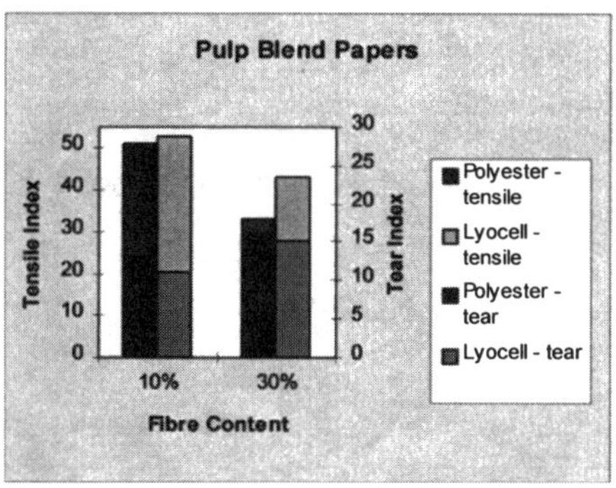

Results show lyocell to give very similar improvements in tear than polyester, when blended with pulp. Tensile results are also similar. Additionally, using lyocell has the advantage of maintaining a furnish composed of 100% cellulose.

ACORDIS LYOCELL can also be used in blend with synthetic fibres, which do not self-bond readily and normally require a chemical bonding agent. Papers can be made from blends of lyocell with microglass without the need for an additional binder - the lyocell adds strength and flexibility to the sheet. Alternatively, inclusion of ACORDIS LYOCELL could allow a reduction in the amount of binder required for a paper.

ACORDIS LYOCELL can also be used as an agent to aid retention of fillers in papers - the fine fibrils allowing particulates to be held in place effectively. This can enable the inclusion of higher levels of filler, or reduce the loss of fillers from the sheet.

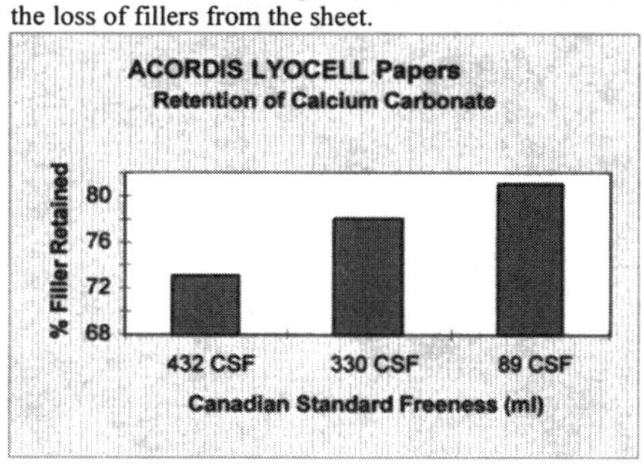

Filtration

The ability for ACORDIS LYOCELL fibre to split into circular, sub-micron fibrils generates the potential to capture small particles, whilst maintaining good air permeability. These properties make ACORDIS LYOCELL an ideal fibre for filtration applications.

One of the many end-uses for which ACORDIS LYOCELL is ideally suited is in the area of filtration. Lyocell fibre can be processed into a fibrillated paper using standard papermaking techniques. The papers produced are strong and have unique properties as the fibrils produced from the fibre are mainly circular in cross-section, with a range of diameters down to sub-micron levels.

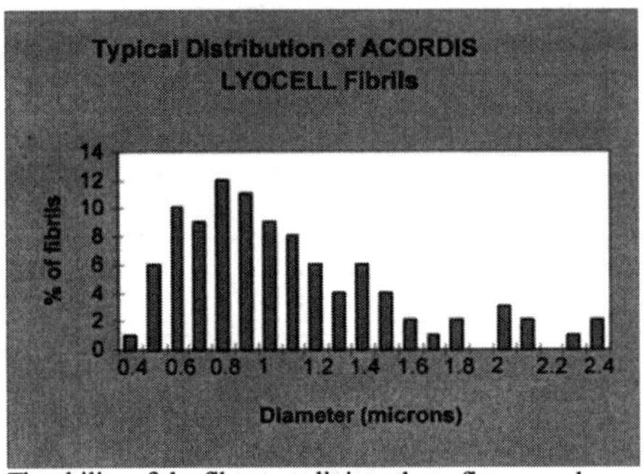

The ability of the fibre to split into these fine, round fibrils means that papers can be made which are ideal for filters - having good permeability coupled with the ability to capture small particles effectively. The pore size of the papers and hence their efficiency can be controlled readily by altering the level of fibrillation generated. This is demonstrated by comparing pore size (measured using a Coulter Porometer) with fibrillation level as measured by CSF (Canadian Standard Freeness).

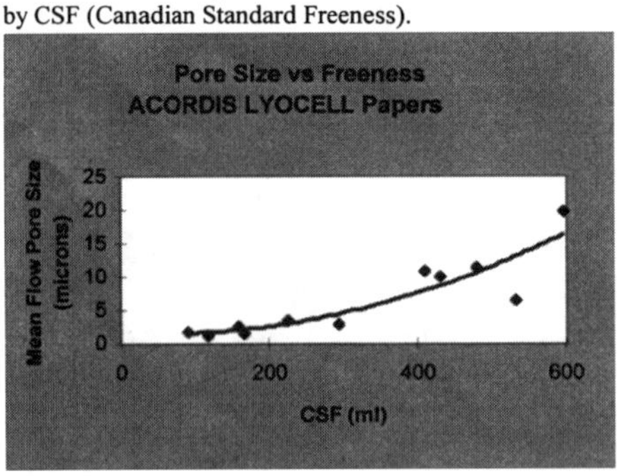

The filtration efficiency of papers made from ACORDIS LYOCELL can be demonstrated using a standard air filtration test. Using a sodium chloride aerosol challenge (BS4400) to assess the performance of flat papers, fibrillated ACORDIS LYOCELL papers were compared with commercial samples of microglass and cellulose (woodpulp) filter papers.

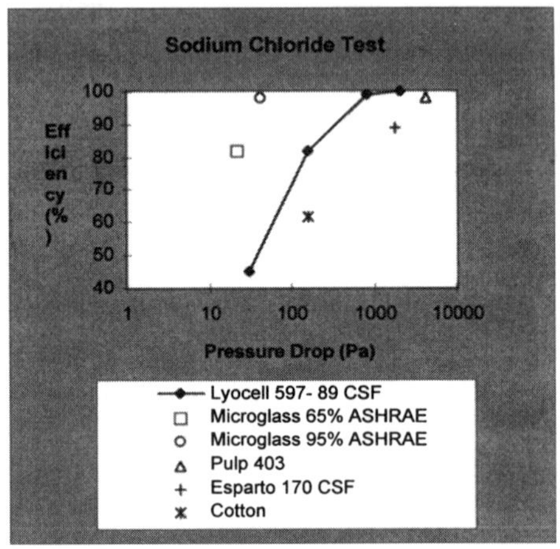

The results demonstrate the performance advantage of ACORDIS LYOCELL papers over woodpulp sheets - higher efficiency at a lower or comparable pressure drop.

Compared to microglass, ACORDIS LYOCELL papers do show higher pressure drops, though the paper properties are significantly better and have the additional advantage of requiring no binder.

	Tensile Index (Nm/g)	Tear Index ($mN.m^2/g$)	Bulk (cc/g)
Microglass	5.1	5.5	7.7
Woodpulp	14.8	4.7	2.2
ACORDIS LYOCELL	10.7	16.0	4.0

Although ACORDIS LYOCELL can be used alone, it is particularly suited to being used in blends. Significant advantages can be achieved using lyocell in combination with other fibres to engineer a product with the correct filtration characteristics. Fibre fibrillation levels can be tailored to suit the application and complement the other fibres being used. Work in blending pulp and lyocell has shown that addition of ACORDIS LYOCELL can improve filtration efficiency without impairing pressure drop.

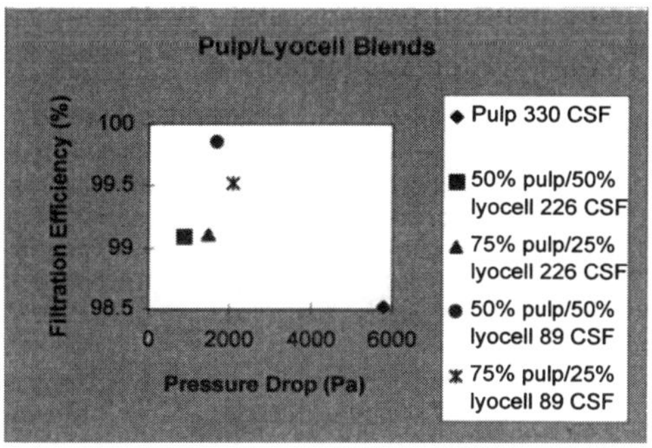

In combination with glass fibre, lyocell offers the opportunity to improve paper strength and integrity, without severely affecting filtration performance. The requirement for binders can also be reduced.

One product which utilises the characteristics of lyocell paper is the CREST™ cigarette filter, developed by ACORDIS in collaboration with Filtrona International Ltd. The Crest material was developed for use in ultra-low tar products (1-2 mg tar) which require a combination of very high tar retentions with acceptable resistance to draw (i.e. pressure drop).

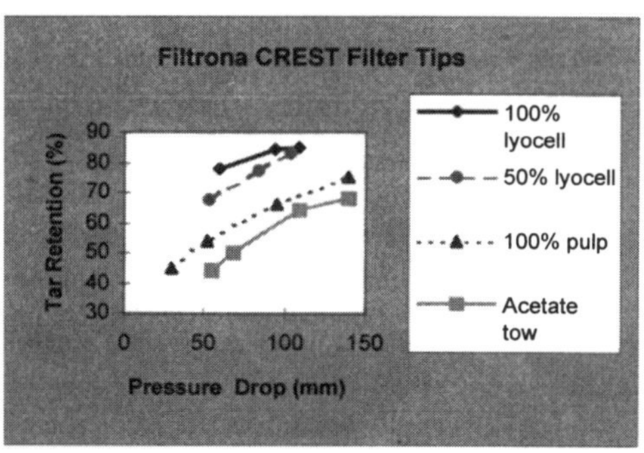

The optimised ACORDIS LYOCELL paper has a intermediate level of fibrillation giving a retentive yet permeable sheet. A blend of pulp and lyocell can also be used. The paper is made into a filter tip in the same way as a normal pulp paper. However, filtration results show the superior performance of both 100% ACORDIS LYOCELL or 50% lyocell 50% pulp, clearly differentiated from acetate or semi-crepe paper tips.

The exceptional performance of ACORDIS LYOCELL paper in filter tips is due to its unique filtration characteristics. The Filtrona CREST™ filter offers design opportunities in the growing ultra-low tar cigarette market, and clearly demonstrates the potential offered by ACORDIS LYOCELL in filtration applications.

Fibrillation of ACORDIS LYOCELL can also be accomplished in structures other than papers for example during hydroentanglement or any wet processes. Such fibrillated fabrics and structures exhibit similar advantages in filtration applications.

Examples of specific filtration applications for ACORDIS LYOCELL include automotive filters (fuel, oil, air), HEPA and ULPA filters, medical filters, vacuum bags and food and beverage filters.

ACORDIS LYOCELL papers have obtained German BGA approval for use in food contact applications.

Appendix B

Archive photographs of regenerated cellulosic fibre processes

The photographs in this appendix are presented courtesy of the Akzo-Nobel (UK) Ltd Archives in Coventry.

Plates 1 to 8 illustrate progressive developments in yarn handling methods from the 1930s' cake-spinning/hank-winding process (Plate 1, but see also Fig. 1.3 in Chapter 1 for the 1905 cake spinning machine) through continuous-warp spinning/washing of tyre yarn (Plates 2 and 3, but see also Fig. 9.5 in Chapter 9 for the self-advancing reel spinning system) to the staple fibre spinning/washing system (Plates 4 to 8).

Plates 9 and 10 are diacetate spinning, a dry process requiring no washing equipment.

Plate 11 shows the mixing of the extremely viscous hot lyocell dope in an early pilotscale mixer.

Plate 12 shows the ease with which the hot lyocell dope can be dry-formed into fibres.

Plate 13 shows extrudable glass-like 'chips' of solid cellulose-in-NMMO.

1 Viscose filament yarn process circa 1930: preparing hanks for washing.

2 Viscose tyre yarn process: tube spinning.

3 Viscose tyre yarn spinning: warp process.

4 Viscose staple fibre process: spinning.

5 Viscose staple fibre process: hot stretching the tows.

6 Viscose staple fibre process: traction unit after hot stretching.

7 Viscose staple fibre process: feeding the tow cutter.

8 Viscose staple fibre process: wash machine.

9 Dry-spinning: diacetate yarn machine spindeck.

10 Diacetate yarn spinning machine.

11 Cellulose in NMMO: premix.

12 Cellulose in NMMO: fibre-forming potential evident.

13 Cellulose in NMMO: chips of solid cellulose.

Index